Lecture Notes in Physics

Springer-Verlag Berlin Heidelberg GmbH

The Editorial Policy for Proceedings

The series Lecture Notes in Physics reports new developments in physical research and teaching – quickly, informally, and at a high level. The proceedings to be considered for publication in this series should be limited to only a few areas of research, and these should be closely related to each other. The contributions should be of a high standard and should avoid lengthy redraftings of papers already published or about to be published elsewhere. As a whole, the proceedings should aim for a balanced presentation of the theme of the conference including a description of the techniques used and enough motivation for a broad readership. It should not be assumed that the published proceedings must reflect the conference in its entirety. (A listing or abstracts of papers presented at the meeting but not included in the proceedings could be added as an appendix.)
When applying for publication in the series Lecture Notes in Physics the volume's editor(s) should submit sufficient material to enable the series editors and their referees to make a fairly accurate evaluation (e.g. a complete list of speakers and titles of papers to be presented and abstracts). If, based on this information, the proceedings are (tentatively) accepted, the volume's editor(s), whose name(s) will appear on the title pages, should select the papers suitable for publication and have them refereed (as for a journal) when appropriate. As a rule discussions will not be accepted. The series editors and Springer-Verlag will normally not interfere with the detailed editing except in fairly obvious cases or on technical matters.
Final acceptance is expressed by the series editor in charge, in consultation with Springer-Verlag only after receiving the complete manuscript. It might help to send a copy of the authors' manuscripts in advance to the editor in charge to discuss possible revisions with him. As a general rule, the series editor will confirm his tentative acceptance if the final manuscript corresponds to the original concept discussed, if the quality of the contribution meets the requirements of the series, and if the final size of the manuscript does not greatly exceed the number of pages originally agreed upon. The manuscript should be forwarded to Springer-Verlag shortly after the meeting. In cases of extreme delay (more than six months after the conference) the series editors will check once more the timeliness of the papers. Therefore, the volume's editor(s) should establish strict deadlines, or collect the articles during the conference and have them revised on the spot. If a delay is unavoidable, one should encourage the authors to update their contributions if appropriate. The editors of proceedings are strongly advised to inform contributors about these points at an early stage.
The final manuscript should contain a table of contents and an informative introduction accessible also to readers not particularly familiar with the topic of the conference. The contributions should be in English. The volume's editor(s) should check the contributions for the correct use of language. At Springer-Verlag only the prefaces will be checked by a copy-editor for language and style. Grave linguistic or technical shortcomings may lead to the rejection of contributions by the series editors. A conference report should not exceed a total of 500 pages. Keeping the size within this bound should be achieved by a stricter selection of articles and not by imposing an upper limit to the length of the individual papers. Editors receive jointly 30 complimentary copies of their book. They are entitled to purchase further copies of their book at a reduced rate. As a rule no reprints of individual contributions can be supplied. No royalty is paid on Lecture Notes in Physics volumes. Commitment to publish is made by letter of interest rather than by signing a formal contract. Springer-Verlag secures the copyright for each volume.

The Production Process

The books are hardbound, and the publisher will select quality paper appropriate to the needs of the author(s). Publication time is about ten weeks. More than twenty years of experience guarantee authors the best possible service. To reach the goal of rapid publication at a low price the technique of photographic reproduction from a camera-ready manuscript was chosen. This process shifts the main responsibility for the technical quality considerably from the publisher to the authors. We therefore urge all authors and editors of proceedings to observe very carefully the essentials for the preparation of camera-ready manuscripts, which we will supply on request. This applies especially to the quality of figures and halftones submitted for publication. In addition, it might be useful to look at some of the volumes already published. As a special service, we offer free of charge LaTeX and TeX macro packages to format the text according to Springer-Verlag's quality requirements. We strongly recommend that you make use of this offer, since the result will be a book of considerably improved technical quality. To avoid mistakes and time-consuming correspondence during the production period the conference editors should request special instructions from the publisher well before the beginning of the conference. Manuscripts not meeting the technical standard of the series will have to be returned for improvement.

For further information please contact Springer-Verlag, Physics Editorial Department II, Tiergartenstrasse 17, D-69121 Heidelberg, Germany

L. Mathelitsch W. Plessas (Eds.)

Broken Symmetries

Proceedings of the
37. Internationale Universitätswochen
für Kern-und Teilchenphysik,
Schladming, Austria,
February 28-March 7, 1998

Springer

Editors

Leopold Mathelitsch
Willibald Plessas
Institut für Theoretische Physik
Karl-Franzens-Universität Graz
Universitätsplatz 5
A-8010 Graz, Austria

Library of Congress Cataloging-in-Publication Data.

Die Deutsche Bibliothek - CIP-Einheitsaufnahme

Broken symmetries : proceedings of the 37. Internationale Universitätswochen für Kern- und Teilchenphysik, Schladming, Austria, February 28 - March 7, 1998 / Leopold Mathelitsch ; Willibald Plessas (ed.).
(Lecture notes in physics ; 521)
ISBN 978-3-662-14240-0 ISBN 978-3-540-49130-9 (eBook)
DOI 10.1007/978-3-540-49130-9

ISSN 0075-8450
ISBN 978-3-662-14240-0

Originally published by Springer-Verlag Berlin Heidelberg New York in 1999
Softcover reprint of the hardcover 1st edition 1999

Typesetting: Camera-ready by the authors/editors
Cover design: *design & production*, Heidelberg

SPIN: 10644301 55/3144 - 5 4 3 2 1 0 – Printed on acid-free paper

Preface

Symmetries have been a part of human life since time immemorial and they have been cultivated in the fine arts, in painting and music, architecture, etc. Symmetries as fundamental principles in physics have a long tradition too, but their eminent role was recognized only in this century, with the Noether theorem as a first milestone. Since then, symmetry considerations have become more and more important, and particle physics especially has taken up these ideas with great success over the past decades.

We know from the examples of arts that perfect symmetries are rather boring and that slight disturbances (asymmetries or broken symmetries) should be added to create excitement. Well-known examples can be found in paintings or musical compositions. It seems that physical nature too wants to make use of this type of excitement: Phenomenological observations lead us to detect and grasp certain symmetries but deeper examinations reveal their breaking to a smaller or larger extent. Symmetry breaking thus introduces a dynamical feature into the rigid construction of a physical theory.

Evidently, due to this dynamical aspect – now in the true physical sense of a force – symmetry breaking has become a vital field of research in particle physics (and other areas of physics). Therefore it appeared timely to devote the 1998 Schladming Winter School to the theme of "Broken Symmetries". It was intended to provide an exhaustive overview of the development of symmetries and symmetry breaking from a few decades ago up to the forefront of current research. We are particularly grateful to the lecturers for having enthusiastically taken up this idea and for preparing the various topics with great effort.

The School started off with R. Peccei giving a detailed overview of discrete and global symmetries in particle physics. W. Bernreuther and H. Wahl concentrated on the subject of CP violation, which at present is a hot topic both experimentally and theoretically, since it may open up perspectives beyond the standard model. Within quantum chromodynamics, the comprehensive description of hadronic processes, including low and intermediate energies, depends decisively on our understanding of the chiral symmetry and its spontaneous breaking; G. Ecker dealt with this subject. H. Fritzsch addressed flavor-symmetry breaking and elaborated on the questions of mass generation. Spontaneous symmetry breaking in general was discussed by W. Thirring on a more mathematical basis. L. Álvarez-Gaumé prepared an introduction to duality in quantum field theory on the basis of low-dimensional gauge theories and paved the way for H.P. Nilles to discuss supersymmetry, string theory and ways of unification.

On behalf of the whole organizing committee we thank the lecturers for preparing the material and for delivering their lectures at the School. We are especially grateful to those lecturers who, in addition, invested much of their scarce time in order to write up their lectures for these proceedings. We are confident that this will be for the benefit of the community of scientists working on symmetries and on symmetry breaking, particularly the younger colleagues. This volume does not contain the lectures of W. Thirring and H. Wahl. The corresponding material may be found in separate publications in the literature. We are also grateful to those participants who gave seminars at the School. We have included one-page abstracts of their presentations in these proceedings.

At this point we also want to express our thanks to the main sponsor of the School, the Austrian Ministry for Science and Transportation. In addition we gratefully acknowledge the generous support by the Government of Styria and the Town of Schladming. Valuable help towards the organisation was received from the Wirtschaftskammer Steiermark (Sektion Industrie), Steyr-Daimler-Puch AG, Mercedes-Benz AG, Minolta-Austria, and Styria Online.

Organizing the 1998 Schladming Winter School, and making it the successful event that it was, would not have been possible without the help of a number of colleagues and graduate students from our institute. Without naming them all we want to acknowledge the traditionally good cooperation within the organizing committee and beyond, which once again guaranteed a smooth running of all organizational, technical, and social matters. Finally we should like to express our sincere thanks to Miss Sabine Fuchs for carrying out the secretarial work and for finalizing the text and layout of these proceedings.

Graz, January 1999

W. Plessas
L. Mathelitsch

Contents

Discrete and Global Symmetries in Particle Physics
Roberto Peccei 1

CP Violation
Werner Bernreuther 51

CP Violation: Experimental Status and Prospects
Heinrich Wahl 79

Spontaneously Broken Symmetries
Heide Narnhofer and Walter Thirring 81

Chiral Symmetry
Gerhard Ecker 83

Quark Mass Hierarchies, Flavour Mixing and Maximal CP Violation
Harald Fritzsch 131

Duality in Quantum Field Theory (and String Theory)
Luis Álvarez-Gaumé 151

Supersymmetry, Strings and Unification
Hans Peter Nilles 225

Abstracts of the Seminars
(Given by the Participants of the School) 279

Discrete and Global Symmetries in Particle Physics

R. D. Peccei

Department of Physics and Astronomy, UCLA, Los Angeles, CA 90095-1547

Abstract. I begin these lectures by examining the transformation properties of quantum fields under the discrete symmetries of Parity, P, Charge Conjugation, C, and Time Reversal, T. With these results in hand, I then show how the structure of the Standard Model helps explain the conservation/violation of these symmetries in various sectors of the theory. This discussion is also used to give a qualitative proof of the CPT Theorem, and some of the stringent tests of this theorem in the neutral Kaon sector are reviewed. In the second part of these lectures, global symmetries are examined. Here, after the distinction between Wigner-Weyl and Nambu-Goldstone realizations of these symmetries is explained, a discussion is given of the various, approximate or real, global symmetries of the Standard Model. Particular attention is paid to the role that chiral anomalies play in altering the classical symmetry patterns of the Standard Model. To understand the differences between anomaly effects in QCD and those in the electroweak theory, a discussion of the nature of the vacuum structure of gauge theories is presented. This naturally raises the issue of the strong CP problem, and I present a brief discussion of the chiral solution to this problem and of its ramifications for astrophysics and cosmology. I also touch briefly on possible constraints on, and prospects for, having real Nambu-Goldstone bosons in nature, concentrating specifically on the simplest example of Majorons. I end these lectures by discussing the compatibility of having global symmetry in the presence of gravitational interactions. Although these interactions, in general, produces small corrections, they can alter significantly the Nambu-Goldstone sector of theories.

1 Discrete Space-Time Symmetries

Lorentz transformations

$$x^\mu \to x'^\mu = \Lambda^\mu{}_\nu x^\nu \tag{1}$$

preserve the invariance of the space-time interval

$$x_\mu x^\mu = \mathbf{r}^2 - c^2t^2 = \mathbf{r}'^2 - c^2t'^2 = x'_\mu x'^\mu \,. \tag{2}$$

This constrains the matrices $\Lambda^\mu{}_\nu$ to obey

$$\eta_{\mu\nu} = \Lambda^\lambda{}_\mu \eta_{\lambda\kappa} \Lambda^\kappa{}_\nu \,, \tag{3}$$

where the matrix tensor $\eta_{\mu\nu}$ is the diagonal matrix

$$\eta_{\mu\nu} = \begin{bmatrix} -1 & & & \\ & 1 & & \\ & & 1 & \\ & & & 1 \end{bmatrix} . \tag{4}$$

The pseudo-orthogonality of the Λ matrices detailed in Eq. (3)

$$\eta = \Lambda^T \eta \Lambda \tag{5}$$

allows the classification of Lorentz transformations depending on whether

$$\det \Lambda = \begin{cases} +1 \\ -1 \end{cases} ; \quad \Lambda^0_0 = \pm \sqrt{1 + \sum_{i=1}^{3} (\Lambda^i{}_0)^2} = \begin{cases} \geq +1 \\ \leq -1 \end{cases} . \tag{6}$$

As a result, the Lorentz group splits into four distinct pieces

$$\begin{aligned} L^{\uparrow}_{+} &: \det \Lambda = +1; \ \Lambda^0_0 \geq 1 \\ L^{\uparrow}_{-} &: \det \Lambda = -1; \ \Lambda^0_0 \geq 1 \\ L^{\downarrow}_{+} &: \det \Lambda = +1; \ \Lambda^0_0 \leq -1 \\ L^{\downarrow}_{-} &: \det \Lambda = -1; \ \Lambda^0_0 \leq -1 . \end{aligned} \tag{7}$$

The transformation matrices Λ in $L^{\uparrow}_{+}$ by themselves form a sub-group of the Lorentz group: the proper orthochronous Lorentz group. All other transformations in the Lorentz group can be obtained from Λ in $L^{\uparrow}_{+}$ by using two discrete transformations, P and T, characterized by the matrices:

$$P^{\mu}_{\ \nu} = \begin{bmatrix} +1 & & & \\ & -1 & & \\ & & -1 & \\ & & & -1 \end{bmatrix} ; T^{\mu}_{\ \nu} = \begin{bmatrix} -1 & & & \\ & +1 & & \\ & & +1 & \\ & & & +1 \end{bmatrix} \tag{8}$$

corresponding to space inversion (Parity) and time reversal. It is clear that if $\Lambda \in L^{\uparrow}_{+}$, then $P\Lambda \in L^{\uparrow}_{-}$; $PT\Lambda \in L^{\downarrow}_{+}$; and $T\Lambda \in L^{\downarrow}_{-}$. Remarkably, nature is invariant only under the proper orthochronous Lorentz transformations. Parity is violated in the weak interactions, something which was first suggested by Lee and Yang (Lee and Yang 1956) in 1956 and soon thereafter observed experimentally (Wu et al 1957). The detection of the decay of K^0_L into pions by Christenson, Cronin, Fitch and Turlay (Christenson et al 1964) in 1964 provided indirect evidence that also time reversal is not a good symmetry of nature.

One can understand why this is so on the basis of the Standard Model of electroweak and strong interactions and of the, so called, CPT theorem, established by Pauli, Schwinger, Lüders and Zumino (Pauli 1955). To appreciate these facts I will need to sketch how quantum fields behave under the discrete space-time transformations of P and T, as well as their behavior under charge conjugation (C) which physically corresponds to reversing the sign of all charges. I will begin with parity.

1.1 Parity

The Parity transformation properties of the electromagnetic fields follow directly from classical considerations.[1] The Lorentz force

$$\mathbf{F} = \frac{d\mathbf{p}}{dt} = q(\mathbf{E} + \mathbf{v} \times \mathbf{B}) \tag{9}$$

obviously changes sign under Parity, since $\mathbf{p} \to -\mathbf{p}$.[2] Hence, it follows that $\mathbf{E}$ is odd and $\mathbf{B}$ is even under Parity:

$$\mathbf{E}(\mathbf{x}, t) \xrightarrow{P} -\mathbf{E}(-\mathbf{x}, t) \; ; \; \mathbf{B}(\mathbf{x}, t) \xrightarrow{P} \mathbf{B}(-\mathbf{x}, t) \; . \tag{10}$$

Formally, the transformation above is induced by a Unitary operator $U(P)$. This operator takes the vector potential $A^\mu(\mathbf{x}, t)$ into a transformed vector potential $A^\mu(-\mathbf{x}, t)$. In view of Eq. (10), it is easy to see that

$$U(P)A^\mu(\mathbf{x}, t)U(P)^{-1} = \eta(\mu)A^\mu(-\mathbf{x}, t) \; , \tag{11}$$

where the symbol $\eta(\mu)$ is a useful notational shorthand, with

$$\eta(\mu) = \begin{cases} -1 & \mu \neq 0 \\ +1 & \mu = 0 \; . \end{cases} \tag{12}$$

Spin-zero scalar, $S(\mathbf{x}, t)$, and pseudoscalar, $P(\mathbf{x}, t)$, fields under parity are, respectively, even and odd. That is,

$$\begin{aligned} U(P)S(\mathbf{x}, t)U(P)^{-1} &= S(-\mathbf{x}, t) \\ U(P)P(\mathbf{x}, t)U(P)^{-1} &= -P(-\mathbf{x}, t) \; . \end{aligned} \tag{13}$$

The behavior of spin-1/2 Dirac fields $\psi(\mathbf{x}, t)$ under Parity is slightly more complex. However, this behavior can be straightforwardly deduced from the requirement that the Dirac equation be invariant under this operation. One finds that

$$U(P)\psi(\mathbf{x}, t)U(P)^{-1} = \eta_P \gamma^0 \psi(-\mathbf{x}, t) \; . \tag{14}$$

Here η_P is a phase factor of unit magnitude ($|\eta_P|^2 = 1$). Because one is always interested in fermion-antifermion bilinears, the phase factor η_P plays no role physically and one can set it to unity ($\eta_P \equiv 1$) without loss of generality.

Given Eq. (14), it is a straightforward exercise to deduce the Parity properties of fermion-antifermion bilinears.[3] Since

$$\gamma^0\gamma^0\gamma^0 = \gamma^0 \; ; \gamma^0\gamma^i\gamma^0 = -\gamma^i \; ; \gamma^0\gamma_5\gamma^0 = -\gamma_5 \; , \tag{15}$$

[1] Henceforth, I shall use natural units where $c = \hbar = 1$.

[2] Since Parity reverses the sign of space coordinates $\mathbf{r} \to -\mathbf{r}$, the velocity also changes sign, $\mathbf{v} \to -\mathbf{v}$.

[3] In my conventions $\{\gamma^\mu, \gamma^\nu\} = -2\eta^{\mu\nu}$, $\gamma^{0\dagger} = \gamma^0$ but $\gamma^{i\dagger} = -\gamma^i$, and $\gamma_5 = i\gamma^0\gamma^1\gamma^2\gamma^3$.

one easily deduces that

$$\begin{aligned} U(P)\bar{\psi}(\mathbf{x},t)\psi(\mathbf{x},t)U(P)^{-1} &= \bar{\psi}(-\mathbf{x},t)\psi(-\mathbf{x},t) \quad \text{(Scalar)} \\ U(P)\bar{\psi}(\mathbf{x},t)i\gamma_5\psi(\mathbf{x},t)U(P)^{-1} &= -\bar{\psi}(-\mathbf{x},t)i\gamma_5\psi(-\mathbf{x},t) \quad \text{(Pseudoscalar)} \\ U(P)\bar{\psi}(\mathbf{x},t)\gamma^\mu\psi(\mathbf{x},t)U(P)^{-1} &= \eta(\mu)\bar{\psi}(-\mathbf{x},t)\gamma^\mu\psi(-\mathbf{x},t) \quad \text{(Vector)} \\ U(P)\bar{\psi}(\mathbf{x},t)\gamma^\mu\gamma_5\psi(\mathbf{x},t)U(P)^{-1} &= -\eta(\mu)\bar{\psi}(-\mathbf{x},t)\gamma^\mu\gamma_5\psi(-\mathbf{x},t) \quad \text{(Pseudovector)} \end{aligned} \tag{16}$$

From the above, one sees immediately that the electromagnetic interaction is parity invariant:

$$W_{\rm int}^{\rm em} = \int d^4x e A^\mu(x)\bar{\psi}(x)\gamma_\mu\psi(x) \xrightarrow{P} W_{\rm int}^{\rm em} . \tag{17}$$

On the other hand, because Parity transforms fields of a given **chirality** into each other[4]

$$\psi_{\rm L}(\mathbf{x},t) \xrightarrow{P} \gamma^0\psi_{\rm R}(-\mathbf{x},t);\ \psi_{\rm R}(\mathbf{x},t) \xrightarrow{P} \gamma^0\psi_{\rm L}(-\mathbf{x},t) , \tag{18}$$

it is obvious that the chirally asymmetric weak interactions will violate parity. Thus, this sector of the Standard Model is Parity violating. The strong interactions, however, are invariant under Parity. These interactions are governed by Quantum Chromodynamics and in QCD both the left-handed and right-handed quarks are triplets under the $SU(3)$ gauge group:

$$q_{\rm L} \sim 3\ ;\ \ q_{\rm R} \sim 3\ . \tag{19}$$

Note the difference here with respect to the weak interactions. Under the weak $SU(2)$ group of the $SU(2)\times U(1)$ theory, the left-handed fields $\psi_{\rm L}$ of both quarks and leptons are doublets, while the right-handed fields $\psi_{\rm R}$ are singlets

$$\psi_{\rm L} \sim 2\ ;\ \ \psi_{\rm R} \sim 1\ . \tag{20}$$

This is the root cause for the violation of Parity in the weak interactions.

1.2 Charge Conjugation

As I alluded to earlier, the process of charge conjugation is connected physically with the reversal of the sign of all electric charges. For the electromagnetic field, therefore, the charge conjugation transformation C brings the vector potential $A^\mu(x)$ into minus itself

$$U(C)A^\mu(x)U(C)^{-1} = -A^\mu(x)\ . \tag{21}$$

For Dirac fields, since charge conjugation should transform particles into antiparticles, this operation essentially corresponds to Hermitian conjugation. That is, one has

$$U(C)\psi(x)U(C)^{-1} = \eta_c C\psi^\dagger(x)\ . \tag{22}$$

[4] Here $\psi_{\rm L}(x) = \frac{1}{2}(1-\gamma_5)\psi(x)$; $\psi_{\rm R}(x) = \frac{1}{2}(1+\gamma_5)\psi_{\rm R}(x)$.

Here η_c is again a phase factor of unit magnitude and, without loss of generality, one can take $\eta_c \equiv 1$. The form of the matrix C can be deduced from the requirement that the transformation (22) should leave the Dirac equation invariant. For this to be the case necessitates that

$$C\gamma_\mu^* C^{-1} = -\gamma_\mu \ . \tag{23}$$

The particular form of C one obtains depends on the form of the γ-matrices used. In the Majorana representation, where the γ-matrices are purely imaginary [Majorana: $\gamma_\mu^* = -\gamma_\mu$] then $C = 1$. On the other hand, in the Dirac representation [Dirac: $\gamma^0 = \begin{bmatrix} 1 & 0 \\ 0 & -1 \end{bmatrix}$; $\gamma^i = \begin{bmatrix} 0 & \sigma^i \\ -\sigma^i & 0 \end{bmatrix}$], then $C = \gamma_2$. Because of the simplicity of C in the Majorana representation, in what follows we shall make use of this representation when dealing with charge conjugation.

Using Eq. (22), it is straightforward to compute the C-conjugation properties of fermion-antifermion bilinears. Let me do this explicitly for the scalar density $\bar\psi\psi$ and then quote the results for the other bilinears. One has

$$\begin{aligned} U(C)\bar\psi(x)\psi(x)U(C)^{-1} &= U(C)\psi_\alpha^\dagger(x)(\gamma^0)_{\alpha\beta}\psi_\beta(x)U(C)^{-1} \\ &= \psi_\alpha(x)(\gamma^0)_{\alpha\beta}\psi_\beta^\dagger(x) \\ &= -\psi_\beta^\dagger(x)(\gamma^0)_{\alpha\beta}\psi_\alpha(x) \\ &= -\psi_\beta^\dagger(x)(\gamma^{0T})_{\beta\alpha}\psi_\alpha(x) \\ &= +\bar\psi(x)\psi(x) \ . \end{aligned} \tag{24}$$

The second line above is the result of using Eq. (22), taking $C = 1$ assuming one is working in the Majorana representation. The third line above follows because fermion fields anticommute (apart from an irrelevant infinite piece which can be subtracted away). Finally, the last line follows since in the Majorana representation γ^0 is an antisymmetric matrix ($\gamma^{0T} = -\gamma^0$).

The full set of results for the behavior of fermion-antifermion bilinears under C is displayed below:

$$\begin{aligned} &U(C)\ \bar\psi(x)\psi(x)U(C)^{-1} = \bar\psi(x)\psi(x) \ \text{(Scalar)} \\ &U(C)\ \bar\psi(x)i\gamma_5\psi(x)U(C)^{-1} = \bar\psi(x)i\gamma_5\psi(x) \ \text{(Pseudoscalar)} \\ &U(C)\ \bar\psi(x)\gamma^\mu\psi(x)U(C)^{-1} = -\bar\psi(x)\gamma^\mu\psi(x) \ \text{(Vector)} \\ &U(C)\ \bar\psi(x)\gamma^\mu\gamma_5\psi(x)U(C)^{-1} = \bar\psi(x)\gamma^\mu\gamma^5\psi(x) \ \text{(Pseudovector)} \ . \end{aligned} \tag{25}$$

These results lead to some immediate consequences. For instance, it follows that electromagnetic interactions are C-invariant. Using Eqs. (21) and (25) it follows that

$$W_{\text{int}}^{\text{em}} = \int d^4x e A^\mu(x)\bar\psi(x)\gamma_\mu\psi(x) \xrightarrow{C} W_{\text{int}}^{\text{em}} \ , \tag{26}$$

since both A^μ and the electromagnetic current $\bar\psi\gamma^\mu\psi$ change sign under C.

The strong interactions are also invariant under charge conjugation. This takes a small discussion, but it is also easy to see. The principal point to note is that the $SU(3)$ currents of QCD do not have the same simple transformation properties as the electromagnetic current, because they involve the non-trivial $SU(3)$ matrices λ_a. Effectively these matrices get transposed in the bilinears, if one makes a charge conjugation transformation. That is, one has

$$U(C)\bar{q}\gamma^\mu \frac{\lambda_a}{2} q U(C)^{-1} = -\bar{q}\gamma^\mu \left(\frac{\lambda_a}{2}\right)^T q \,. \tag{27}$$

Because $\lambda_1, \lambda_3, \lambda_4, \lambda_6$, and λ_8 are symmetric, while λ_2, λ_5, and λ_7 are antisymmetric, it follows that

$$J_a^\mu \to -\eta(a) J_a^\mu \,, \tag{28}$$

where

$$\eta(a) = \begin{cases} +1 \text{ for } a = 1,3,4,6 \text{ and } 8 \\ -1 \text{ for } a = 2,5 \text{ and } 7 \end{cases} \tag{29}$$

To guarantee invariance of the quark gluon interaction terms

$$W_{\text{int}} = \int d^4x g_3 A_a^\mu J_{\mu a} \tag{30}$$

under charge conjugation it is necessary to assume that the charge conjugation properties of the gluon fields themselves vary according to which component one is dealing with. Namely, for invariance of Eq. (30) under C one needs

$$U(C)A_a^\mu(x)U(C)^{-1} = -\eta(a)A_a^\mu(x) \,. \tag{31}$$

It is easy to check that the above transformation property is precisely what is needed to have the nonlinear gluon field strengths have well defined C-properties. Recall that

$$G_a^{\mu\nu} = \partial^\mu A_a^\nu - \partial^\nu A_a^\mu + g f_{abc} A_b^\mu A_c^\nu \,. \tag{32}$$

Now, for $SU(3)$, the only non-vanishing structure constants f_{abc} are (Slansky 1981)

$$f_{abc} \neq 0 \text{ for } abc = \{123, 147, 156, 246, 257, 345, 367, 458, 678\} \,. \tag{33}$$

One sees that $f_{abc} \neq 0$ only for cases in which there is an odd number of indices which themselves are odd (i.e. the indices: 2,5, and 7). This assures that, indeed, $G_a^{\mu\nu}$ transforms in the same way as A_a^μ does under C:

$$U(C)G_a^{\mu\nu}(x)U(C)^{-1} = -\eta(a)G_a^{\mu\nu}(x) \,. \tag{34}$$

This last property then insures that

$$W^{\text{QCD}} = \int d^4x \left[-\bar{q}\left(\gamma^\mu \frac{1}{i} D_\mu + m_q\right) q - \frac{1}{4} G_a^{\mu\nu} G_{a\mu\nu}\right] \xrightarrow{C} W^{\text{QCD}} \,. \tag{35}$$

The situation is different for the weak interactions since these involve both vector and pseudovector interactions. Let us focus, for example, on the $SU(2)$ current for leptons of the first generation

$$J_i^\mu = (\bar{\nu}_e\ \bar{e})_{\rm L}\gamma^\mu \frac{\tau_i}{2}\begin{pmatrix}\nu_e\\ e\end{pmatrix}_{\rm L} = \frac{1}{4}(\bar{\nu}_e\ \bar{e})\gamma^\mu(1-\gamma_5)\tau_i\begin{pmatrix}\nu_e\\ e\end{pmatrix} . \tag{36}$$

This current transforms differently in its vector and pseudovector pieces as well as in its 1, 3 and 2 components:

$$\begin{aligned} U(C)J^\mu_{1,3}U(C)^{-1} &= -\frac{1}{4}(\bar{\nu}_e\ \bar{e})\gamma^\mu(1+\gamma_5)\tau_{1,3}\begin{pmatrix}\nu_e\\ e\end{pmatrix} \\ U(C)J^\mu_2U(C)^{-1} &= +\frac{1}{4}(\bar{\nu}_e\ \bar{e})\gamma^\mu(1+\gamma_5)\tau_2\begin{pmatrix}\nu_e\\ e\end{pmatrix} . \end{aligned} \tag{37}$$

The difference in behavior in the 1,3 and 2 components is absorbed by postulating the following C-transformation properties for the W_i^μ fields.[5]

$$U(C)W_i^\mu(x)U(C)^{-1} = -\eta(i)W_i^\mu(x) , \tag{38}$$

with

$$\eta(i) = \begin{cases} +1 & i=1,3 \\ -1 & i=2 \end{cases} \tag{39}$$

Note that these properties are what one might expect since they imply that

$$W_\pm^\mu = \frac{1}{\sqrt{2}}(W_1^\mu \mp iW_2^\mu) \stackrel{C}{\longrightarrow} -W_\mp^\mu . \tag{40}$$

However, even so, the simultaneous presence of vector and pseudovector pieces in the currents which enter the weak interactions forces one to conclude that

$$W_{\text{weak interactions}} \stackrel{C}{\not\longrightarrow} W_{\text{weak interactions}} , \tag{41}$$

as is observed experimentally.

1.3 Time Reversal

Classically, T-invariance corresponds to the fact that the equations of motion describing a particle going from A to B along some path also allow, as a permitted motion, the time reversed motion. That is, a motion where the particle follows the same path, but is now going from B to A. Clearly, in this time reversed motion all momenta are reflected, but the coordinates remain the same. So, classically, under a T-transformation

$$\mathbf{p} \stackrel{T}{\longrightarrow} -\mathbf{p} ; \quad \mathbf{F} = \frac{d\mathbf{p}}{dt} \stackrel{T}{\longrightarrow} \mathbf{F} . \tag{42}$$

[5] These transformation properties guarantee that $F_i^{\mu\nu}$ and W_i^μ have the same C-properties.

Quantum mechanically, the interchange of initial and final states is implemented by having the operator $U(T)$, corresponding to time reversal, be an **anti-unitary** operator (Wigner 1932), with

$$U(T) = V(T)K \ . \tag{43}$$

In the above, $V(T)$ is a unitary operator while K complex conjugates any c-number quantity it acts on. The operation of complex conjugation as part of $U(T)$ is what renders this operator anti-unitary. The need for complex conjugation, in connection with time reversal, is already seen at the level of the Schrödinger equation. From

$$i\frac{\partial}{\partial t}\psi(\mathbf{x},t) = H\psi(\mathbf{x},t) \tag{44}$$

one deduces that $\psi^*(\mathbf{x},-t)$ obeys the equation

$$i\frac{\partial}{\partial t}\psi^*(\mathbf{x},-t) = H^*\psi^*(\mathbf{x},-t) \ . \tag{45}$$

So, provided that the Hamiltonian is real ($H^* = H$), then one sees that $\psi^*(\mathbf{x},-t)$ is also a solution of the Schrödinger equation. Therefore, in quantum mechanics, complex conjugation of the wave function (along with the reality of the Hamiltonian) accompanies the reversal in the direction of time.

The association of complex conjugation with time reversal effectively interchanges incoming and outgoing states (Low 1967)

$$\langle U(T)\phi|U(T)\psi\rangle = \langle\psi|\phi\rangle \ . \tag{46}$$

Thus, if T is a good symmetry of the theory, one relates processes to their time reversed process (e.g. the decay $A \to BC$ to the formation of A from the coalescence of B and C, $BC \to A$). More precisely, if time reversal is a good symmetry, then one relates the S-matrix element S_{fi} to that for $S_{\tilde{i}\tilde{f}}$, where the states, $\tilde{i}, \tilde{f}$ have all the momentum directions $\{\mathbf{p}\}$ reversed in comparison to the states i, f. That is

$$S_{fi} = {}_{\rm out}\langle f|i\rangle_{\rm in} = {}_{\rm in}\langle U(T)i|U(T)f\rangle_{\rm out} = {}_{\rm out}\langle\tilde{i}|\tilde{f}\rangle_{\rm in} = S_{\tilde{i}\tilde{f}} \ . \tag{47}$$

The next to last step above is only valid if time reversal is a good symmetry of the theory, since in this case it follows that

$$U(T)|f\rangle_{\rm out} = |\tilde{f}\rangle_{\rm in} \ ; \ \ U(T)|i\rangle_{\rm in} = |\tilde{i}\rangle_{\rm out} \ . \tag{48}$$

I should add a comment here about the issue of the reality of the Hamiltonian needed for time reversal to hold at the Schrödinger equation level. This is not quite the case when spin is involved and is the reason for the possible additional operator $V(T)$ in the definition of $U(T)$ in Eq. (43). More correctly, in general, what is needed is that

$$V(T)H^*V(T)^{-1} = H \ . \tag{49}$$

When there is no spin $V(T)$ is just the unit matrix, but with spin its presence allows for T-invariance. The simplest example of this is provided by the ordinary spin-orbit interaction of atomic physics

$$H_{\rm s-o} = \lambda \boldsymbol{\sigma} \cdot \mathbf{L} \ , \tag{50}$$

with λ some real constant. Since $\mathbf{L} = \mathbf{r} \times \frac{1}{i} \boldsymbol{\nabla}$, it follows that

$$H^*_{\rm s-o} = \lambda \boldsymbol{\sigma}^* \cdot \mathbf{L}^* = -\lambda \boldsymbol{\sigma}^* \cdot \mathbf{L} \ , \tag{51}$$

which is not the same as Eq. (50) because $\sigma_2^* = -\sigma_2$ but $\sigma_{1,3}^* = \sigma_{1,3}$. However, since $\sigma_2 \boldsymbol{\sigma}^* \sigma_2 = -\boldsymbol{\sigma}$, using $V(T) = \sigma_2$ guarantees that

$$V(T) H^*_{\rm s-o} V(T)^{-1} = H_{\rm s-o} \ , \tag{52}$$

reflecting physically that, indeed, time reversal not only changes $\mathbf{L} \to -\mathbf{L}$, but also, effectively, $\boldsymbol{\sigma} \to -\boldsymbol{\sigma}$.

In field theory, it is again straightforward to deduce what is the effect of a time-reversal transformation on the electromagnetic fields by focusing on what happens classically. Since the Lorentz force is invariant under T

$$\mathbf{F} = \frac{d\mathbf{p}}{dt} = q(\mathbf{E} + \mathbf{v} \times \mathbf{B}) \xrightarrow{T} \mathbf{F} \ , \tag{53}$$

it follows that $\mathbf{E}$ is even and $\mathbf{B}$ is odd under time-reversal. In terms of the vector potential, therefore, one has

$$U(T) A^\mu(\mathbf{x},t) U(T)^{-1} = \eta(\mu) A^\mu(\mathbf{x},-t) \ . \tag{54}$$

For spin-1/2 fields one can deduce the transformation properties of $\psi(\mathbf{x},t)$ under T-transformations by again asking that the action of $U(T)$ on $\psi(\mathbf{x},t)$ produce another solution of the Dirac equation. Writing

$$U(T)\psi(\mathbf{x},t) U(T)^{-1} = \eta_T T \psi(\mathbf{x},-t) \ , \tag{55}$$

with η_T a phase of unit magnitude (which we shall take, without loss of generality, to be unity, $\eta_T \equiv 1$), and remembering that $U(T)$ complex conjugates all c-numbers, one finds that for invariance of the Dirac equation the matrix T must obey

$$\begin{aligned} T\gamma^{0*}T^{-1} &= \gamma^0 \\ T\gamma^{i*}T^{-1} &= -\gamma^i \ . \end{aligned} \tag{56}$$

As was the case for the charge conjugation matrix C, the form of the matrix T also depends on which representation of the γ-matrices one uses. In the convenient Majorana representation, where $\gamma^{\mu *} = -\gamma^\mu$, one finds that

$$T = \gamma^0 \gamma_5 \ . \tag{57}$$

Armed with Eqs. (55) and (57), a simple calculation then produces the following transformation properties for the familiar fermion-antifermion bilinears:[6]

$$\begin{aligned}
&U(T)\,\bar{\psi}(\mathbf{x},t)\psi(\mathbf{x},t)U(T)^{-1} = \bar{\psi}(\mathbf{x},-t)\psi(\mathbf{x},-t) \quad \text{(Scalar)}\\
&U(T)\,\bar{\psi}(\mathbf{x},t)i\gamma_5\psi(\mathbf{x},t)U(T)^{-1} = -\bar{\psi}(\mathbf{x},-t)i\gamma_5\psi(\mathbf{x},-t) \quad \text{(Pseudoscalar)}\\
&U(T)\,\bar{\psi}(\mathbf{x},t)\gamma^\mu\psi(\mathbf{x},t)U(T)^{-1} = \eta(\mu)\bar{\psi}(\mathbf{x},-t)\gamma^\mu\psi(\mathbf{x},-t) \quad \text{(Vector)}\\
&U(T)\,\bar{\psi}(\mathbf{x},t)\gamma^\mu\gamma_5\psi(\mathbf{x},t)U(T)^{-1} = \eta(\mu)\bar{\psi}(\mathbf{x},-t)\gamma^\mu\gamma_5\psi(\mathbf{x},-t) \quad \text{(Pseudovector)}
\end{aligned} \tag{58}$$

It is obvious from the above and Eq. (54), as well from the reality of the electromagnetic coupling constant e, that the electromagnetic interactions are T-invariant

$$W_{\rm int}^{\rm em} = \int d^4x eA^\mu(x)\bar{\psi}(x)\gamma_\mu\psi(x) \xrightarrow{T} W_{\rm int}^{\rm em}\,. \tag{59}$$

It is easy to check also that the gauge interactions in both QCD and the $SU(2)\times U(1)$ electroweak theory are also T-invariant, provided one properly defines how the gauge fields transform. Since for $SU(3)$ only λ_2, λ_5 and λ_7 are imaginary, and for $SU(2)$ only σ_2 is imaginary, it is easy to check that the desired T-transformation properties are:[7]

$$\begin{aligned}
U(T)A_a^\mu(\mathbf{x},t)U(T)^{-1} &= \eta(\mu)\eta(a)A_a^\mu(\mathbf{x},-t) \quad (SU(3))\\
U(T)W_i^\mu(\mathbf{x},t)U(T)^{-1} &= \eta(\mu)\eta(i)W_i^\mu(\mathbf{x},-t) \quad (SU(2))\\
U(T)Y^\mu(\mathbf{x},t)U(T)^{-1} &= \eta(\mu)Y^\mu(\mathbf{x},-t) \qquad\quad (U(1))\,.
\end{aligned} \tag{60}$$

Note that in contrast to C, T-transformations affect vector and pseudovector currents in the same way. Thus, using (58) and (60), it follows immediately that

$$W^{\rm SM}_{\text{gauge interactions}} \xrightarrow{T} W^{\rm SM}_{\text{gauge interactions}}\,. \tag{61}$$

The Standard Model can have, however, T-violating interactions in the electroweak sector involving the scalar Higgs field. The couplings of the Higgs field, in contrast to the gauge couplings, do not need to be real. These complex couplings then provide the possibility of having T-violating interactions. I examine this point in the simplest case where one has only one complex Higgs doublet

$$\Phi = \begin{pmatrix}\phi^0\\ \phi^-\end{pmatrix} \tag{62}$$

in the theory. The scalar Higgs self-interactions, which trigger the breakdown of $SU(2)\times U(1)$, only involve real coefficients since one must require the Higgs potential to be Hermitian. That is

$$V = \lambda\left(\Phi^\dagger\Phi - \frac{v^2}{2}\right)^2 = V^\dagger \tag{63}$$

[6] In deducing Eq. (58), care must be taken to remember that $U(T)$ complex conjugates c-numbers.

[7] Of course, the gauge coupling constants, just like e, are real.

implies that both λ and v are real parameters.

The Yukawa interactions of Φ with the quark fields, however, can have complex coefficients.[8] With i, j being family indices, one can write, in general, these interactions as

$$\mathcal{L}_{\text{Yukawa}} = -\Gamma^u_{ij}(\bar{u}, \bar{d})_{\text{L}i}\Phi u_{\text{R}j} - \Gamma^d_{ij}(\bar{u}, \bar{d})_{\text{L}i}\tilde{\Phi} d_{\text{R}j} + \text{h.c.} \,. \quad (64)$$

Here $\tilde{\Phi} = i\sigma_2\Phi^*$ and the coefficient matrices Γ^u_{ij} Γ^d_{ij} are arbitrary complex matrices. After the electroweak interactions are spontaneously broken $(SU(2) \times U(1) \to U(1)_{\text{em}})$, effectively all that remains of the doublet field Φ is one scalar excitation—the Higgs boson H—and the vacuum expectation value v:

$$\Phi \to \frac{1}{\sqrt{2}}\begin{pmatrix} v + H \\ 0 \end{pmatrix} \quad (65)$$

Thus the Yukawa interactions (64) generate mass terms for the charge 2/3 and charge -1/3 quarks

$$M^{u,d}_{ij} = \frac{1}{\sqrt{2}}\Gamma^{u,d}_{ij} v \,. \quad (66)$$

As is well known, these mass matrices can be diagonalized by a bi-unitary transformation

$$(U^{u,d}_{\text{L}})^\dagger M^{u,d} U^{u,d}_{\text{R}} = \mathcal{M}^{u,d} \,. \quad (67)$$

The diagonal matrices $\mathcal{M}^{u,d}$ have real eigenvalues m_i, corresponding to the physical quark masses. Further, the bi-unitary transformations on the quark fields diagonalizes the Yukawa coupling matrices, since M and Γ are linearly related. Whence, all that remains of the Yukawa sector after these transformations is the simple interaction

$$\mathcal{L}^{\text{eff}}_{\text{Yukawa}} = -\sum_i m_i \bar{q}_i(x) q_i(x)\left[1 + \frac{H(x)}{v}\right] \,. \quad (68)$$

Provided $H(\mathbf{x}, t)$ has the canonical T-transformation one expects for a scalar field,

$$U(T)H(\mathbf{x}, t)U(T)^{-1} = H(\mathbf{x}, -t) \,. \quad (69)$$

Eq. (68) is a T-conserving interaction also. Nevertheless, the complex nature of the original Yukawa couplings does end up by producing some T-violating interactions.

It is easy to understand this last point. The bi-unitary transformations performed on the quarks to diagonalize the quark mass matrices alter the form of the charged current weak interactions. Before these transformations, these interactions had the form

$$\mathcal{L}^{\text{cc}} = \frac{e}{2\sqrt{2}\,\sin\theta_W}\left[W^\mu_+ J^0_{-\mu} + W^\mu_- J^0_{+\mu}\right] \,, \quad (70)$$

[8] I concentrate here only on the quark sector, because if one does not introduce right-handed neutrinos in the theory—so that neutrinos are effectively massless—then all the phases in the Yukawa couplings in the lepton sector can be rotated away.

with

$$J^0_{-\mu} = (\bar{u}_1, \bar{u}_2, \bar{u}_3)\gamma_\mu(1-\gamma_5)\,\mathbf{1}\begin{pmatrix} d_1 \\ d_2 \\ d_3 \end{pmatrix} \tag{71}$$

and

$$J^0_{+\mu} = (J^0_{-\mu})^\dagger\ . \tag{72}$$

Clearly, this interaction is T-invariant. However, after the bi-unitary transformation on the quark fields to diagonalize M [Eq. (67)], the charged current $J^0_{-\mu}$ is altered to

$$J_{-\mu} = (\bar{u}, \bar{c}, \bar{t})\gamma_\mu(1-\gamma_5)\mathbf{V}_{\mathrm{CKM}}\begin{pmatrix} d \\ s \\ b \end{pmatrix}\ , \tag{73}$$

where the Cabibbo-Kobayashi-Maskawa quark mixing matrix (Cabibbo 1963, and Kobayashi and Maskawa 1973)

$$\mathbf{V}_{\mathrm{CKM}} = U^{u\dagger}_{\mathrm{L}} U^d_{\mathrm{L}} \tag{74}$$

is a unitary matrix, since U^u_{L} and U^d_{L} are. Because, in general, $\mathbf{V}_{\mathrm{CKM}}$ is complex, its presence in the currents J^μ_- (and J^μ_+) can lead to T-violation.

For three families of quarks and leptons, as we apparently have, it is not difficult to show that the matrix $\mathbf{V}_{\mathrm{CKM}}$ has only one physical phase, δ. All the other phases can be rotated away through further harmless redefinitions of the quark fields. If $\delta \neq 0$, then the charged current weak interactions are not T-invariant

$$\mathcal{L}^{\mathrm{cc}}(\mathbf{x}, t) = \frac{e}{2\sqrt{2}\ \sin\theta_{\mathrm{W}}}\left[W^\mu_+ J_{-\mu} + W^\mu_- J_{+\mu}\right] \stackrel{T}{\not\to} \mathcal{L}^{\mathrm{cc}}(\mathbf{x}, -t) \tag{75}$$

and the standard model can give rise to observable manifestations of T-violation. We return to this point in more detail in the next subsection, after we discuss the CPT theorem.

1.4 The CPT Theorem

If nature is described by a local Lorentz invariant field theory, where there is the usual connection between spin and statistics, then one can prove a deep theorem, now known as the CPT Theorem (Pauli 1955). Namely, in these circumstances, one can show that the action of the theory is **always invariant** under the combined application of a C-, a P-, and a T-transformation. That is

$$W \stackrel{CPT}{\longrightarrow} W\ . \tag{76}$$

I will not attempt here to establish the CPT theorem with rigor. The interested reader can turn, for example, to the erudite manuscript of Streater and Wightman (Streater and Wightman 1964) for this. Rather, I want to show why and how the CPT Theorem works, based on the preceding discussion of the C, P, and T transformation properties of quantum fields.

To get started, let us look at the effect of a CPT transformation on the electromagnetic interactions. Using Eqs. (11), (16), (21), (25), (54), and (58), one has

$$\begin{aligned} A^{\mu}(\mathbf{x},t) &\xrightarrow{CPT} [-1][\eta(\mu)][\eta(\mu)]A^{\mu}(-\mathbf{x},-t) = -A^{\mu}(-\mathbf{x},-t) \\ J^{\mu}_{\rm em}(x,t) &= \bar{\psi}(\mathbf{x},t)\gamma^{\mu}\psi(\mathbf{x},t) \\ &\xrightarrow{CPT} [-1][\eta(\mu)][\eta(\mu)]J^{\mu}_{\rm em}(-\mathbf{x},-t) = -J^{\mu}_{\rm em}(-\mathbf{x},-t)\,. \end{aligned} \tag{77}$$

Obviously, therefore, under a CPT transformation

$$W^{\rm em}_{\rm int} = \int d^4x e A^{\mu}(x) J^{\rm em}_{\mu}(x) \xrightarrow{CPT} W^{\rm em}_{\rm int}\,. \tag{78}$$

This, however, is a trivial case, since $W^{\rm em}_{\rm int}$ was **separately** invariant under C-, P-, and T-transformations!

CPT invariance, if it is a general property, must hold also when there is violation of the individual symmetries. A more significant test is provided by the electroweak theory. There, for example, both C and P are violated in the neutral current interactions, while T and CPT are conserved. Let us check this. The action for the neutral current interactions is given by

$$W^{\rm NC}_{\rm int} = \frac{e}{2\cos\theta_W \sin\theta_W}\int d^4x Z_{\mu} J^{\mu}_{\rm NC}\,. \tag{79}$$

The neutral current

$$J^{\mu}_{\rm NC} = 2[J^{\mu}_3 - \sin^2\theta_W J^{\mu}_{\rm em}] = V^{\mu} + A^{\mu} \tag{80}$$

contains both vector and pseudovector pieces, since these latter components are present in the $SU(2)$ current J^{μ}_3. Parity and Charge Conjugation are violated in Eq. (79) because the vector and pseudovector currents transform in opposite ways under each of these transformations. That is, one has, under Parity

$$\begin{aligned} &Z^{\mu}(\mathbf{x},t) \xrightarrow{P} \eta(\mu)Z^{\mu}(-\mathbf{x},t) \;\; ; \;\; V^{\mu}(\mathbf{x},t) \xrightarrow{P} \eta(\mu)V^{\mu}(-\mathbf{x},t); \\ &A^{\mu}(\mathbf{x},t) \xrightarrow{P} -\eta(\mu)A^{\mu}(-\mathbf{x},t) \end{aligned} \tag{81}$$

while, under Charge Conjugation,

$$Z^{\mu}(\mathbf{x},t) \xrightarrow{C} -Z^{\mu}(\mathbf{x},t);\; V^{\mu}(\mathbf{x},t) \xrightarrow{C} -V^{\mu}(\mathbf{x},t); A^{\mu}(\mathbf{x},t) \xrightarrow{C} A^{\mu}(\mathbf{x},t) \tag{82}$$

On the other hand, T is conserved by Eq. (79), since under time reversal

$$\begin{aligned} &Z^{\mu}(\mathbf{x},t) \xrightarrow{T} \eta(\mu)Z^{\mu}(\mathbf{x},-t);\; V^{\mu}(\mathbf{x},t) \xrightarrow{T} \eta(\mu)V^{\mu}(\mathbf{x},-t); \\ &A^{\mu}(\mathbf{x},t) \xrightarrow{T} \eta(\mu)\mathbf{A}(\mathbf{x},-t)\;\;. \end{aligned} \tag{83}$$

Using the above three equations, it is easy to see that the neutral current interactions conserve CPT. One has

$$Z^\mu(x,t) \xrightarrow{CPT} -Z^\mu(-\mathbf{x},-t)\ ;\ V^\mu(\mathbf{x},t) \xrightarrow{CPT} -V^\mu(-\mathbf{x},-t);$$
$$A^\mu(\mathbf{x},t) \xrightarrow{CPT} -A^\mu(-\mathbf{x},-t)\ . \tag{84}$$

From the above, it is also clear that CP and T are **equivalent** transformations for the neutral current action

$$W_{\rm int}^{\rm NC} \xrightarrow{CP} W_{\rm int}^{\rm NC} \xrightarrow{T} W_{\rm int}^{\rm NC}\ . \tag{85}$$

The equivalence between a T-transformation and a CP-transformation also holds when both of these potential symmetries are violated. Hence, even in this case, the combined CPT-transformation is indeed an invariance of the action. This is the essence of the CPT Theorem. To appreciate this point let me examine, specifically, the T-violating charged current interaction between the u and b quarks, typified by the complex CKM matrix element V_{ub}.[9] One has

$$W_{ub}^{\rm cc} = \frac{e}{2\sqrt{2}\ \sin\theta_W}\int d^4x\,\left\{V_{ub}W_+^\mu\bar{u}\gamma_\mu(1-\gamma_5)b + V_{ub}^* W_-^\mu\bar{b}\gamma_\mu(1-\gamma_5)u\right\}\ , \tag{86}$$

where

$$W_\pm^\mu = \frac{1}{\sqrt{2}}(W_1^\mu \mp iW_2^\mu)\ . \tag{87}$$

Because under T

$$W_1^\mu(\mathbf{x},t) \xrightarrow{T} \eta(\mu)W_1^\mu(\mathbf{x},-t);\ W_2^\mu(\mathbf{x},t) \xrightarrow{T} -\eta(\mu)W_2^\mu(\mathbf{x},-t) \tag{88}$$

and remembering the i factor in Eq. (87), it follows that

$$W_\pm^\mu(\mathbf{x},t) \xrightarrow{T} \eta(\mu)W_\pm^\mu(\mathbf{x},-t)\ . \tag{89}$$

On the other hand, under T, the $u-b$ currents behave as

$$\bar{u}(\mathbf{x},t)\gamma_\mu(1-\gamma_5)b(\mathbf{x},t) \xrightarrow{T} \eta(\mu)\bar{u}(\mathbf{x},-t)\gamma_\mu(1-\gamma_5)b(\mathbf{x},-t)$$
$$\bar{b}(\mathbf{x},t)\gamma_\mu(1-\gamma_5)u(\mathbf{x},t) \xrightarrow{T} \eta(\mu)\bar{b}(\mathbf{x},-t)\gamma_\mu(1-\gamma_5)u(\mathbf{x},-t)\ . \tag{90}$$

Hence, one sees, indeed, that the action $W_{ub}^{\rm cc}$ is not T-invariant

$$W_{ub}^{\rm cc} \xrightarrow{T} \tilde{W}_{ub}^{\rm cc} = \frac{e}{2\sqrt{2}\sin\theta_W}\int d^4x\,\left\{V_{ub}^* W_+^\mu\bar{u}\gamma_\mu(1-\gamma_5)b\right.$$
$$\left. + V_{ub}W_-^\mu\bar{b}\gamma_\mu(1-\gamma_5)u\right\}\ . \tag{91}$$

[9] One can pick phase conventions where V_{ub} is real. In this case, however, other pieces in the charged current Lagrangian give rise to T-violation. The final result for physically measured parameters must be phase-convention independent. I focus here on the V_{ub} term for definitiveness, since in the standard convention for the CKM matrix (Cabibbo 1963, and Kobayashi and Maskawa 1973) V_{ub} is complex and its phase is precisely $-\delta$.

The behavior of the various ingredients in W_{ub}^{cc} under CP is individually different than it is under T. For instance, one has

$$W_1^\mu(\mathbf{x},t) \xrightarrow{CP} -\eta(\mu)W_1^\mu(-\mathbf{x},t);\; W_2^\mu(\mathbf{x},t) \xrightarrow{CP} \eta(\mu)W_2^\mu(-\mathbf{x},t)\,. \tag{92}$$

Hence, since one also does not complex conjugate the i in $W_\pm^\mu$ in this case, one has

$$W_\pm^\mu(\mathbf{x},t) \xrightarrow{CP} -\eta(\mu)W_\mp(-\mathbf{x},t)\,. \tag{93}$$

Similarly, one finds, that under CP the $u-b$ currents transform as

$$\begin{aligned}\bar{u}(\mathbf{x},t)\gamma_\mu(1-\gamma_5)b(\mathbf{x},t) &\xrightarrow{CP} -\eta(\mu)\bar{b}(-\mathbf{x},t)\gamma_\mu(1-\gamma_5)u(-\mathbf{x},t)\\ \bar{b}(\mathbf{x},t)\gamma_\mu(1-\gamma_5)u(\mathbf{x},t) &\xrightarrow{CP} -\eta(\mu)\bar{u}(-\mathbf{x},t)\gamma_\mu(1-\gamma_5)b(-\mathbf{x},t)\,.\end{aligned} \tag{94}$$

The net effect, however, on W_{ub}^{cc} is the same as that of a T-transformation. One finds

$$\begin{aligned}W_{ub}^{cc} \xrightarrow{CP} \tilde{W}_{ub}^{cc} = \frac{e}{2\sqrt{2}\,\sin\theta_W}\int d^4x\,\{&V_{ub}W_-^\mu\bar{b}\gamma_\mu(1-\gamma_5)u\\ &+V_{ub}^*W_+^\mu\bar{u}\gamma_\mu(1-\gamma_5)b\}\,.\end{aligned} \tag{95}$$

One can extract from this example the underlying reason why the CPT theorem holds. It results really from a combination of the needed Hermiticity of the Lagrangian and the complementary role that T and CP play on the operators and c-numbers that enter in the Lagrangian. Hermiticity means that a given term in the Lagrangian, containing some operator $O(x)$ and some c-number a, has the form

$$\mathcal{L}(x) = aO(x) + a^*O^\dagger(x)\,. \tag{96}$$

Under T, the operator is unchanged (except for replacing t by $-t$), but the c-number is complex conjugated

$$O(\mathbf{x},t) \xrightarrow{T} O(\mathbf{x},-t)\;;\; a \xrightarrow{T} a^*\,. \tag{97}$$

Under CP, on the other hand, the operator O gets essentially replaced by its Hermitian adjoint, but the c-number a stays the same:

$$O(\mathbf{x},t) \xrightarrow{CP} O^\dagger(-\mathbf{x},t)\;;\; a \xrightarrow{CP} a\,. \tag{98}$$

Combining the operations of T and CP changes, effectively, the first term in Eq. (96) into the second term and *vice versa*

$$\mathcal{L} = aO(x) + a^*O^\dagger(x) \xrightarrow{CPT} \mathcal{L}(-x) = a^*O^\dagger(-x) + aO(-x) \tag{99}$$

leaving the action invariant

$$W = \int d^4x\mathcal{L}(x) \xrightarrow{CPT} W\,. \tag{100}$$

1.5 CP and CPT Tests in the Neutral Kaon Complex

The $K^0 \sim \bar{s}d$ and $\bar{K}^0 \sim s\bar{d}$ states provide an excellent laboratory to test CP and CPT. These states are unstable, decaying into particles with no strangeness through a first-order weak process. In addition, second order weak processes, giving rise to the transition $\bar{s}d \leftrightarrow s\bar{d}$, allow the K^0 to mix with the $\bar{K}^0$. The quantum mechanical evolution of this two-state system leads to the physical eigenstates K_L^0 and K_S^0, characterized by their, respective, long and short lifetimes.

The physical eigenstates K_L^0 and K_S^0 are obtained by diagonalizing the 2×2 effective Hamiltonian

$$H_{\rm eff} = M - \frac{i}{2}\Gamma \,. \tag{101}$$

Here M and Γ are Hermitian matrices describing the mass mixing and decay properties of the neutral Kaon complex. If CPT is a good symmetry of nature, then the diagonal matrix elements of M and Γ are equal, since this symmetry changes effectively K^0 into $\bar{K}^0$.

$$M_{11} = M_{22} \ ; \quad \Gamma_{11} = \Gamma_{22} \ \text{[CPT Conservation]} \,. \tag{102}$$

CP conservation, on the other hand, guarantees the reality of the mass and decay matrices. It provides therefore a constraint on the off-diagonal matrix elements of M and Γ. Namely:

$$M_{12} = M_{12}^* \ ; \quad \Gamma_{12} = \Gamma_{12}^* \ \text{[CP Conservation]} \,. \tag{103}$$

If one does not impose the above constraints of CPT and CP conservation on M and Γ, the eigenstates of the Schrödinger equation

$$H_{\rm eff} \begin{pmatrix} |K^0\rangle \\ |\bar{K}^0\rangle \end{pmatrix} = i\frac{\partial}{\partial t} \begin{pmatrix} |K^0\rangle \\ |\bar{K}^0\rangle \end{pmatrix} \tag{104}$$

are linear superpositions of the $|K^0\rangle$ and $|\bar{K}^0\rangle$ states, involving parameters δ_K and ϵ_K which reflect the breaking of these symmetries. The physical $|K_L^0\rangle$ and $|K_S^0\rangle$ eigenstates have the standard time evolution

$$|K_{L,S}(t)\rangle = \exp[-im_{L,S}t] \exp\left[-\frac{1}{2}\Gamma_{L,S}t\right] |K_{L,S}(0)\rangle \,, \tag{105}$$

characterized by the mass and width of these particles. The states $|K_{L,S}(0)\rangle$ involve the following superposition of the $|K^0\rangle$ and $|\bar{K}^0\rangle$ states:

$$\begin{aligned} |K_L(0)\rangle &= \frac{1}{\sqrt{2}} \left\{ (1+\epsilon_K+\delta_K)|K^0\rangle + (1-\epsilon_K-\delta_K)|\bar{K}^0\rangle \right\} \\ |K_S(0)\rangle &= \frac{1}{\sqrt{2}} \left\{ (1+\epsilon_K-\delta_K)|K^0\rangle - (1-\epsilon_K+\delta_K)|\bar{K}^0\rangle \right\} \,. \end{aligned} \tag{106}$$

In the above

$$\epsilon_K = e^{i\phi_{SW}} \left[\frac{-\text{Im } M_{12} + \frac{i}{2} \text{ Im } \Gamma_{12}}{\sqrt{2}\, \Delta m} \right]$$

$$\delta_K = ie^{i\phi_{SW}} \left[\frac{(M_{11} - M_{22}) - \frac{i}{2}(\Gamma_{11} - \Gamma_{22})}{2\sqrt{2}\, \Delta m} \right] , \tag{107}$$

where

$$\phi_{SW} = \tan^{-1} \frac{2\Delta m}{\Gamma_S - \Gamma_L} \ ; \ \ \Delta m = m_L - m_S \ . \tag{108}$$

Experimentally, one finds (Particle Data Group 1996)

$$\phi_{SW} = (43.49 \pm 0.08)^o \ ; \ \ \Delta m = (3.491 \pm 0.009) \times 10^{-12} \text{ MeV} \ . \tag{109}$$

Note that $\epsilon_K = 0$, if CP is conserved and $\delta_K = 0$, if CPT is conserved. Only if both ϵ_K and δ_K vanish are the eigenstates $|K_L^0\rangle$ and $|K_S^0\rangle$ CP eigenstates. If both these symmetries hold then

$$CP|K_{L,S}^0\rangle = \mp|K_{L,S}^0\rangle \ \ \text{[CP, CPT Conservation]} \ . \tag{110}$$

What is measured experimentally are the CP violating ratios of the amplitude of the K_L and K_S to go into two pions

$$\eta_{+-} = \frac{A(K_L \to \pi^+\pi^-)}{A(K_S \to \pi^+\pi^-)} = |\eta_{+-}|e^{i\phi_{+-}} = \epsilon + \epsilon'$$

$$\eta_{00} = \frac{A(K_L \to \pi^0\pi^0)}{A(K_S \to \pi^0\pi^0)} = |\eta_{00}|e^{i\phi_{00}} = \epsilon - 2\epsilon' \ . \tag{111}$$

Experimentally, one finds that $\eta_{+-} \simeq \eta_{00}$ (so $\epsilon \gg \epsilon'$), with (Particle Data Group 1996)

$$|\eta_{+-}| = (2.285 \pm 0.019) \times 10^{-3} \ ; \ \ \phi_{+-} = (43.7 \pm 0.6)^o \ . \tag{112}$$

Neglecting the contribution of the widths compared to the masses, which is a very good approximation, one finds that the parameter ϵ above is simply (Buchanan et al 1992)

$$\epsilon \simeq \epsilon_K - \delta_K \simeq e^{i\phi_{SW}} \left[\frac{-\text{Im } M_{12}}{\sqrt{2}\, \Delta m} \right] + ie^{i\phi_{SW}} \left[\frac{M_{22} - M_{11}}{2\sqrt{2}\, \Delta m} \right] \ . \tag{113}$$

Note that the CPT violating contribution in the above is 90° **out of phase** from the CP violating contribution. Because $\phi_{SW} = (43.49 \pm 0.08)^o$ is consistent with $\phi_{+-} = (43.7 \pm 0.6)^o$, one deduces immediately that the non-zero value for η_{+-} observed is mostly a signal of CP-violation [Im $M_{12} \neq 0$] rather than of CPT violation [$M_{11} \neq M_{22}$].

If one neglects altogether the possibility that there is any CPT violation in the neutral Kaon decay amplitudes—something one would eventually need to check—then one can write approximately

$$M_{22} - M_{11} \simeq |\eta_{+-}| 2\sqrt{2}\, \Delta m \tan(\phi_{+-} - \phi_{SW}) \ . \tag{114}$$

This equation, given the values of the experimental parameters involved, provides a spectacularly strong bound on CPT violation, because the $K_L - K_S$ mass difference Δm is so small. One finds, at the 90% CL,

$$\left|\frac{m_{\bar{K}^0} - m_{K^0}}{m_{K^0}}\right| < 9 \times 10^{-19} , \tag{115}$$

which is an incredibly stringent test of CPT.

Experiments at the just completed Frascati Phi Factory will be able to directly measure δ_K, without further assumptions, to an accuracy similar to the present accuracy for ϵ. This will be accomplished by studying the difference in relative time decay patterns of the doubly semileptonic decays of the $K_L K_S$ states produced in the Φ decay. If one studies the relative time dependence of the process $\Phi \to K_L K_S \to \pi^- e^+ \nu_e(t_1) \pi^+ e^- \bar{\nu}_e(t_2)$, then one can show that the pattern at large $\Delta t = t_1 - t_2$ is sensitive to Re δ_K, while the pattern at small Δt is sensitive to Im δ_K (Buchanan et al 1992).

2 Continuous Global Symmetries

In the Standard Model there are a variety of global symmetries, both exact and approximate. Some of these symmetries are manifest [Wigner-Weyl realized], while others are spontaneously broken [Nambu-Goldstone realized]. I wish here to examine these matters in some detail.

An important distinction exists for a continuous global symmetry depending on whether or not the vacuum state respects the symmetry. Let us denote the global symmetry group for the theory by G. This group, in general, will have generators g_i which obey an algebra

$$[g_i, g_j] = i c_{ijk} g_k , \tag{116}$$

where c_{ijk} are the structure constants for the group. If the generators g_i, for all i, annihilate the vacuum

$$g_i|0\rangle = 0 , \tag{117}$$

then the symmetry group is realized in a Wigner-Weyl way, with degenerate multiplets of states in the spectrum (Wigner 1952 and Weyl 1929). If, on the other hand, for some generators g_i

$$g_i|0\rangle \neq 0 \tag{118}$$

then the symmetry group G is spontaneously broken to a subgroup H $(G \to H)$ and $n = \dim G/H$ massless scalars appear in the spectrum of the theory. This is the Nambu-Goldstone realization of the symmetry G and the massless scalars are known as Nambu-Goldstone bosons (Nambu 1980 and Goldstone 1981).

Physically, approximate global symmetries are easy to understand. These symmetries result from being able to neglect dynamically certain parameters in the theory. A well known example is provided by Quantum Chromodynamics (QCD). The Lagrangian of QCD

$$\mathcal{L}_{\rm QCD} = -\sum_i \bar{q}_i \left(\gamma^\mu \frac{1}{i} D_\mu + m_i\right) q_i - \frac{1}{4} G_a^{\mu\nu} G_{a\mu\nu} \tag{119}$$

has an approximate global symmetry, connected to the fact that the lightest quark masses m_u and m_d are much smaller than the dynamical scale of the theory, $\Lambda_{\rm QCD}$.[10] Neglecting the light quark masses, one sees that the QCD Lagrangian is invariant under a large global symmetry transformation

$$\mathcal{L}_{\rm QCD} \stackrel{U(n_f)_{\rm L}\times U(n_f)_{\rm R}}{\longrightarrow} \mathcal{L}_{\rm QCD} , \tag{120}$$

where n_f is the number of flavors whose masses are neglected. Under this group of transformations the n_f light quarks go into each other. For example, for $n_f = 2$, neglecting m_u and m_d in the QCD Lagrangian allows the symmetry transformation

$$\begin{pmatrix} u \\ d \end{pmatrix}_{\rm L} \to e^{i a_{i\rm L} T_i} \begin{pmatrix} u \\ d \end{pmatrix}_{\rm L} ; \quad \begin{pmatrix} u \\ d \end{pmatrix}_{\rm R} \to e^{i a_{i\rm R} T_i} \begin{pmatrix} u \\ d \end{pmatrix}_{\rm R} ; \tag{121}$$

where $T_i = (\tau_i, 1)$.

The global $U(2)_{\rm L} \times U(2)_{\rm R}$ approximate symmetry of QCD, arising from the fact that $m_u, m_d \ll \Lambda_{\rm QCD}$, is actually only a symmetry at the classical level. At the quantum level, there is an Adler-Bell-Jackiw (Adler, Bell and Jackiw 1969) anomaly in a $U(1)_{\rm R-L}$ subgroup of this symmetry and the real approximate global symmetry of QCD is reduced to

$$G = SU(2)_{\rm R+L} \times SU(2)_{\rm R-L} \times U(1)_{\rm R+L} \equiv SU(2)_V \times SU(2)_A \times U(1)_B . \tag{122}$$

Only $SU(2)_V$ and $U(1)_B$, however, are manifest symmetries of nature. The $SU(2)_A$ symmetry is spontaneously broken by the formation of u and d quark condensates, due to the QCD dynamics (see, for example, Donoghue et al 1992)

$$\langle \bar{u}u \rangle = \langle \bar{d}d \rangle \neq 0 . \tag{123}$$

The manifest $SU(2)_V$ symmetry is the well-known isospin symmetry of the strong interactions (Heisenberg 1932), leading to the approximate nucleon $N = (p, n)$ and pion $\pi = (\pi^\pm, \pi^0)$ multiplets. $U(1)_B$ corresponds to baryon number and its existence as a good symmetry guarantees that nucleons and antinucleons have the same mass. The spontaneously broken $SU(2)_A$ symmetry leads to the appearance of three Nambu-Goldstone bosons, which are identified as the pions. Indeed, one can show that (see, for example, Peccei 1987)

$$m_\pi^2 \to 0 \ \text{ as } \ m_{u,d} \to 0 . \tag{124}$$

[10] The strange quark mass $m_s \sim \Lambda_{\rm QCD}$ may also be neglected in some circumstances, leading to a larger $SU(3) \times SU(3)$ global symmetry.

Although $SU(2)_V \times SU(2)_A$ are only **approximate** symmetries of QCD, valid if we neglect m_u and m_d in the QCD Lagrangian, $U(1)_B$ is actually an exact global symmetry of the theory corresponding to the transformation

$$q_i \to \exp\left[\frac{i}{3}\alpha_B\right] q_i \ . \tag{125}$$

This transformation, since it affects all quarks equally, is also clearly a symmetry of the electroweak theory. Indeed, since all interactions always involve $q-\bar{q}$ pairs, it follows immediately that

$$\mathcal{L}_{\rm SM} \stackrel{U(1)_B}{\longrightarrow} \mathcal{L}_{\rm SM} \ , \tag{126}$$

with the associated conserved current being given by

$$J_B^\mu = \frac{1}{3}\sum_i \bar{q}_i \gamma^\mu q_i \ . \tag{127}$$

Precisely the same argument can be made for leptons, since again all interactions in the Standard Model always involve a lepton-antilepton pair. Whence, one has

$$\mathcal{L}_{\rm SM} \stackrel{U(1)_L}{\longrightarrow} \mathcal{L}_{\rm SM} \ , \tag{128}$$

with

$$J_{\rm L}^\mu = \sum_i \bar{\ell}_i \gamma^\mu \ell_i \tag{129}$$

being the corresponding conserved current.

At the quantum level, however, it turns out that neither $U(1)_{\rm L}$ or $U(1)_{\rm B}$ are good symmetries, because of the chiral nature of the weak interactions. Because the left-handed fields under the $SU(2) \times U(1)$ Standard Model group behave differently than the right-handed fields, effectively in the electroweak theory both $J_{\rm B}^\mu$ and $J_{\rm L}^\mu$ feel corresponding ABJ anomalies ('t Hooft 1976a). As we shall see, the breaking of $U(1)_{\rm B}$ and $U(1)_{\rm L}$ by these anomalies is the same. Hence, in the electroweak theory, at the quantum level, there remains only one true global quantum symmetry, $U(1)_{\rm B-L}$:

$$\mathcal{L}_{\rm SM} \stackrel{U(1)_{\rm B-L}}{\longrightarrow} \mathcal{L}_{\rm SM} \ . \tag{130}$$

We shall soon discuss these matters in some detail. However, before doing so, let me remark that the electroweak theory has actually a larger set of global symmetries if the neutrino masses vanish ($m_{\nu_i} = 0$).[11] In this case, each **individual** lepton number (L_e, L_μ and L_τ) is separately conserved at the classical level, while, say, $3L_e - B$, $3L_\mu - B$, $3L_\tau - B$ are conserved at the quantum level.

If one includes right-handed neutrinos in the standard model, so that $m_{\nu_i} \neq 0$, then one expects in general neutrino mixing, much as in the quark case.

[11] Theoretically, this is simply achieved by not including any right-handed neutrino fields $\nu_{\rm R_i}$ in the theory.

One knows, however, experimentally that neutrino masses, if they exist at all are very light (Particle Data Group 1996)—typically with masses in the eV range. With such light neutrino masses, effectively the Standard Model produces extremely small lepton flavor violations. For instance, one knows experimentally that (Particle Data Group 1996)

$$BR(\mu \to e\gamma) < 5 \times 10^{-11} \, . \tag{131}$$

Such a transition can occur at the one-loop level in the SM, but its ratio is extremely suppressed due to the tiny neutrino masses (Pal and Wolfenstein 1982). Typically, one finds

$$BR(\mu \to e\gamma) \sim \frac{\alpha G_F \sin\theta_\nu (m_{\nu_1}^2 - m_{\nu_2}^2)}{M_W^2} \sim 10^{-24} \, . \tag{132}$$

Here θ_ν is a neutrino mixing angle and the numerical result corresponds to taking $\sin\theta_\nu \sim 10^{-1}$ and $\Delta m_\nu^2 \sim (\mathrm{eV})^2$.

2.1 Chiral Anomalies

The existence of chiral anomalies (Adler, Bell and Jackiw 1969) has important consequences for the Standard Model. Anomalies, as we shall see, alter the classical global symmetry structure of the model. In addition, they bring into play the gauge field strength structure

$$F_a^{\mu\nu} \tilde{F}_{a\mu\nu} = \frac{1}{2} \epsilon^{\mu\nu\alpha\beta} F_{a\alpha\beta} F_{a\mu\nu} \, . \tag{133}$$

This structure is C even, but is both P and T odd. Hence, it can provide additional sources of CP violation. In the Standard Model, it does so through the, so-called, $\bar{\theta}$-term effective interaction

$$\mathcal{L}_{\text{CP viol.}} = \bar{\theta} \frac{\alpha_3}{8\pi} G_a^{\mu\nu} \tilde{G}_{a\mu\nu} \, , \tag{134}$$

where $G_a^{\mu\nu}$ is the gluon field strength for QCD and α_3 is the corresponding (squared) coupling constant $[\alpha_3 = g_3^2/4\pi]$.

For pedagogical reasons, it is important to sketch the *raison d'etre* for chiral anomalies. This is done best in the simple example provided by a theory which has a single fermion field ψ and a $U(1)_V \times U(1)_A$ global symmetry. In such a theory, at the classical (Lagrangian) level there are two conserved currents

$$J_V^\mu = \bar{\psi}\gamma^\mu\psi \quad \text{with} \quad \partial_\mu J_V^\mu = 0 \tag{135}$$

and

$$J_A^\mu = \bar{\psi}\gamma^\mu\gamma_5\psi \quad \text{with} \quad \partial_\mu J_A^\mu = 2m\bar{\psi}i\gamma_5\psi \stackrel{m\to 0}{\longrightarrow} 0 \, . \tag{136}$$

That is, the chiral $U(1)_A$ symmetry obtains if the fermion ψ is massless. At the quantum level, however, it is not possible to preserve **both** the conservation laws

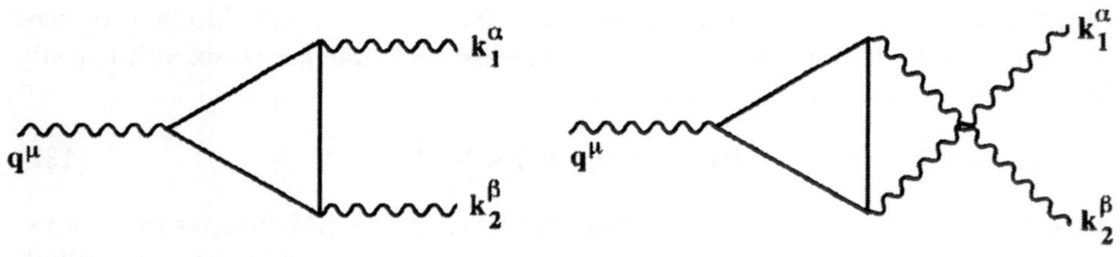

Fig. 1. Triangle graphs contributing to the AVV anomaly

for J_A^μ and J_V^μ. This is the origin of the chiral anomaly (Adler, Bell and Jackiw 1969).

More specifically, the source of the anomaly is the singular behavior of the triangle graph (shown in Fig. 1) involving one axial current J_A^μ and two vector currents J_V^μ. The individual graphs in Fig. 1 are each logarithmic divergent. However, their sum is finite. One can write the Green's function for two vector currents J_V^μ and an axial current as (Adler 1970)

$$T^{\mu\alpha\beta} = F(q^2, k_1^2, k_2^2) P^{\mu\alpha\beta}(k_1, k_2) \; . \tag{137}$$

The pseudotensor $P^{\mu\alpha\beta}(k_1, k_2)$ by Bose symmetry obeys

$$P^{\mu\alpha\beta}(k_1, k_2) = P^{\mu\beta\alpha}(k_2, k_1) \; . \tag{138}$$

Further, the conservation of the vector currents imposes the constraints

$$k_{1\alpha} P^{\mu\alpha\beta}(k_1, k_2) = k_{2\beta} P^{\mu\alpha\beta}(k_1, k_2) = 0 \; . \tag{139}$$

The above equations imply a unique structure for the pseudotensor $P^{\mu\alpha\beta}(k_1, k_2)$, namely

$$P^{\mu\alpha\beta}(k_1, k_2) = \epsilon^{\alpha\beta\rho\sigma} k_{1\rho} k_{2\sigma} q^\mu \; . \tag{140}$$

Because of the momentum factors in $P^{\mu\alpha\beta}(k_1, k_2)$, it follows that the invariant function $F(q^2, k_1^2, k_2^2)$ is indeed finite.

Given the above, imagine regularizing the triangle graphs in Fig. 1 via a Pauli-Villars regularization, to make each of the individual graphs finite (Adler 1970). Denoting the graphs in Fig. 1, respectively, by $t^{\mu\alpha\beta}(k_1, k_2)$ and $t^{\mu\beta\alpha}(k_2, k_1)$ this procedure yields for $T^{\mu\alpha\beta}$ the expression

$$\begin{aligned} T^{\mu\alpha\beta} &= \epsilon^{\alpha\beta\rho\sigma} k_{1\rho} k_{2\sigma} q^\mu F(q^2, k_1^2, k_2^2) \\ &= \left[t^{\mu\alpha\beta}(k_1, k_2) \big|_m - t^{\mu\alpha\beta}(k_1, k_2) \big|_M \right] \\ &\quad + \left[t^{\mu\beta\alpha}(k_2, k_1) \big|_m - t^{\mu\beta\alpha}(k_2, k_1) \big|_M \right] \; . \end{aligned} \tag{141}$$

Here M is the Pauli-Villars regularization mass. Taking the divergence of the above and setting the fermion mass $m \to 0$ yields the expression

$$q_\mu T^{\mu\alpha\beta} = -2iMP^{\alpha\beta}(M) \; . \tag{142}$$

Here the pseudoscalar structure $P^{\alpha\beta}(M)$ involves similar graphs to those in Fig. 1, except that the axial vertex is proportional to γ_5 and not $\gamma^\mu\gamma_5$.

Because the function $F(q^2, k_1^2, k_2^2)$ is finite, one knows that the Pauli-Villars regularization is really irrelevant and that one can therefore let $M \to \infty$. By straightforward calculation (Adler 1970) one finds that

$$\lim_{M\to\infty} -2iMP^{\alpha\beta}(M) = \frac{i}{2\pi^2}\epsilon^{\alpha\beta\rho\sigma}k_{1\rho}k_{2\sigma} \; . \tag{143}$$

Hence, one deduces the Adler-Bell-Jackiw anomalous divergence equation (Adler, Bell and Jackiw 1969)

$$q_\mu T^{\mu\alpha\beta} = \frac{i}{2\pi^2}\epsilon^{\alpha\beta\rho\sigma}k_{1\rho}k_{2\sigma} \; . \tag{144}$$

The anomalous Ward identity for $T^{\mu\alpha\beta}$ above can be interpreted in terms of an effective violation of the conservation equation for the axial current J^μ_A. Because the $U(1)_V$ gauge bosons – "photons" – couple to J^α_V and J^β_V, it is easy to show that Eq. (144) is equivalent to the anomalous divergence equation

$$\partial_\mu J^\mu_A = \frac{e^2}{8\pi^2}F_{\alpha\beta}\tilde{F}^{\alpha\beta} = \frac{\alpha}{2\pi}F_{\alpha\beta}\tilde{F}^{\alpha\beta} \; , \tag{145}$$

where e is the $U(1)_V$ coupling constant. The above is the famous Adler-Bell-Jackiw chiral anomaly (Adler, Bell and Jackiw 1969).

The above result, whose derivation we sketched for the $U(1)_V \times U(1)_A$ theory, can easily be generalized to the case where the fields in the current J^μ_A carry some non-Abelian charge. In this case the fermions in the anomalous triangle graphs carry some non-Abelian index and the graph, instead of simply involving e^2, now contains a factor of

$$g^2 \; \mathrm{Tr}\, \frac{\lambda_a}{2}\frac{\lambda_a}{2} = \frac{1}{2}g^2\delta_{ab} \; . \tag{146}$$

Here g is the coupling constant associated to the non-Abelian group and $\lambda_a/2$ is the appropriate generator matrix for the fermion fields, assuming they transform according to the fundamental representation of the non-Abelian group. It follows, therefore, that in the non-Abelian case the chiral anomaly (145) is replaced by

$$\partial_\mu J^\mu_A = \frac{g^2}{16\pi^2}F^{\alpha\beta}_a\tilde{F}_{a\alpha\beta} = \frac{\alpha_g^2}{4\pi}F^{\alpha\beta}_a\tilde{F}_{a\alpha\beta} \; , \tag{147}$$

where $F^{\alpha\beta}_a$ are the field strengths for the non-Abelian gauge bosons.

One can use the above results to analyze the Baryon (B) and Lepton (L) number currents in the Standard Model ('t Hooft 1976a). These currents, as we mentioned earlier, are conserved at the Lagrangian level. Decomposing these currents into chiral components, one has

$$J^\mu_{\rm B} = \frac{1}{3}\sum_i \bar{q}_i\gamma^\mu q_i = \frac{1}{3}\sum_i (\bar{q}_{i\rm L}\gamma^\mu q_{i\rm L} + \bar{q}_{i\rm R}\gamma^\mu q_{i\rm R})$$
$$J^\mu_{\rm L} = \sum_i \bar{\ell}_i\gamma^\mu \ell_i = \sum_i (\bar{\ell}_{i\rm L}\gamma^\mu \ell_{i\rm L} + \bar{\ell}_{i\rm R}\gamma^\mu \ell_{i\rm R}) \ . \tag{148}$$

Because the quarks and leptons interact with the $SU(2)\times U(1)$ electroweak fields the divergence of $J^\mu_{\rm B}$ and $J^\mu_{\rm L}$ will not vanish, as a result of the chiral anomalies. A straightforward computation of the relevant triangle graphs gives

$$\partial_\mu J^\mu_{\rm B} = -\frac{\alpha_2}{8\pi}N_g W_i^{\mu\nu}\tilde{W}_{i\mu\nu} + \frac{\alpha_1}{8\pi}N_g\left(\frac{4}{9}+\frac{1}{9}-\frac{1}{18}\right)Y^{\alpha\beta}\tilde{Y}_{\alpha\beta} \tag{149}$$

and

$$\partial_\mu J^\mu_{\rm L} = -\frac{\alpha_2 N_g}{8\pi}W_i^{\mu\nu}\tilde{W}_{i\mu\nu} + \frac{\alpha_1}{8\pi}N_g\left(1-\frac{1}{2}\right)Y^{\alpha\beta}\tilde{Y}_{\alpha\beta} \ . \tag{150}$$

In the above, N_g is the number of generations. The various numbers in front of the contributions involving the $U(1)$ gauge bosons contain the squares of the appropriate hypercharges, multiplied by the corresponding number of states [e.g. $u_{\rm R}$ contributes a factor of 4/9, while the doublet $(u,d)_{\rm L}$ contributes a factor of $2\times 1/36$]. Note that for the Baryon number current and for the Lepton number current, not only the $SU(2)$ but also the $U(1)$ factors are the same [(4/9 + 1/9-1/18) = (1-1/2) = 1/2]. It follows therefore that, as advertised, the total fermion number B+L is broken at the quantum level, but B-L is conserved:

$$\partial_\mu J^\mu_{\rm B+L} = \frac{\alpha_1^2}{8\pi}N_g Y^{\alpha\beta}\tilde{Y}_{\alpha\beta} - \frac{\alpha_2^2}{4\pi}N_g W_i^{\alpha\beta}\tilde{W}_{i\alpha\beta}$$
$$\partial_\mu J^\mu_{\rm B-L} = 0 \ . \tag{151}$$

A similar situation obtains in QCD. In the limit as $m_u, m_d \to 0$, this theory has a global symmetry at the classical level of $SU(2)_V\times SU(2)_A\times U(1)_V\times U(1)_A$. However, the $U(1)_A$ current

$$J^\mu_5 = \frac{1}{2}[\bar{u}\gamma^\mu\gamma_5 u + \bar{d}\gamma^\mu\gamma_5 d] \tag{152}$$

has a chiral anomaly, since the quarks carry color and interact with the gluons. Taking into account the contribution of both the u and d quarks in the triangle graph, one finds

$$\partial_\mu J^\mu_5 = \frac{\alpha_3^2}{4\pi}G_a^{\alpha\beta}\tilde{G}_{a\alpha\beta} \ . \tag{153}$$

The violation of the (B+L)-current in the electroweak theory and of the $U(1)_A$ current in QCD, codified by Eqs. (151) and (153), have a similar aspect. Nevertheless, these quantum corrections are quite different physically in their import. As we shall see, the current J^μ_5 is really badly broken by the above quantum QCD effects. As a result, as we mentioned earlier, the classical $U(1)_A$ symmetry is never a good (approximate) symmetry of the strong interactions. In contrast, $J^\mu_{\rm B+L}$ is extraordinarily weakly broken by the quantum corrections, except in the early Universe where temperature-dependent effects enhance these contributions. Thus, at zero temperature, the total fermion number (B+L) is essentially conserved.

Physically, these two results are what is needed. The formation of u and d-quark condensates

$$\langle \bar{u}u \rangle = \langle \bar{d}d \rangle \neq 0 \tag{154}$$

in QCD clearly breaks both the $SU(2)_A$ and $U(1)_A$ symmetries spontaneously. If $U(1)_A$ were really a symmetry, one would expect to have an associated Nambu-Goldstone boson—the η—with similar properties to the $SU(2)_A$ Nambu-Goldstone bosons—the π mesons. Although these states are supposed to be massless when the respective global symmetries are exact, both states should get similar masses once one includes quark mass terms for the u and d quarks (Weinberg 1975). However, experimentally, one finds $m_\eta^2 \gg m_\pi^2$ and one concludes that $U(1)_A$ cannot really be a true symmetry of QCD. Thus the strong breaking of J^μ_A by the anomaly is a welcome result.

In contrast, for the electroweak theory it is important that the anomalous breaking of (B+L) should not physically lead to large effects, since one has very strong experimental bounds on baryon number violation. For instance the B-violating decay $p \to e^+\pi^0$ has a bound (Particle Data Group 1996)

$$\tau(p \to e^+\pi^0) > 5.5 \times 10^{32} \text{ years} . \tag{155}$$

To understand why the anomaly contribution in Eq. (153) connected to the $U(1)_A$ current is important, while the anomaly contribution in Eq. (151) connected to the (B+L) current is irrelevant, requires an examination of the properties of the gauge theory vacuum. We turn to this next.

2.2 The Gauge Theory Vacuum

The resolution of the above issues came through a better understanding of the vacuum structure of gauge theories ('t Hooft 1976b and Polyakov 1977). The vacuum state is, by definition, a state where all fields vanish. For gauge fields, this needs to be slightly extended since these fields themselves are not physical. So, in the case of gauge fields, the vacuum state is one where either $A^\mu_a = 0$ or the gauge fields are a gauge transformation of $A^\mu_a = 0$. For our purposes it suffices to examine an $SU(2)$ gauge theory, since this example serves to exemplify what happens in a more general case.

It proves particularly convenient (Callan, Dashen and Gross 1976) to study the $SU(2)$ gauge theory in a temporal gauge where $A^0_a = 0$ $\{a = 1, 2, 3\}$. In this gauge the space components of the gauge fields are time-independent $A^i_a(\mathbf{r}, t) = A^i_a(\mathbf{r})$. Even so, there is still some residual gauge freedom. Defining a gauge matrix $A^i(\mathbf{r})$ by contracting the gauge fields with the Pauli matrices, $A^i(\mathbf{r}) =$

$\frac{\tau_a}{2}A^i_a(\mathbf{r})$, in the $A^0_a = 0$ gauge one is left with the freedom to perform the following gauge transformations

$$A^i(\mathbf{r}) \to \Omega(\mathbf{r})A^i(\mathbf{r})\Omega(\mathbf{r})^{-1} + \frac{i}{g}\Omega(\mathbf{r})\nabla^i\Omega(\mathbf{r})^{-1} , \tag{156}$$

where g is the gauge coupling for the $SU(2)$ theory. In view of the above, one concludes that in the $A^0_a = 0$ gauge, pure gauge fields corresponding to the vacuum configuration are simply the set $\{0, \frac{i}{g}\Omega(\mathbf{r})\nabla^i\Omega(\mathbf{r})^{-1}\}$.

The behavior of $\Omega(\mathbf{r})$ as $\mathbf{r} \to \infty$ distinguishes classes of pure gauge fields. In particular, the requirement that (Callan, Dashen and Gross 1976)

$$\Omega(\mathbf{r}) \overset{\mathbf{r}\to\infty}{\longrightarrow} 1, \tag{157}$$

provides a map of physical space $[S_3]$ onto the group space $[SU(2) \sim S_3]$. This $S_3 \to S_3$ map splits the matrices $\Omega(\mathbf{r})$ into different homotopy classes $\{\Omega_n(\mathbf{r})\}$, characterized by an integer n—the winding number—specifying how $\Omega(\mathbf{r})$ goes to unity at spatial infinity:

$$\Omega_n(\mathbf{r}) \overset{\mathbf{r}\to\infty}{\longrightarrow} e^{2\pi i n} . \tag{158}$$

Thus the set of pure gauge fields is $\{0, A^i_n(\mathbf{r})\}$, where

$$A^i_n(\mathbf{r}) = \frac{i}{g}\Omega_n(\mathbf{r})\nabla^i\Omega_n(\mathbf{r})^{-1} . \tag{159}$$

The winding number n is just the Jacobian of the $S_3 \to S_3$ transformation (Crewther 1978) and one can show that

$$n = \frac{ig^3}{24\pi^2}\int d^3r \,\mathrm{Tr}\, \epsilon_{ijk}A^i_n(\mathbf{r})A^j_n(\mathbf{r})A^k_n(\mathbf{r}) . \tag{160}$$

Furthermore, one can construct the transformation matrix $\Omega_n(\mathbf{r})$ with winding number n by compounding n-times the transformation matrix of unit winding

$$\Omega_n(\mathbf{r}) = [\Omega_1(\mathbf{r})]^n . \tag{161}$$

A representative $n = 1$ matrix, giving rise to a, so called, **large gauge transformation** is given by

$$\Omega_1(\mathbf{r}) = \frac{\mathbf{r}^2 - \lambda^2}{\mathbf{r}^2 + \lambda^2} + \frac{2i\lambda\boldsymbol{\tau}\cdot\mathbf{r}}{\mathbf{r}^2 + \lambda^2} , \tag{162}$$

with λ an arbitrary scale parameter.

Using the above properties, it is clear that the n-vacuum state—corresponding to the pure gauge field configuration $A^i_n(\mathbf{r})$—is not fully gauge invariant. Indeed, a large gauge transformation can change the gauge field $A^i_n(\mathbf{r})$ into that of $A^i_{n+1}(\mathbf{r})$

$$A^i_{n+1}(\mathbf{r}) = \Omega_1(\mathbf{r})A^i_n(\mathbf{r})\Omega_1^{-1}(\mathbf{r}) + \frac{i}{g}\Omega_1(\mathbf{r})\nabla^i\Omega_1^{-1}(\mathbf{r}) , \tag{163}$$

or

$$\Omega_1|n\rangle = |n+1\rangle \ . \tag{164}$$

The correct vacuum state for a gauge theory must be gauge invariant. As such it must be a linear superposition of these n-vacuum states. This is the, so-called, θ-vacuum ('t Hooft 1976b and Polyakov 1977)

$$|\theta\rangle = \sum_n e^{-in\theta}|n\rangle \ . \tag{165}$$

Clearly, since

$$\Omega_1|\theta\rangle = \sum_n e^{-in\theta}\Omega_1|n\rangle = \sum_n e^{-in\theta}|n+1\rangle = e^{i\theta}|\theta\rangle \ , \tag{166}$$

the $|\theta\rangle$ vacuum is gauge invariant.

Using the θ-vacuum as the correct vacuum state for gauge theories, it is clear that the vacuum functional for these theories splits into distinct sectors (Callan, Dashen and Gross 1976). If $|\theta\rangle_\pm$ are the θ-vacuum states at $t = \pm\infty$, then the vacuum functional for a gauge theory takes the form

$$\begin{aligned} {}_+\langle\theta|\theta\rangle_- &= \sum_{n,m} e^{im\theta}e^{-in\theta}\ {}_+\langle m|n\rangle_- \\ &= \sum_\nu e^{i\nu\theta}\left[\sum_n {}_+\langle n+\nu|n\rangle_-\right] \ . \end{aligned} \tag{167}$$

That is, the vacuum functional sums over vacuum to vacuum amplitudes in which the winding numbers at $t = \pm\infty$ differ by ν, weighing each by a factor $e^{i\nu\theta}$. We anticipate here that the superposition of amplitudes with **different** phases $e^{i\nu\theta}$ will lead to CP-violating effects. Recalling that the vacuum functional is given by a path integral over gauge field configurations, each weighted by the classical action, one arrives at the formula

$${}_+\langle\theta|\theta\rangle_- = \int_{\text{Paths}} \delta A_\mu e^{iS[A]} = \sum_\nu e^{i\nu\theta}\left[\sum_\nu {}_+\langle n+\nu|n\rangle_-\right] \ . \tag{168}$$

Although the formula for ${}_+\langle\theta|\theta\rangle_-$ above was derived in the A_a^0 gauge, the parameter ν entering in this formula has actually a gauge invariant meaning. One finds ('t Hooft 1976b and Polyakov 1977)

$$\nu = n_+ - n_- = \frac{g^2}{32\pi^2}\int d^4x G_a^{\mu\nu}\tilde{G}_{a\mu\nu} \ . \tag{169}$$

To prove this result requires using Bardeen's identity (Bardeen 1972) which expresses the product of $G\tilde{G}$ as a total derivative:

$$G_a^{\mu\nu}\tilde{G}_{a\mu\nu} = \partial_\mu K^\mu \ , \tag{170}$$

where the "current" K^μ is given by

$$K^\mu = \epsilon^{\mu\alpha\beta\gamma} A_{a\alpha} \left[G_{a\beta\gamma} - \frac{g}{3} \epsilon_{abc} A_{b\beta} A_{c\gamma} \right] . \tag{171}$$

For pure gauge fields $[G_{a\beta\gamma} = 0]$ and in the $A_a^0 = 0$ gauge this current has only a temporal component:

$$K^i = 0; \; K^0 = -\frac{g}{3} \epsilon_{ijk} \epsilon_{abc} A_a^i A_b^j A_c^k = \frac{4}{3} i g \epsilon_{ijk} \operatorname{Tr} A^i A^j A^k . \tag{172}$$

Using these relations, in this gauge one can write the winding numbers $n_\pm$ as

$$n_\pm = \frac{ig^3}{24\pi^2} \int d^3 r \epsilon_{ijk} \operatorname{Tr} A^i A^j A^k = \frac{g^2}{32\pi^2} \int d^3 r K^0 |_{t=\pm\infty} . \tag{173}$$

The above formula allows one to express the winding number difference $\nu = n_+ - n_-$ as

$$\nu = n_+ - n_- = \frac{g^3}{32\pi^2} \int d^3 r K^0 |_{t=-\infty}^{t=+\infty} = \frac{g^2}{32\pi^2} \int d\sigma_\mu K^\mu . \tag{174}$$

Whence, Eq. (169) follows by using Gauss's theorem and Bardeen's identity.

Having identified ν as an integral over $G\tilde{G}$, one can rewrite the formula for the vacuum functional in terms of an effective action. Defining

$$S_{\text{eff}}[A] = S[A] + \theta \frac{g^2}{32\pi^2} \int d^4 x G_a^{\mu\nu} \tilde{G}_{a\mu\nu} , \tag{175}$$

one sees that

$$_+\langle\theta|\theta\rangle_- = \sum_\nu \int_{\text{Paths}} \delta A_\mu e^{iS_{\text{eff}}[A]} \delta \left[\nu - \frac{g^2}{32\pi^2} \int d^4 x G_a^{\mu\nu} \tilde{G}_{a\mu\nu} \right] . \tag{176}$$

The more complicated structure of the gauge theory vacuum [θ-vacuum] effectively adds an additional term to the gauge theory Lagrangian:

$$\mathcal{L}_{\text{eff}} = \mathcal{L}_{\text{gauge theory}} + \theta \frac{g^2}{32\pi^2} G_a^{\mu\nu} \tilde{G}_{a\mu\nu} . \tag{177}$$

Perturbation theory is connected to the $\nu = 0$ sector, since $\int d^4 x G\tilde{G} = 0$. Effects of non-zero winding number differences ($\nu \neq 0$) involve **non-perturbative** contributions. These are naturally selected by the connection of the pseudoscalar density $G\tilde{G}$ with the divergence of chiral currents, through the chiral anomaly (Adler, Bell and Jackiw 1969).

Let me examine this first for QCD. Assuming there are n_f flavors whose mass can be neglected ($m_f = 0$), the axial current in QCD

$$J_5^\mu = \frac{1}{2} \sum_{i=1}^{n_f} \bar{q}_i \gamma^\mu \gamma_5 q_i \tag{178}$$

is still not conserved as a result of the chiral anomaly. One has

$$\partial_\mu J_5^\mu = n_f \frac{g_3^2}{32\pi^2} G_a^{\mu\nu} \tilde{G}_{a\mu\nu} \,. \tag{179}$$

In view of the above, chirality changes ΔQ_5, are simply related to ν:

$$\Delta Q_5 = \int d^4x \partial_\mu J_5^\mu = n_f \frac{g_3^2}{32\pi^2} \int d^4x G_a^{\mu\nu} \tilde{G}_{a\mu\nu} = n_f \nu \,. \tag{180}$$

Clearly, if $\nu \neq 0$ sectors are important in QCD, then the above changes are important and the corresponding $U(1)_A$ symmetry is **never** a symmetry of the theory. This then is the physical explanation why (in the relevant $n_f = 2$ case) the η does not have the properties of a Goldstone boson.

't Hooft ('t Hooft 1976c), by using semiclassical methods, provided an estimate of the likelihood of the occurrence of processes involving $\nu \neq 0$ transitions. Basically, he viewed the transition from an n-vacuum at $t = -\infty$ to an $(n + \nu)$-vacuum at $t = +\infty$ as a tunneling process and estimated the tunneling probability by WKB methods. 't Hooft's result ('t Hooft 1976c)

$$A[\nu] \sim e^{-S_E[\nu]} \tag{181}$$

uses as the WKB factor in the exponent the minimal Euclidean action for the gauge theory. Such a minimal action obtains if the gauge field configurations are those provided by instantons (Belavin et al 1975). These are self-dual solutions of the field equations in Euclidean space $[G_a^{\mu\nu} = \tilde{G}_a^{\mu\nu}]$ and their action is simply related to ν. For these solutions

$$S_E[\nu] = \frac{1}{4} \int d^4x_E G_a^{\mu\nu} G_a^{\mu\nu} = \frac{1}{4} \int d^4x_E G_a^{\mu\nu} \tilde{G}_a^{\mu\nu} = \frac{8\pi^2}{g_3^2} \nu \,. \tag{182}$$

What 't Hooft showed in his careful calculation ('t Hooft 1976c) is that the coupling constant that enters in $S_E[\nu]$ is actually a running coupling, with its scale set by the scale of the instanton solution involved. Further, to evaluate the amplitude in question one must integrate over all such scales. Thus, schematically, 't Hooft's result is

$$A[\nu] \sim \int d\rho \, \exp\left[-\frac{2\pi\nu}{\alpha_3(\rho^{-1})}\right] \,. \tag{183}$$

In QCD, since the gauge coupling squared $\alpha_3(\rho^{-1})$ grows for large distances, there is no particular suppression due to the tunneling factor for large size instantons. Because of this, although one cannot really calculate $A[\nu]$, one expects that

$$A[\nu \neq 0] \sim A[0] \,. \tag{184}$$

Thus, as advertized, $U(1)_A$ is not really a symmetry of QCD.

Much of the above discussion applies to the electroweak theory. However, as we shall see, there is a crucial difference. Since the electroweak theory is based on the group $SU(2) \times U(1)$, because of the $SU(2)$ factor there is also here a non-trivial vacuum structure. The $W\tilde{W}$ density connected to the index difference in this case is directly related to the divergence of the B+L current. Focusing on this contribution, one has

$$\partial_\mu J^\mu_{\rm B+L} = -\frac{g_2^2}{16\pi^2} N_g W_i^{\mu\nu} \tilde{W}_{i\mu\nu} \,. \tag{185}$$

Hence, the change in (B+L) in the electroweak theory is also simply connected to the (weak) index ν ('t Hooft 1976a)

$$\Delta(B+L) = \int d^4x \partial_\mu J^\mu_{\rm B+L} = -\frac{g_2^2}{16\pi^2} N_g \int d^4x W_i^{\mu\nu} \tilde{W}_{i\mu\nu} = -2N_g\nu \,. \tag{186}$$

I note that for three generations [$N_g = 3$] the minimal violation of the (B+L)-current is $|\Delta(\mathrm{B+L})| = 6$. So, even though baryon number is violated in the Standard Model the process $p \to e^+\pi^0$, which involves $\Delta(\mathrm{B+L}) = 2$, is still forbidden! More importantly, however, the amplitude for (B+L)-violation itself is totally negligible. This amplitude, at least semiclassically, will again be given by a result similar to what was obtained in QCD (except with $\alpha_3 \to \alpha_2$). However, because the electroweak symmetry is broken, the integration over instanton sizes cuts off at sizes of order $1/v$ (or momentum scales of order M_Z). Hence, one estimates ('t Hooft 1976a)[12]

$$A[\nu]_{\rm (B+L)-violation} \sim \exp\left[-\frac{2\pi\nu}{\alpha_2(M_Z)}\right] \sim 10^{-80\nu} \,. \tag{187}$$

I want to remark that, although the above result is negligibly small, in the early Universe (B+L)-violation in the electroweak theory can be important. This was first observed by Kuzmin, Rubakov, and Shaposhnikov (Kuzmin et al 1985), who pointed out that in a thermal bath the semiclassical estimate of 't Hooft ceases to be accurate. Effectively, in these circumstances, the gauge configurations associated with (B+L)-violating processes are not governed by a tunnelling factor, but by a Boltzman factor. As one nears the electroweak phase transitions, furthermore, this Boltzman factor tends to unity and the (B+L)-violating processes proceed essentially unsuppressed.

2.3 The Strong CP Problem

The θ-vacuum of QCD is a new source of CP-violation,[13] as a result of the effective interaction

$$\mathcal{L}_{\rm CP-violation} = \theta \frac{\alpha_3}{8\pi} G_a^{\mu\nu} \tilde{G}_{a\mu\nu} \,, \tag{188}$$

---

[12] Here we use $\alpha_2(M_Z) = \frac{\alpha(M_Z)}{\sin^2\theta_W} \sim \frac{1}{30}$.

[13] One can show that the equivalent θ-parameter in the electroweak theory can be rotated away as a result of the chiral nature of these interactions (Krasnikov et al 1978).

which reflects the presence of the vacuum angle. It turns out, in fact, that the situation is a little bit more complicated, because of the electroweak interactions. Recall that the quark mass matrices arising as a result of the spontaneous breakdown of $SU(2) \times U(1)$ are, in general, neither Hermitian nor diagonal

$$\mathcal{L}_{\text{mass}} = -\bar{q}_{\text{L}i} M_{ij} q_{\text{R}j} - \bar{q}_{\text{R}i} (M^\dagger)_{ij} q_{Lj} \ . \tag{189}$$

These matrices can, however, be diagonalized by performing appropriate unitary transformations on the quark fields

$$q_{\text{R}} \to q'_{\text{R}} = U_{\text{R}} q_{\text{R}} \ ; \quad q_{\text{L}} \to q'_{\text{L}} = U_{\text{L}} q_{\text{L}} \ . \tag{190}$$

It is easy to check that part of the above transformations involves a $U(1)_A$ transformation. In fact, the $U(1)_A$ piece of these transformations is just

$$\begin{aligned} q_{\text{R}} \to q'_{\text{R}} &= \exp\left[\frac{i}{2n_f} \text{ Arg det } M\right] q_{\text{R}} \equiv \exp\left[\frac{i}{2}\alpha\right] q_{\text{R}} \\ q_{\text{L}} \to q'_{\text{L}} &= \exp\left[-\frac{i}{2n_f} \text{ Arg det } M\right] q_{\text{L}} \equiv \exp\left[-\frac{i}{2}\alpha\right] q_{\text{L}} \ . \end{aligned} \tag{191}$$

It turns out that such $U(1)_A$ transformations engender a change in the vacuum angle (Jackiw and Rebbi 1976). Thus they effectively add a contribution to Eq. (188), beyond that of the QCD angle θ.

To prove this contention (Jackiw and Rebbi 1976), one has to examine carefully what is the result of a chiral $U(1)_A$ transformation. Although the current J_5^μ connected to $U(1)_A$ has an anomaly, it is always possible to construct a conserved current by using the current K^μ which enters in Bardeen's identity (Bardeen 1972). Recalling Eqs. (170) and (179), it is obvious that the desired conserved chiral current $\tilde{J}_5^\mu$ is

$$\tilde{J}_5^\mu = J_5^\mu - \frac{n_f \alpha_3}{4\pi} K^\mu \ . \tag{192}$$

The charge which generates chiral transformations, $\tilde{Q}_5$, needs to be time-independent. By necessity, it must therefore be related to $\tilde{J}_5^\mu$—the conserved current:

$$\tilde{Q}_5 = \int d^3x \tilde{J}_5^0 \ . \tag{193}$$

Although $\tilde{Q}_5$ is time-independent, this charge is not invariant under large gauge transformations, since K^μ is itself not a gauge-invariant current like J_5^μ. One finds

$$\Omega_1 \tilde{Q}_5 \Omega_1 = \Omega_1 \left[Q_5 - \frac{n_f \alpha_3}{4\pi} \int d^3x K^0\right] \Omega_1 = \tilde{Q}_5 + n_f \ . \tag{194}$$

Consider the action of a large gauge transformation Ω_1 on a chirally rotated θ-vacuum state $e^{i\alpha\tilde{Q}_5}|\theta\rangle$. One has

$$\begin{aligned} \Omega_1 \left[e^{i\alpha\tilde{Q}_5}|\theta\rangle\right] &= \Omega_1 e^{i\alpha\tilde{Q}_5} \Omega_1^{-1} \Omega_1 |\theta\rangle \\ &= e^{i(\alpha n_f + \theta)} \left[e^{i\alpha\tilde{Q}_5}|\theta\rangle\right] \ . \end{aligned} \tag{195}$$

It follows from the above, immediately, that a chiral $U(1)_A$ rotation indeed shifts the vacuum angle (Jackiw and Rebbi 1976):

$$e^{i\alpha\tilde{Q}_5}|\theta\rangle = |\theta + \alpha n_f\rangle \ . \tag{196}$$

For the electroweak theory, the chiral rotation one needs to perform to diagonalize the quark mass matrices has a parameter $\alpha = \frac{1}{n_f}\det M$. Whence, it follows that the effective CP-violating Lagrangian term arising from the structure of the gauge theory vacuum is

$$\mathcal{L}^{\text{eff}}_{\text{CP-violation}} = \bar{\theta}\frac{\alpha_3}{8\pi}G^{\mu\nu}_a\tilde{G}_{a\mu\nu} \ , \tag{197}$$

where

$$\bar{\theta} = \theta + \text{Arg det } M \ . \tag{198}$$

The effective CP-violating parameter $\bar{\theta}$ is the sum of a QCD contribution—the vacuum angle θ—and an electroweak piece–Arg det M—related to the phase structure of the quark mass matrix.

The interaction (197) is C even, and T and P odd. Thus it violates CP also. It turns out, as we shall see below, that unless $\bar{\theta}$ is very small [$\bar{\theta} \leq 10^{-10}$] this interaction produces an electric dipole moment for the neutron which is beyond the present experimental bound for this quantity. It is difficult to understand why a parameter like $\bar{\theta}$, which is a sum of two very different contributions, should be so small. This conundrum is known as the strong CP problem.

Before discussing the strong CP problem further, let me first indicate how to calculate the contribution of the effective Lagrangian (197) to the electric dipole moment of the neutron. This is most easily done by transforming the $\bar{\theta}$ interaction from an interaction involving gluons to one involving quarks. For simplicity, let me concentrate on the two-flavor case ($n_f = 2$) and take, again for simplicity, $m_u = m_d = m_q$. In this case, it is easy to see that the chiral $U(1)_A$ transformation

$$\begin{pmatrix} u \\ d \end{pmatrix} \to \exp\left[i\frac{\bar{\theta}\gamma_5}{4}\right]\begin{pmatrix} u \\ d \end{pmatrix} \tag{199}$$

will get rid of the $\bar{\theta}G\tilde{G}$ term. However, the above transformation will, at the same time, generate a CP-violating γ_5-dependent mass term for the u and d quarks:

$$\mathcal{L}^{\text{eff}}_{\text{CP-violation}} = i\bar{\theta}m_q\left[\bar{u}\frac{\gamma_5}{2}u + \bar{d}\frac{\gamma_5}{2}d\right] \ . \tag{200}$$

One can use the above effective Lagrangian directly to calculate the neutron electric dipole moment. One has, in general

$$d_n\bar{n}\sigma_{\mu\nu}k^\nu\gamma_5 n = \langle n|T(J^{\text{em}}_\mu i\int d^4x\mathcal{L}^{\text{eff}}_{\text{CP-violation}})|n\rangle \ . \tag{201}$$

To arrive at a result for d_n one inserts a complete set of states $|X\rangle$ in the matrix element above and tries to estimate which set of states $|X\rangle$ dominates. In the literature there are two calculations along these lines. Baluni (Baluni 1979) uses for $|X\rangle$ the odd parity $|N^-_{1/2}\rangle$ states which are coupled to the neutron by $\mathcal{L}^{\rm eff}_{\rm CP-violation}$. Crewther *et al.* (Crewther et al 1979), instead, do a soft pion calculation (effectively $|X\rangle \sim |N\pi_{\rm soft}\rangle$). The result of these calculations are rather similar and lead to an expression for d_n whose form could have been guessed. Namely

$$d_n \sim \frac{e}{M_n}\left(\frac{m_q}{M_n}\right)\bar{\theta} \sim \begin{cases} 2.7 \times 10^{-16}\,\bar{\theta}\ \text{ecm (Baluni 1979)} \\ 5.2 \times 10^{-16}\,\bar{\theta}\ \text{ecm (Crewther et al 1979)} \end{cases} \tag{202}$$

The present bound on d_n (Particle Data Group 1996) is, at 95% C.L.,

$$d_n < 1.1 \times 10^{-25}\ \text{ecm}\ . \tag{203}$$

Whence, to avoid contradictions with experiment, the parameter $\bar{\theta}$ must be less than 2×10^{-10}. Why this should be so is a mystery. This is the strong CP problem.

2.4 The Chiral Solution to the Strong CP Problem

About twenty years ago, Helen Quinn and I (Peccei and Quinn 1977) suggested a possible dynamical solution to the strong CP problem. If our mechanism holds in nature then $\bar{\theta}$ actually vanishes, and there is no need to explain a small number like 10^{-10} cropping up in the theory.[14] To "solve" the strong CP problem, Quinn and I postulated that the Lagrangian of the Standard Model was invariant under an additional global $U(1)$ chiral symmetry—$U(1)_{\rm PQ}$. This required imposing certain constraints on the Higgs sector of the theory, but otherwise appeared perfectly possible. Because the $U(1)_{\rm PQ}$ symmetry is a chiral symmetry, if this symmetry were exact, it is trivial to see that the $\bar{\theta}G\tilde{G}$ term can be eliminated, since the chiral rotation $\exp\left[-i\frac{\bar{\theta}}{n_f}\tilde{Q}_5^{\rm PQ}\right]$ gives

$$\exp\left[-i\frac{\bar{\theta}}{n_f}\tilde{Q}_5^{\rm PQ}\right]|\bar{\theta}\rangle = |0\rangle\ . \tag{204}$$

That is, by a $U(1)_{\rm PQ}$ transformation the effective vacuum angle $\bar{\theta}$ is set to zero and this parameter is no longer present in the theory. Physically, however, if $U(1)_{\rm PQ}$ is an extra global symmetry of the Standard Model, it is not possible for this symmetry to remain unbroken. What Quinn and I showed (Peccei and Quinn 1977) was that, even if $U(1)_{\rm PQ}$ is spontaneously broken, one still is able to eliminate the $\bar{\theta}G\tilde{G}$ term.

To see this, it is useful to focus on the associated Nambu-Goldstone boson resulting from the spontaneous breakdown of the $U(1)_{\rm PQ}$ symmetry. This excitation is the axion, first discussed by Weinberg and Wilczek (Weinberg and

[14] Even incorporating a $U(1)_{\rm PQ}$ symmetry into the theory it turns out that CP violating effects in the electroweak interactions do not allow $\bar{\theta}$ to totally vanish. However, the effective $\bar{\theta}$ induced back through weak CP-violation is tiny ($\bar{\theta} \sim 10^{-15}$) (Georgi et al 1986) and well within the bound provided by the neutron electric dipole moment.

Wilczek 1978) in connection with the $U(1)_{PQ}$ symmetry. It turns out that the axion is not quite massless, so it is really a pseudo-Goldstone boson (Weinberg 1972). This is a consequence of the $U(1)_{PQ}$ symmetry having an anomaly due to QCD interactions. One finds (Weinberg and Wilczek 1978) that the axion mass is of order

$$m_a \sim \frac{\Lambda_{QCD}^2}{f} , \tag{205}$$

where Λ_{QCD} typifies the scale of the QCD interactions, while f is the scale of the $U(1)_{PQ}$ breakdown. If $f \gg \Lambda_{QCD}$, then axions turn out to be very much lighter than ordinary hadrons.

If we denote the axion field by $a(x)$, it turns out that imposing a $U(1)_{PQ}$ symmetry on the standard model effectively serves to replace the CP-violating $\bar{\theta}$ parameter by the dynamical CP-conserving axion field:

$$\bar{\theta} \to \frac{a(x)}{f} . \tag{206}$$

To understand why this is so, recall that since the axion is the Nambu-Goldstone boson of the broken $U(1)_{PQ}$ symmetry, this field translates under a $U(1)_{PQ}$ transformation:

$$a(x) \overset{U(1)_{PQ}}{\longrightarrow} a(x) + \alpha f , \tag{207}$$

where α is the parameter associated with the $U(1)_{PQ}$ transformation. Because of Eq. (207), the axion field can only enter in the Lagrangian of the theory through derivative terms. Even though the detailed axion interactions are somewhat model-dependent, this property allows one to understand how to augment the Lagrangian of the Standard Model so that it becomes $U(1)_{PQ}$ invariant.

Focusing only on the possible additional contributions due to the inclusion of the axion field, one is led to the following effective Lagrangian for the theory

$$\begin{aligned} \mathcal{L}_{SM}^{eff} = \mathcal{L}_{SM} + \bar{\theta}\frac{\alpha_3}{8\pi}G_a^{\mu\nu}\tilde{G}_{a\mu\nu} - \frac{1}{2}\partial_\mu a \partial^\mu a \\ + \mathcal{L}_{axion}^{int}\left[\frac{\partial_\mu a}{f} ; \psi\right] + \frac{a}{f}\xi\frac{\alpha_3}{8\pi}G_a^{\mu\nu}\tilde{G}_{a\mu\nu} . \end{aligned} \tag{208}$$

The third term above is the kinetic energy term for the axion field, while the fourth term in Eq. (208) schematically indicates the kind of interactions the axion field can participate in with the other fields $[\psi]$ in the theory. The last term above, as can be noticed, does not involve a derivative of the axion field, thereby violating the usual expectations for Nambu-Goldstone fields. The reason why this term is included, however, is clear. The $U(1)_{PQ}$ symmetry is anomalous[15]

$$\partial_\mu J_{PQ}^\mu = \xi\frac{\alpha_3}{8\pi}G_a^{\mu\nu}\tilde{G}_{a\mu\nu} . \tag{209}$$

This anomaly must be reflected in the effective Lagrangian (208) when one performs a chiral $U(1)_{PQ}$ transformation. This is guaranteed by having the last

[15] Here ξ is a model-independent number of $O(1)$ (see, for example, Peccei 1989).

term in Eq. (208), since it precisely reproduces the anomaly when the axion field undergoes the $U(1)_{\rm PQ}$ transformation (207).

The last term in Eq. (208), whose origin is intimately connected to the chiral anomaly, because it contains the axion field directly (and not its derivative) provides a potential for the axion field. As a result, it is not true anymore that all values of the vacuum expectation value (VEV) of $a(x)$ are allowed.[16] The minimum of $V_{\rm eff}$ in the vacuum is simply

$$\left\langle \frac{\partial V_{\rm eff}}{\partial_a} \right\rangle = -\frac{\xi}{f}\frac{\alpha_3}{8\pi}\langle G_a^{\mu\nu}\tilde{G}_{a\mu\nu}\rangle\,|_{\langle a\rangle\neq 0} \ . \tag{210}$$

What Quinn and I showed (Peccei and Quinn 1977), in essence, is that the periodicity of $\langle G\tilde{G}\rangle$ in the effective vacuum angle $\theta_{\rm eff}$ for the Lagrangian of Eq. (208)

$$\theta_{\rm eff} = \bar{\theta} + \frac{\xi}{f}\langle a(x)\rangle \ , \tag{211}$$

requires that $\theta_{\rm eff} = 0$, or

$$\langle a(x)\rangle = -\frac{f}{\xi}\bar{\theta} \ . \tag{212}$$

As a result of Eq. (212), only the physical axion field

$$a(x)_{\rm phy} = a(x) - \langle a(x)\rangle \tag{213}$$

interacts with the gluon field strengths, eliminating altogether the $\theta G\tilde{G}$ term. Thus, indeed, imposing an additional $U(1)_{\rm PQ}$ symmetry in the Standard Model, even in the case this symmetry is spontaneously broken, solves the strong CP problem.

As we remarked earlier, the axion is actually massive because of the anomaly in the $U(1)_{\rm PQ}$ current. This follows readily from the effective Lagrangian (208). The second derivative of the effective potential $V_{\rm eff}$, which arose precisely because of the chiral anomaly in the $U(1)_{\rm PQ}$ symmetry, when evaluated at its minimum value $\langle a(x)\rangle$ gives for the axion mass squared the value

$$m_a^2 = \left\langle \frac{\partial^2 V_{\rm eff}}{\partial a^2} \right\rangle\Big|_{\langle a\rangle} = -\frac{\xi}{f}\frac{\alpha_3}{8\pi}\frac{\partial}{\partial a}\,\langle G_a^{\mu\nu}\tilde{G}_{a\mu\nu}\rangle\Big|_{\langle a\rangle} \sim \frac{\Lambda_{\rm QCD}^2}{f} \ . \tag{214}$$

Using the above results, it is clear that the effective theory incorporating $U(1)_{\rm PQ}$ and axions no longer suffers from the strong CP problem. All that remains as a signal of this erstwhile problem is the direct interaction of the (massive) axion field with the gluonic pseudoscalar density.

$$\begin{aligned}\mathcal{L}_{\rm SM}^{\rm eff} &= \mathcal{L}_{\rm SM} + \mathcal{L}_{\rm axion}^{\rm int}\left[\frac{\partial_\mu a_{\rm phys}}{f}\ ;\ \psi\right] - \frac{1}{2}\partial_\mu a_{\rm phys}\partial^\mu a_{\rm phys} \\ &\quad - \frac{1}{2}m_a^2 a_{\rm phys}^2 + \frac{a_{\rm phys}}{f}\xi\frac{\alpha_3}{8\pi}G_a^{\mu\nu}\tilde{G}_{a\mu\nu} \ .\end{aligned} \tag{215}$$

[16] This would be true if $\mathcal{L}_{\rm SM}^{\rm eff}$ only contained interactions involving $\partial_\mu a$, since these cannot fix a value for the VEV of a, $\langle a\rangle$.

As is obvious from the above equation, the physics of axions depends on the scale of $U(1)_{\rm PQ}$ breaking f. In the original model Helen Quinn and I put forth (Peccei and Quinn 1977), we associated f quite naturally with the scale of electroweak symmetry breaking $v = (\sqrt{2}\,G_F)^{-1/2}$. To impose the $U(1)_{\rm PQ}$ symmetry on the Standard Model we had to have two distinct Higgs doublets, Φ_1 and Φ_2, with different $U(1)_{\rm PQ}$ charges. The axion field then turns out to be the common phase field of Φ_1 and Φ_2 which is orthogonal to the weak hypercharge (Peccei 1989). Isolating just this contribution in Φ_1 and Φ_2, one has

$$\Phi_1 = \frac{v_1}{\sqrt{2}} \exp\left[ix\frac{a}{f}\right]\begin{pmatrix}1\\0\end{pmatrix} \; ; \; \Phi_2 = \frac{v_2}{\sqrt{2}} \exp\left[i\frac{a}{xf}\right]\begin{pmatrix}0\\1\end{pmatrix} . \tag{216}$$

Here $x = v_2/v_1$, is the ratio of the two Higgs VEV's and the $U(1)_{\rm PQ}$ symmetry breaking scale f is given by

$$f = \sqrt{v_1^2 + v_2^2} = (\sqrt{2}\,G_F)^{-1/2} \simeq 250 \text{ GeV} . \tag{217}$$

The Φ_1 field has weak hypercharge of $-1/2$, while the Φ_2 field has weak hypercharge of $+1/2$. Hence, in the Yukawa interactions Φ_1 couples the $u_{{\rm R}j}$ fields to the left-handed quark doublets, while Φ_2 couples $d_{{\rm R}j}$ to these same fields

$$\mathcal{L}_{\rm Yukawa} = -\Gamma^u_{ij}(\bar{u},\bar{d})_{{\rm L}i}\Phi_1 u_{{\rm R}j} - \Gamma^d_{ij}(\bar{u},\bar{d})_{{\rm L}i}\Phi_2 d_{{\rm R}j} + \text{h.c.} \tag{218}$$

In view of Eq. (216), it is clear that the above interaction is $U(1)_{\rm PQ}$ invariant. The shift of the axion field by αf [cf Eq. (207)] under a $U(1)_{\rm PQ}$ transformation is compensated by an appropriate rotation of the right-handed quark fields. Specifically, under a $U(1)_{\rm PQ}$ transformation one has

$$\begin{aligned} a_{\rm phys} &\xrightarrow{PQ} a_{\rm phys} + \alpha f \\ u_{{\rm R}j} &\xrightarrow{PQ} \exp[-i\alpha x]\, u_{{\rm R}j} \\ d_{{\rm R}j} &\xrightarrow{PQ} \exp\left[-i\frac{\alpha}{x}\right] d_{{\rm R}j} . \end{aligned} \tag{219}$$

It is clear from the above that this $U(1)_{\rm PQ}$ transformation encompasses also a $U(1)_A$ transformation. As a result, one can use $U(1)_{\rm PQ}$ to send $\bar{\theta} \to 0$, as advertized.

Unfortunately, weak interaction scale axions [with $f \sim 250$ GeV; $m_a \sim 100$ keV] of the type which ensue in the model suggested by Helen Quinn and myself, or in variations thereof, have been ruled out experimentally. I do not want to review all the relevant data here, as this is done already fully elsewhere (Peccei 1989). An example, however, will give a sense of the strength of this assertion. If weak scale axions were to exist, one expects a rather sizable branching ratio for the decay $K^\pm \to \pi^\pm a$ (Bardeen et al 1987)

$$BR(K^\pm \to \pi^\pm a) \sim 3 \times 10^{-5} . \tag{220}$$

Experimentally, however, the process $K^+ \to \pi^+$ "Nothing", which would reflect the axion decay of the K^+ meson, has a bound roughly three orders of magnitude lower (Asano et al 1981)

$$BR(K^+ \to \pi^+ + \text{Nothing}) < 3.8 \times 10^{-8} . \tag{221}$$

One can bypass this bound by modifying the $U(1)_{\rm PQ}$ properties of the Higgs fields involved. However, these variant model themselves run into other experimental troubles (Peccei 1989).

Although weak scale axions do not exist, it is still possible that the strong CP problem is solved because of the existence of a $U(1)_{\rm PQ}$ symmetry. The dynamical adjustment of $\bar{\theta} \to 0$ works **independently** of what is the scale, f, of the spontaneous symmetry breaking of $U(1)_{\rm PQ}$. Obviously if $f \gg (\sqrt{2}\, G_F)^{-1/2}$, the resulting axions are extremely light ($m_a \sim \Lambda^2_{\rm QCD}/f$), extremely weakly coupled (couplings $\sim f^{-1}$) and very long lived ($\tau_a \sim f^5$) and thus are essentially **invisible**. A variety of invisible axion models have been suggested in the literature (Kim et al 1979) and they offer an interesting, if perhaps unconventional, resolution of the strong CP problem. Fortunately, as we shall see, these models are actually testable.

If $f \gg (\sqrt{2}\, G_F)^{-1/2}$, it is clear that the spontaneous breakdown of $U(1)_{\rm PQ}$ must occur through a VEV of a field which is an $SU(2) \times U(1)$ singlet. Thus, in invisible axion models, the axion is essentially the phase associated with an $SU(2) \times U(1)$ singlet field σ.[17] Keeping only the axion degrees of freedom, one has

$$\sigma = \frac{f}{\sqrt{2}} e^{ia/f} . \tag{222}$$

It turns out that astrophysics and cosmology give important constraints on the $U(1)_{\rm PQ}$ breaking scale f, or equivalently the axion mass (Peccei 1989)

$$m_a \simeq 6 \left[\frac{10^6 \text{ GeV}}{f}\right] \text{eV} . \tag{223}$$

These constraints restrict the available parameter space for invisible axion models and suggest ways in which these excitations, if they exist, could be detected. Let me briefly discuss these matters.

The astrophysical bounds on axions arise because, if f is not large enough, axion emission removes energy from stars, altering their evolution. These bounds are reviewed in great details in a recent monograph by Raffelt (Raffelt 1996). Although these bounds are somewhat dependent on the type of invisible axion model one is considering, typically invisible axions avoid all astrophysical constraints if

$$f \geq 5 \times 10^9 \text{ GeV} ; \quad m_a \leq 10^{-3} \text{ eV} . \tag{224}$$

Cosmology, on the other hand, provides an upper bound on f (Preskill et al 1983). At the $U(1)_{\rm PQ}$ phase transition in the early Universe, at temperatures

[17] The field σ need not necessarily be an elementary scalar field (Kim 1979).

$T \sim f$, the effects of the QCD anomaly are not yet felt and the axion vacuum expectation value $\langle a \rangle$ is not aligned dynamically to cancel the $\bar{\theta}$ term. This cancellation only occurs as the Universe cools towards temperatures T of order $T \sim \Lambda_{\rm QCD}$. The axion VEV $\langle a \rangle$, as the temperature decreases, is driven to the correct minimum in an oscillatory fashion. These coherent, zero momentum, axion oscillations contribute to the Universe's energy density. If f is too large, in fact, the energy density due to axions can overclose the Universe. Demanding that this not happen gives a bound (Preskill et al 1983):

$$f \leq 10^{12}\ \mathrm{GeV}\ ;\quad m_a \geq 6 \times 10^{-6}\ \mathrm{eV}\ . \tag{225}$$

This bound has some uncertainties, related to cosmology (for a discussion see, for example, Peccei 1996), but otherwise is not very dependent on the properties of the invisible axions themselves.

If axions contribute substantially to the Universe's energy density, the value of f (or m_a) will be close to the above bound. If this is the case, axions could be the source for the dark matter in the Universe. Remarkably, then, it may be actually possible, experimentally, to detect signals for these invisible axions. The basic idea, due to Sikivie (Sikivie 1983), is to try to convert axions, trapped in the galactic halo, into photons in a laboratory magnetic field.

If invisible axions constitute the dark matter of our galactic halo, they would have a velocity typical of the virial velocity in the galaxy, $v_a \sim 10^{-3}$c. Further, as the dominant components of the energy density of the Universe, axions would have a typical energy density in the halo of order

$$\rho_a^{\rm halo} \sim 5 \times 10^{-25}\ \mathrm{g/cm}^3 \sim 300\ \mathrm{MeV/cm}^3\ . \tag{226}$$

As a result of the (electromagnetic) anomaly, axions have an interaction with the electromagnetic field given by the effective Lagrangian (Peccei 1989)

$$\mathcal{L}^{\rm eff}_{a\gamma\gamma} = \frac{\alpha}{\pi f} K_{a\gamma\gamma}\, a\, \mathbf{E}\cdot\mathbf{B}\ . \tag{227}$$

Here $K_{a\gamma\gamma}$ is a model dependent parameter of $O(1)$. As a result of the above interaction, in the presence of an external magnetic field a galactic axion can convert into a photon.

Specifically, the electric field produced by an axion of energy $E_a \simeq m_a$ in the presence of a magnetic field $\mathbf{B}_0$ can be deduced from the modified wave equation

$$\left(\boldsymbol{\nabla}^2 - \frac{\partial^2}{\partial t^2}\right)\mathbf{E} = \frac{\alpha}{\pi f} K_{a\gamma\gamma} \mathbf{B}_o \frac{\partial^2 a}{\partial t^2}\ . \tag{228}$$

Experimentally, the generated electromagnetic energy can be detected by means of a resonant cavity. When the cavity is tuned to the axion frequency $w_a \simeq m_a$, one should get a narrow line on top of the noise spectrum. On resonance, the axion to photon conversion power is given by the expression (Sikivie 1983)

$$P_{\rm axion} = \frac{\rho_a}{m_a}\cdot V B_o^2 \cdot \left[\frac{\alpha}{\pi f} K_{a\gamma\gamma}\right]^2 C_{\rm overlap}\cdot Q_{\rm eff}\ . \tag{229}$$

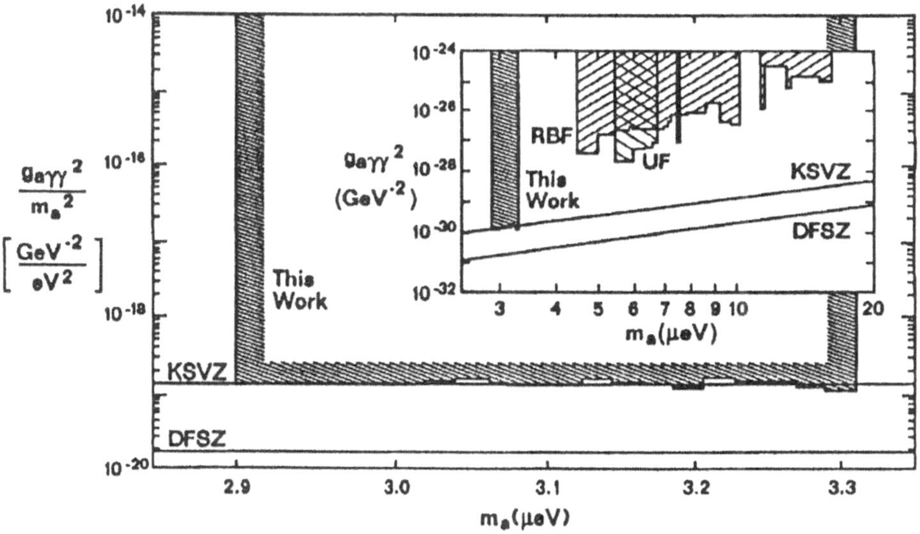

Fig. 2. Result of the Livermore experiment, along with limits from some previous axion searches

In the above, the first factor gives the expected number of axions per unit volume, the second details the magnetic energy stored in the cavity, the third contains the coupling strength squared, $g^2_{a\gamma\gamma} = \left[\frac{\alpha}{\pi f} K_{a\gamma\gamma}\right]^2$. Finally, $C_{\text{overlap}} \simeq 0.7$ is an effectiveness factor for the cavity and Q_{eff} is the least value between the Q of the cavity itself $[Q \sim 10^6]$ and the Q due to the energy spread in the spectrum of halo axion, $Q_a \simeq [v_a^2/c^2]^{-1} \sim 10^6$.

Halo axions produce microwave photons, since 4×10^{-6} eV $\equiv 1$ GHz. Two pilot experiments carried out in the late 80's (de Panfilis et al 1987) had limited magnetic energy $[VB_o^2 \simeq 0.5 \text{ m}^3 \text{ (Tesla)}^2]$ and relatively noisy amplifiers. These experiments set limits for $g^2_{a\gamma\gamma}$ about 2 to 3 orders of magnitude above the theoretical expectations. Presently, there are two second generation experiments underway, one at the Lawrence Livermore National Laboratory and the other in Kyoto. The Livermore experiment uses a very large $VB_o^2 \sim 12 \text{ m}^3 \text{ (Tesla)}^2$ and low noise "state-of-the-art" amplifiers. Although the signal expected at 1 GHz is tiny, $P_{\text{axion}} \sim 5 \times 10^{-22}$ Watts, this experiment has already excluded a set of invisible axion masses, at the level of strength expected theoretically. These recent results (Hagmann et al 1998), along with some of the older data are shown in Fig. 2. The Kyoto experiment (Matsuki et al 1991) uses a moderate $VB_o^2 \sim 0.2 \, m^3 \text{ (Tesla)}^2$. However, it utilizes an extremely clever technique for counting the number of photons converted from axions—using Rydberg atoms—

which makes up for the small VB_o^2. The Kyoto experiment is presently in a testing phase. One hopes that when both the Livermore and Kyoto experiments are completed, in 3-5 years time, they will have settled the important question of whether axions exist or not.

2.5 Do Real Nambu-Goldstone Bosons Exist?

We have known for almost 40 years that when a global symmetry group G breaks down spontaneously to a subgroup H $[G \to H]$, dim G/H massless Nambu-Goldstone bosons (Nambu 1980 and Goldstone 1981) appear in the spectrum of the theory. However, we have no **real** physical examples still of this phenomenon. To be fair, pions are an excellent example of states which are **nearly** Nambu-Goldstone bosons. However, although there is no question that pions are the Nambu-Goldstone excitations associated with the breakdown of the $SU(2)_V \times SU(2)_A$ approximate global symmetry of QCD to $SU(2)_V$, pions have a small mass since the u and d quarks are not exactly massless.

For a while, it was believed that it was impossible for real physical Nambu-Goldstone bosons to exist in nature. The argument was simple. Because these particles are massless, their existence seemed to be precluded by the fact that the only long-range forces we know in nature are gravity and electromagnetism. However, in the early 1980's it was realized that the existence of $m = 0$ Nambu-Goldstone bosons does not pose a contradiction, so that one can actually contemplate the interesting possibility that such states may actually exist.

This idea came up first as a result of studying the possibility that lepton number may be spontaneously violated. Chikashige, Mohapatra, and I (Chikashige, Mohapatra and Peccei 1981a) dubbed the Nambu-Goldstone boson associated with the spontaneous breakdown of lepton number a Majoron. Soon thereafter, others (Wilczek 1982) considered theories where one had a global family number which could also be spontaneously broken, resulting in other types of real Nambu-Goldstone bosons, given the name of Familons.

In this subsection I want to explain briefly why, in general, real Nambu-Goldstone bosons, are not dangerous excitations to have in a theory. After having done so, I then want to discuss briefly a specific type of Majoron model, to illustrate some of the consequences of these kind of models. Succinctly, the reason why Nambu-Goldstone bosons do not run afoul with present limits on possible additional long-range forces is due to a little theorem of Gelmini, Nussinov, and Yanagida (Gelmini, Nussinov and Yanagida 1983), which shows that the exchange of Nambu-Goldstone bosons leads only to a long-range tensor force.

The proof of the Gelmini-Nussinov-Yanagida theorem is very simple. One is interested in the potential produced by the exchange of a Nambu-Goldstone boson between two fermions. Recall that Nambu-Goldstone boson fields, π, always shift under a broken symmetry transformation [cf Eq. (207) for the axion]. Therefore, one has

$$\pi(x) \xrightarrow{\xi} \pi(x) + v_\pi \xi \,. \tag{230}$$

Here ξ is a parameter in G/H and v_π is a scale parameter associated with the symmetry breakdown in question. As a result of Eq. (230), clearly Nambu-Goldstone fields must always be derivatively coupled. Hence, the most general coupling of a Nambu-Goldstone boson π to two fermions f_1 and f_2 takes the form

$$\mathcal{L}_{\rm NGB}^{\rm fermion} = i\frac{\partial_\mu \pi}{v_\pi}\bar{f}_1[a\gamma_\mu + b\gamma_\mu\gamma_5]f_2 + {\rm h.c.}\ , \tag{231}$$

where a and b are numerical coefficients. If one uses the fermion equations of motion, one can reduce the above to a more useful form involving the π field directly

$$\mathcal{L}_{\rm NGB}^{\rm fermion} = \frac{\pi}{v_\pi}\bar{f}_1[a(m_1 - m_2) + b(m_1 + m_2)\gamma_5]f_2 + {\rm h.c.}\ , \tag{232}$$

where m_1 and m_2 are the masses of the fermion fields f_1 and f_2, respectively.

In calculating the potential due to π-exchange between two fermions one needs, at each vertex, to use the interaction Lagrangian above with $f_1 = f_2$. Obviously, for two equal fermions, the effective coupling of a Nambu-Goldstone boson is always a **pseudoscalar** coupling. Thus Nambu-Goldstone boson exchange cannot really generate coherent long-range forces, since a pseudoscalar coupling in the non-relativistic limit reduces to a $\boldsymbol{\sigma}\cdot\mathbf{p}$ coupling. More precisely, the effective diagonal coupling of a Nambu-Goldstone boson, π, to a fermion, f, is given by

$$\mathcal{L}_{\rm NGB}^{\rm diag} = ig_\pi\frac{m_f}{v_\pi}\bar{f}\gamma_5 f\pi\ , \tag{233}$$

where g_π is a, dimensionless, coupling constant. In the non-relativistic limit, the above reduces to

$$\mathcal{L}_{\rm NGB}^{\rm diag} \to g_\pi\chi_f^*\frac{\boldsymbol{\sigma}\cdot\boldsymbol{\nabla}}{v_\pi}\chi_f\pi\ , \tag{234}$$

where χ_f is a Pauli spinor. Such an interaction gives an exchange potential between two fermions which is spin-dependent and tensorial, with an $1/r^3$ not an $1/r$ fall off

$$V_{\rm NGB-exchange}^{\rm eff} = \frac{g_\pi^2/4\pi}{v_\pi^2}\left\{\frac{\boldsymbol{\sigma}_1\cdot\boldsymbol{\sigma}_2 - 3(\boldsymbol{\sigma}_1\cdot\hat{r})(\boldsymbol{\sigma}_2\cdot\hat{r})}{r^3} + \frac{4\pi}{3}\delta^3(r)\boldsymbol{\sigma}_1\cdot\boldsymbol{\sigma}_2\right\}\ . \tag{235}$$

There have been analyses in the literature (Feinberg and Sucher 1979) of the size of possible non-magnetic dipole-dipole interactions in matter, precisely of the type one would obtain from the exchange of a real Nambu-Goldstone boson. These bounds effectively limit how small the scale v_π can be. One finds no contradiction with experiment (Chikashige, Mohapatra and Peccei 1981a) provided that

$$\frac{v_\pi}{g_\pi} \geq {\rm TeV}\ . \tag{236}$$

Thus, one can contemplate having real Nambu-Goldstone bosons of global symmetries which are broken down at scales not much bigger than the weak scale! If

v_π/g_π is much above the bound (236), clearly one expects no measurable effects in matter. Furthermore, if v_π/g_π is large, these Nambu-Goldstone bosons are also hard to directly produce, since the effective coupling for producing them from a fermion f scales like $g_\pi m_f/v_\pi$.

2.6 Majorons

I want to illustrate the above discussion by briefly considering the simplest example of spontaneously broken Lepton number and its associated Majoron. As we discussed earlier, Lepton number is a classical global symmetry of the Standard Model. Even at the quantum level, because the $\nu \neq 0$ amplitudes are highly suppressed, this remains an almost exact symmetry. However, there is no reason why Lepton number should remain a symmetry of the theory, once one considers extensions of the Standard Model. Indeed, the simplest extension of the Standard Model introduces right-handed neutrino fields ν_{Ri} for each family. Because these fields are $SU(2) \times U(1)$ singlets, one can write a Majorana (fermion-fermion) mass term involving these fields of the form

$$\mathcal{L}_{\text{mass}} = -\frac{(M_{\text{R}})_{ij}}{2} \nu_{\text{R}i}^T C \nu_{\text{R}j} + \text{h.c.} , \tag{237}$$

with C being the charge conjugation matrix introduced earlier ($C = 1$ in the Majorana representation). Obviously $\mathcal{L}_{\text{mass}}$ does not respect Lepton number, since its two terms carry Lepton number +2 and -2, respectively.

One can restore Lepton number as a symmetry in the above example by introducing an appropriately transforming Higgs field. In this case, what one needs is a complex $SU(2) \times U(1)$ singlet field σ, which carries Lepton number -2 (Chikashige, Mohapatra and Peccei 1981a). Clearly the interaction Lagrangian

$$\mathcal{L}_{\text{CMP}} = -\frac{h_{ij}}{\sqrt{2}} \left[\nu_{\text{R}i}^T C \nu_{\text{R}j} \sigma + \text{h.c.}\right] \tag{238}$$

is L invariant by construction. If the dynamics of the theory forces σ to acquire a VEV, $\langle\sigma\rangle = \frac{1}{\sqrt{2}} V$, then the above Lagrangian reproduces the effect of having an explicit Majorana mass term for the right-handed neutrino fields. In this case, one has

$$(M_{\text{R}})_{ij} = h_{ij} V , \tag{239}$$

and Lepton number is spontaneously broken. Hence this theory must also contain an explicit Nambu-Goldstone boson—the Majoron. This is the model which I first studied with Chikashige and Mohapatra (Chikashige, Mohapatra and Peccei 1981a).

As was the case for the axion, the Majoron can also be identified here as the phase field associated with the complex field σ. Focusing only on the Majoron, χ, degrees of freedom, one can write

$$\sigma \simeq \frac{V}{\sqrt{2}} e^{i\chi/V} . \tag{240}$$

If the interaction (238) was the only interaction that the $\nu_{\mathrm{R}i}$ fields had, then clearly χ would couple only to these fields. However, once one introduces right-handed neutrino fields, one cannot avoid coupling $\nu_{\mathrm{R}i}$ to the usual leptonic doublet fields $(\nu, e)_{\mathrm{L}i}$ via the ordinary Higgs doublet field. As a result of these couplings, the Majoron field χ also ends up by having a (small) interaction with the left-handed neutrino fields. However, if the right-handed Majorana mass M_{R} (or, equivalently, the VEV of the σ-field V) is large, the Majoron still predominantly couples to the right-handed neutrinos.

Let us see how this goes in detail. As a result of the spontaneous breaking of both Lepton number and $SU(2) \times U(1)$, the neutrino fields have both a Dirac (fermion-antifermion) and a Majorana mass term:

$$\mathcal{L}_{\text{mass}} = -\frac{1}{2}(M_{\mathrm{R}})_{ij}\left[\nu_{\mathrm{R}i}^T \nu_{\mathrm{R}j}\right] - (M_D)_{ij}[\bar{\nu}_{\mathrm{L}i}\nu_{\mathrm{R}j}] + \text{h.c.}\,, \tag{241}$$

with the Dirac mass matrix M_D being proportional to the doublet Higgs VEV.[18] If the eigenvalues of M_R are much greater than those of M_D, then the neutrino mass matrix

$$\mathcal{M} = \begin{pmatrix} 0 & M_{\mathrm{D}} \\ M_{\mathrm{D}} & M_{\mathrm{R}} \end{pmatrix} \tag{242}$$

has a set of large eigenvalues, corresponding to the eigenvalues of M_R, and a set of extremely small eigenvalues, associated with the matrix M_D^2/M_R. This is the famous see-saw mechanism (Yanagida et al 1979). As a result, one ends up with a spectrum of neutrinos with both superheavy states and superlight states:

$$\mathcal{L}_{\text{mass}} \simeq -\frac{1}{2}\left[\frac{M_{\mathrm{D}}^2}{M_{\mathrm{R}}}\right]_{ij} \bar{\eta}_{1i}\eta_{1j} - \frac{1}{2}[M_{\mathrm{R}}]_{ij}\bar{\eta}_{2i}\eta_{2j}\,. \tag{243}$$

The light neutrinos η_{1i} are mostly left-handed, while the heavy neutrinos η_{2i} are mostly right-handed.

The mass mixing discussed above, has a counterpart in the interactions of the Majoron. Although the field χ mostly couples to the heavy fields η_{2i}, there will also be a small coupling of χ to η_{1i}. That is, the Majoron χ as a result of the neutrino mass mixing actually has also a small coupling to the ordinary left-handed neutrinos. Specifically (Chikashige, Mohapatra and Peccei 1981a), one finds

$$\mathcal{L}_{\text{Majoron}}^{\text{int}} = -\frac{h_{ij}}{2}\bar{\eta}_{2i}i\gamma_5\eta_{2j}\chi - \left(\frac{hM_{\mathrm{D}}^2}{M_{\mathrm{R}}^2}\right)_{ij}\bar{\eta}_{1i}i\gamma_5\eta_{2j}\chi\,. \tag{244}$$

It follows that the Majoron coupling to the light neutrinos is of order $M_{\mathrm{D}}^2/M_{\mathrm{R}}^2 \sim m_\nu/M_{\mathrm{R}}$, where m_ν is the mass (matrix) for the light neutrinos. The Majoron has an even weaker coupling to ordinary matter, which is induced at one-loop

[18] Naively, one would expect M_D to be similar to the mass matrix M_ℓ for the charged leptons.

order via mixing of the χ with the Z^o. One finds (Chikashige, Mohapatra and Peccei 1981a)

$$\mathcal{L}^{\rm eff}_{\rm matter} = i\frac{m_f}{v_\chi}\bar{f}\gamma_5 f\chi \tag{245}$$

with the scale v_χ of order $v_\chi \sim (G_F m_\nu)^{-1} \gg$ TeV. So clearly the Majoron in this model easily satisfies the constraints imposed on additional dipole-dipole interactions in matter (Feinberg and Sucher 1979).

If Majorons exist, it is possible for the heaviest of the light neutrinos to decay into the other neutrinos by Majoron emission. The process $\nu_i \to \nu_j\chi$, if it were fast enough, would serve to open up a region of neutrino masses forbidden by cosmology. For stable neutrinos, one knows that neutrinos in the mass range from a few eV to a few GeV (Kolb and Turner 1990) overclose the Universe. However, these bounds cease to apply for unstable neutrinos. If the lifetime τ for the neutrino decay $\nu_i \to \nu_j\chi$ is much shorter than the Universe's lifetime T_o, then effectively one can redshift the ν_i energy beyond its mass m_{ν_i}. Hence the contribution of these neutrinos to the energy density of the Universe is reduced to (Chikashige, Mohapatra and Peccei 1981b)

$$\rho_{\nu_i} \sim m_{\nu_i}\left[\frac{\tau}{T_o}\right]^{1/2} T_\nu^3 \tag{246}$$

where T_ν is the neutrino temperature now, $T_\nu \sim T_\gamma \sim 3^o$ K.

The lifetime τ for the process $\nu_i \to \nu_j\chi$ naively scales as (Chikashige, Mohapatra and Peccei 1981b)

$$\tau(\nu_i \to \nu_j\chi) \sim \frac{1}{m_{\nu_i}}\left[\frac{M_{\rm R}}{M_{\rm D}}\right]^4 \tag{247}$$

and can be made short enough if $M_{\rm R}$ is not too large. However, in the simplest Majoron model discussed here (Chikashige, Mohapatra and Peccei 1981a) this lifetime is lengthened by a further factor of $\left[\frac{M_{\rm R}}{M_{\rm D}}\right]^4$ (Schechter and Valle 1984), making it very doubtful that $\tau < T_o$. More elaborate models (Gelmini and Roulet 1995) restore the simple formula (247) and the possibility that, through Majoron decay, neutrinos with masses in the "forbidden" cosmological range could exist. This is not entirely an academic exercise, as the existing bounds on m_{ν_μ} and m_{ν_τ} [$m_{\nu_\mu} \leq 170$ keV; $m_{\nu_\tau} < 24$ MeV (Particle Data Group 1996)] allow these particles to have masses precisely in this range.

2.7 Global Symmetries and Gravity

In the preceding subsections I have discussed various interesting global symmetries, which may be associated with the interactions of the Standard Model, and have explored a bit the consequences of these symmetries. There are, however, some arguments one can adduce from the analysis of gravitational interactions which bring into question the whole notion of having theories with exact global symmetries. I want to end my lectures by discussing this point briefly.

Perhaps the simplest way to see why gravitational interactions may cause trouble is to focus on the "No Hair" theorem for black holes. Basically this theorem (see, for example, Banks 1990) asserts that black holes can be characterized only by a few fundamental quantities, like mass and spin, but possess otherwise no other quantum numbers. Because black holes can absorb particles which carry global charge, while carrying no global charge themselves, it appears that through these processes one can get an explicit violation of whatever symmetry is associated with the global charge. That is, global charge can be lost when particles carrying this charge are swallowed by a black hole.

One can parametrize the effect of the breaking of global symmetries by gravitational interactions by adding to the low-energy Lagrangian non-renormalizable terms, scaled by inverse powers of the Planck mass $M_{\mathrm{P}} \sim 10^{19}$ GeV. These terms, of course, should be constructed so as to explicitly violate the symmetries in question. Schematically, therefore, the full Lagrangian of the theory, besides containing the usual Standard Model terms, should also include some effective non-renormalizable interactions containing various operators O_n, breaking explicitly the Standard Model global symmetries:

$$\mathcal{L}^{\mathrm{eff}}_{\mathrm{grav.\ int.}} = \sum_m \frac{1}{M_{\mathrm{P}}^n} O_n \ . \qquad (248)$$

Here the dimension of the operators O_n, which explicitly breaks some of the Standard Model global symmetries, is $n+4$.

Let me make two remarks. First, the Lagrangian (248) can often be augmented by other effective interactions which themselves break certain global symmetries even more strongly than gravity. For instance, an explicit mass term for the right-handed neutrinos [cf Eq. (237)] does break L directly and more strongly than the operators in Eq. (248) do. This said, however, in what follows I will concentrate only on the gravitational effects embodied in Eq. (248).

Because $M_{\mathrm{P}} \sim 10^{19}$ GeV $\gg (\sqrt{2}\, G_F)^{-1/2} \sim 250$ GeV, the naive expectation is that Eq. (248) cannot be that important, except at superheavy scales. This turns out to be true for the interactions themselves, but fails when one considers the effect of Eq. (248) on the Nambu-Goldstone sector. To demonstrate the first point, let me consider the example of (B+L)-violation. The dominant, $d=6$, (B+L)-violating interaction induced by gravity schematically has the form (Weinberg, Wilczek and Zee 1979)

$$\mathcal{L}_{(B+L)-violation} \sim \frac{1}{(M_{\mathrm{P}})^2} u_i^c d_j u_k^c e f_{ijk} \ . \qquad (249)$$

Such a term leads to a proton lifetime, for the process $p \to e^+\pi^0$, of order

$$\tau(p \to e^+\pi^0) \sim (M_{\mathrm{P}})^4 \sim 10^{46} \text{ years} \ , \qquad (250)$$

much greater than the present experimental bound on this process discussed earlier [Eq. (155)]. So the breaking of B+L provided through gravitational effects is indeed irrelevant.

The situation is, however, different when one considers the Nambu-Goldstone sector. Let us consider again the simple example of spontaneously broken Lepton number with its associated Majoron. To the Lepton number conserving potential, which forces the $SU(2)\times U(1)$ singlet field σ to acquire a VEV, one must now add non-renormalizable Lepton number violating terms induced by the gravitational interactions. The simplest such term involves a dimension 5 operator. Thus, one is invited to study the potential

$$V_{\text{total}} = \lambda\left(\sigma^\dagger\sigma - \frac{V^2}{2}\right)^2 - \frac{\lambda'}{M_P}(\sigma^\dagger\sigma)^2[\sigma+\sigma^\dagger]\,. \tag{251}$$

The first term above is clearly invariant under the Lepton number transformation $\sigma \to e^{-2ia}\sigma$. This is not so for the term which scales as M_P^{-1}. Writing, as before,

$$\sigma \simeq \frac{V}{\sqrt{2}}\exp\left[i\frac{\chi}{V}\right]\,, \tag{252}$$

one sees that the effect of including the gravitational corrections is to produce a mass term for the erstwhile Nambu-Goldstone field χ. One finds, for the Majoron, a mass

$$m_\chi^2 = \frac{\lambda'}{2\sqrt{2}}V^2\left(\frac{V}{M}\right)\,. \tag{253}$$

Note that the size of the Majoron mass depends on the value of V, the scale of the spontaneous breakdown of Lepton number. For example, if we took $V \sim$ TeV—the lowest it can be according to the bound of Eq. (236)—then $m_\chi \sim V(V/M_P)^{1/2} \sim 10^{-8}$ V $\simeq 10$ keV. If V is larger, the mass of the Majoron grows as

$$m_\chi \sim 10\left[\frac{V}{\text{TeV}}\right]^{3/2}\ \text{keV}\,. \tag{254}$$

Clearly, if the Majoron is massive, some of its physical properties are altered substantially. For instance, it could happen that the decay $\nu_i \to \nu_j\chi$ is actually kinematically forbidden! Of course, the above results are predicated on the assumption that the global Lepton number symmetry is violated explicitly by a dim 5 interaction. If the violation were due to a higher dimensional operator of dimension d, then one finds for the Majoron mass the formula

$$m_\chi \sim V\left(\frac{V}{M_P}\right)^{\frac{d-4}{2}} \tag{255}$$

which leads to masses which become smaller the larger d is.

These considerations are particularly troubling for the $U(1)_{PQ}$ solution to the strong CP problem (Holman et al 1992). Not only potentially do gravitational effects give an additional contribution to the axion mass, but they can also alter the QCD potential so that $\bar{\theta}$ does not finally adjust to zero! One can understand what is going on by schematically sketching the form of the effective axion potential in the absence and in the presence of the $U(1)_{\rm PQ}$ breaking gravitational interactions (Barr and Seckel 1992). Without gravity, a useful parametrization for the physical axion effective potential, which follows from examining the contributions of instantons (Peccei and Quinn 1977), is

$$V_{\rm axion} = -\Lambda^4_{\rm QCD} \cos a_{\rm phys}/f \ . \tag{256}$$

This potential displays the necessary periodicity in $a_{\rm phys}/f$, has a minimum at $\langle a_{\rm phys}\rangle = \bar{\theta}_{\rm eff} = 0$, and leads to an axion mass $m_a = \Lambda^2_{\rm QCD}/f$.

Including gravitational effects changes the above potential by adding a sequence of terms involving operators of different dimensions. Let us just consider one such term and examine the potential (Barr and Seckel 1992)

$$\tilde{V}_{\rm axion} = -\Lambda^4_{\rm QCD} \cos \frac{a_{\rm phys}}{f} - \frac{cf^d}{M_{\rm P}^{d-4}} \cos\left[\frac{a_{\rm phys}}{f} - \delta\right] \ . \tag{257}$$

Here c is some dimensionless constant and δ is a CP-violating phase which enters through the gravitational interactions. This potential modifies the formula for the axion mass, giving now

$$m_a^2 \simeq \frac{\Lambda^4_{\rm QCD}}{f^2} + c\frac{f^{d-2}}{M_{\rm P}^{d-4}} \ . \tag{258}$$

For f in the range of interest for invisible axions, the second term above coming from the gravitational effects dominates the QCD mass estimate for the axion, unless c is extraordinarily small and/or the dimension d is rather large. More troublesome still, $\tilde{V}_{\rm axion}$ now no larger has a minimum at $\langle a_{\rm phys}\rangle = 0$. Rather one finds a minimum of $\tilde{V}_{\rm axion}$ for values of

$$\bar{\theta}_{\rm eff} = \frac{\langle a_{\rm phys}\rangle}{f} \simeq c\sin\delta \frac{f^d}{M_{\rm P}^{d-4}\Lambda^4_{\rm QCD}} \ . \tag{259}$$

That is, the gravitational effects (provided there is a CP violating phase associated with them) induce a non-zero $\bar{\theta}$, even in the presence of a $U(1)_{\rm PQ}$ symmetry! To satisfy the bound $\bar{\theta} \leq 10^{-10}$ again necessitates that d be large and/or that the constant c be extraordinarily small.

To date there is no clear resolution to this problem and it could be that these considerations actually vitiate the chiral solution to the strong CP problem. Since this is the most appealing solution to this conundrum, this is somewhat troubling. Nevertheless, it is worth noting a number of points. First, one does not really understand quantum gravity. Thus it is possible that when matters are better understood the effective global symmetry breaking interactions we introduced may in fact not be there at all, or be tremendously suppressed. Second, there are some encouraging results in this direction coming from string theory. Axions associated with broken chiral symmetries arise very naturally in string theory (Witten 1984). Furthermore, CP is conserved, at least in higher dimensions in string theory (Choi, Kaplan and Nelson 1993), so perhaps it is possible that $\sin\delta = 0$. Finally, there are arguments that for large compactification radii, the effective $U(1)_{PQ}$ symmetries are broken very little in strings, so that the tiny number needed for c [$c \leq 10^{-51}$] may not be out of the question (see, for example, Choi 1997).

Irrespective of the above considerations, one should note that if the gravitational effects induce values of $\bar{\theta} < 10^{-10}$, so that the strong CP problem is still solved by imposing a $U(1)_{PQ}$ symmetry, then also the axion mass is approximately given by its QCD form. Thus, perhaps the best way to resolve these thorny theoretical questions is to find experimental evidence for the existence of invisible axions, with the canonical properties!

Acknowledgments

I am grateful to Professor W. Plessas for the very nice hospitality shown to me at Schladming. This work is supported in part by the Department of Energy under Grant No. FG03-91ER40662, Task C.

References

Adler, S.(1970): in *Lectures on Elementary Particles and Quantum Field Theory*, 1970 Brandeis Summer Institute, eds. S. Deser, M. Grisaru and H. Pendelton (MIT Press, Cambridge, MA).

Adler, S. (1969): Phys. Rev. **177**, 2426; Bell, J. S., Jackiw, R. (1969): Nuovo Cimento **51**, 47.

Asano, Y., *et al.* (1981): Phys. Lett. **107B**, 159.

Baluni, V. (1979): Phys. Rev. D**19**, 2227.

Banks, T. (1990): Physicalia **12**, 19.

Bardeen, W. A. (1974): Nucl. Phys. B**75**, 246.

Bardeen, W. A., Peccei, R. D., Yanagida, T. (1987): Nucl. Phys. B**279**, 401.

Barr S. M., Seckel, D. (1992): Phys. Rev. D**46**, 539.

Belavin, A. A., Polyakov, A. M., Schwartz, A. A., Tyupkin, Y. S. (1975): Phys. Lett. **59B**, 85.

Buchanan, C. D., Cousins, R., Dib, C. O., Peccei, R. D., Quackenbush, J. (1992): Phys. Rev. D**45**, 4088; Dib, C. O., Peccei, R. D. (1992): Phys. Rev. D**46**, 2265.

Cabibbo, N. (1963): Phys. Rev. Lett. **12**, 531; Kobayashi, M., Maskawa, T. (1973): Prog. Theor. Phys. **49**, 652.

Callan, C. G., Dashen, R., Gross, D. (1976): Phys. Lett. **63B**, 172.

Chikashige, Y., Mohapatra, R. N., Peccei, R. D. (1981a): Phys. Lett. **96B**, 265.

Chikashige, Y., Mohapatra, R. N., Peccei, R. D. (1981b): Phys. Rev. Lett. **45**, 1926.

Choi, K. W. (1997): Phys. Rev. D**56**. 6588.

Choi, K. W., Kaplan D. B., Nelson, A. E. (1993): Nucl. Phys. B**391**, 515.

Christenson, J. H., Cronin, J. W., Fitch, V. L., Turlay, R. (1964): Phys. Rev. Lett. **13**, 138.

Crewther, R. J. (1978):, Acta Phys. Austria Suppl. **19**, 47, Proceedings of the XVII Int. Universität Wochen für Kernphysik, Schladming, Austria, 1978.

Crewther, R., di Vecchia, P., Veneziano, G., Witten, E. (1979): Phys. Lett. **88B**, 123; (1980): **91B**, 487(E).

de Panfilis, S., *et al.* (1987): Phys. Rev. Lett. **59**, 839; Wuensch, W. U., *et al.* (1989): Phys. Rev. D**40**, 3153; Hagmann, C. *et al.* (1990): Phys. Rev. D**42**, 1297.

Donoghue, J. F., Golowich, E., Holstein, B. R. (1992): *Dynamics of the Standard Model* (Cambridge University Press, Cambridge, UK).

Feinberg, G., Sucher, J. (1979): Phys. Rev. D**20**, 1717.

Gelmini, G., Nussinov, S., Yanagida, T. (1983): Nucl. Phys. B**219**, 31.

Gelmini, G., Roulet, E. (1995): Rept. Prog. Phys. **58**, 1207.

Georgi, H., Kaplan, D. B., Randall, L. (1986): Phys. Lett. **169B**, 73.

Hagmann, C. *et al.* (1998): Phys. Rev. Lett. **80**, 2043.

Heisenberg, W., Z. (1932): Phys. **77**, 1; see also Cassen, B., Condon, E. U. (1936): Phys. Rev. **60**, 846; Wigner, E. P. (1937): Phys. Rev. **51**, 106.

Holman R., *et al.* (1992): Phys. Lett. B**282**, 132; Kamionkowski, M., March Russel, J. (1992): Phys. Lett. B**282**, 137; Barr, S. M., Seckel, D. (1992): Phys. Rev. D**46**, 539.

Jackiw. R., Rebbi, C. (1976): Phys. Rev. Lett. **37**, 2.

Kim, J. (1979): Phys. Rev. Lett. **43**, 103.

Kim, J. (1979): Phys. Rev. Lett. **43**, 103; Shifman, M. A., Vainshtein, A. I., Zakharov, V. I. (1980): Nucl. Phys. B**166**, 453; Dine, M., Fischler, W., and Srednicki, M. (1981): Phys. Lett. **104B**, 199: Zhitnisky, A. P. (1980): Sov. Jour. Nucl. Phys. **31**, 260.

Kolb, E. W., Turner, M. (1990): *The Early Universe* (Addison Wesley, Redwood City, California).

Krasnikov, N., Rubakov, V., Tokarev, V. (1978): Phys. Lett. **79B**, 423; see also, Anselm A. A., Johansen, A. A. (1994): Nucl. Phys. B**412**, 553.

Kuzmin, V., Rubakov, V., Shaposhnikov, M. (1985): Phys. Lett. **55B**, 36.

Lee, T. D., Yang, C. N. (1956): Phys. Rev. **104**, 254.

Low, F. E.(1967): *Symmetries and Elementary Particles* (Gordon and Breach, New York).

Matsuki, S., Yamamoto, K. (1991): Phys. Lett. **263B**, 523; Ogawa, I., Matsuki, S., Yamamoto, K. (1996): Phys. Rev. D**53**, R1740.

Nambu, Y. (1960): Phys. Rev. Lett. **4**, 380; Goldstone, J. (1961): Nuovo Cimento **19**, 1425; see also, Goldstone, J., Salam, A., Weinberg, S. (1962): Phys. Rev. **127**, 965.

Pal, P. B., Wolfenstein, L. (1982): Phys. Rev. D**25**, 788.

Particle Data Group, Barnett, R. M. *et al.* (1996): Phys. Rev. D**54**, 1.

Pauli, W. (1955): in *Niels Bohr and the Development of Physics*, ed. W. Pauli (Pergamon Press, New York); Schwinger, J. (1951): Phys. Rev. **82**, 914; Lüders, G. (1954): Dansk Mat. Fys. Medd **28**, 5; Lüders, G., Zumino, B. (1958): Phys. Rev. **110**, 1450.

Peccei, R. D. (1987): in *Concepts and Trends in Particle Physics*, eds. H. Latal and H. Mitter, in Proceedings of the XXV Int. Universität Wochen für Kernphysik, Schladming, Austria, 1986 (Springer-Verlag, Berlin).

Peccei, R. D. (1989): in *CP Violation*, ed. C. Jarlskog (World Scientific, Singapore).

Peccei, R. D. (1996): Jour. Korean Phys. Soc. (Proc. Suppl.) **29**, S199.

Peccei, R. D., Quinn, H. R. (1977): Phys. Rev. Lett. **38**, 1440; Phys. Rev. D**16**, 1791.

Preskill, J., Wise, M., Wilczek, F. (1983): Phys. Lett. **120B**, 127; Abbott, L., Sikivie, P. (1983): Phys. Lett. **120B**, 133; Dine, M., Fischler, W. (1983): Phys. Lett. **120B**, 137.

Raffelt, G. (1996): *Stars as Laboratories for Fundamental Physics* (University of Chicago Press, Chicago).

Schechter J., Valle, J. (1984): Phys. Rev. D**25**, 774.

Sikivie, P. (1983): Phys. Rev. Lett. **51**, 1415; (1985): Phys. Rev. D**32**, 2988; Krauss, L. *et al.* (1985): Phys. Rev. Lett. **55**, 1797.

Slansky, R. (1981): Phys. Reports **79C**, 1.

Streater, R. F., Wightman, A. S. (1964): *PCT, Spin and Statistics, and All That*, (W. A. Benjamin, New York).

't Hooft, G. (1976a): Phys. Rev. Lett. **37**, 8.

't Hooft, G. (1976b): Phys. Rev. Lett. **37**, 8; Phys. Rev. D**14**, 3432; Polyakov, A. (1977): Nucl. Phys. B**120**, 429; for a review, see, for example, Schäfer, T., and Shuryak, E. V. (1998): Rev. Mod. Phys. **70**, 323.

't Hooft, G. (1976c): Phys. Rev. D**14**, 3432.

Weinberg, S. (1972): Phys. Rev. Lett. **29**, 1698.

Weinberg, S. (1975): Phys. Rev. D**11**, 3583.

Weinberg, S. (1978): Phys. Rev. Lett. **40**, 223; Wilczek, F. (1978): Phys. Rev. Lett. **40**, 271.

Weinberg, S. (1979): Phys. Rev. Lett. **43**, 1556; Wilczek, F., Zee, A. (1979): Phys. Rev. Lett. **43**, 1571.

Wigner, E. P. (1932): Nach. der Gess. Wiss. Göttingen **32**, 35; see also, *Group Theory and its Applications to the Quantum Mechanics of Atomic Spectra* (Academic Press, New York, 1954).

Wigner, E. P. (1952): Proc. Nat. Acad. Sci. (US) **38**, 449; Weyl, H. (1929): Z. Phys. **56**, 330.

Wilczek, F. (1982): Phys. Rev. Lett. **49**, 1549; Reiss, D. B. (1982): Phys. Lett. **115B**, 217.

Witten, E. (1984): Phys. Lett. **149B**, 551.

Wu, C. S. *et al.* (1957): Phys. Rev. **105**, 1413; **106**, 1361; Freedman, J. I., Telegdi, V. L. (1957): Phys. Rev. **105**, 1681; Garwin, R. L., Lederman, L. M., Weinrich, M. (1957): Phys. Rev. **105**, 1425.

Yanagida, T. (1979): in Proceedings of the Workshop on Unified Theory and the Baryon Number of the Universe, KEK, Japan ; Gell-Mann, M., Ramond, P., Slansky, R. (1979): in *Supergravity*, ed. by P. Van Neuwenhuisen (North Holland, Amsterdam).

CP Violation

Werner Bernreuther

Institut für Theoretische Physik, RWTH Aachen,
D-52056 Aachen, Germany

Abstract. The salient features of CP-violating interactions in the standard electroweak theory and in a few of its popular extensions are discussed. Moreover a brief overview is given on the status and prospects of searches for CP non-conservation effects in low- and high-energy experiments.

1 Introduction

More than 30 years after the discovery of CP non-conservation in the neutral K meson system neither the physics that causes this phenomenon has been clarified nor have other CP- or time-reversal-violating effects been found in laboratory experiments. The standard theory of electroweak interactions can explain the experimental findings by a complex phase in the coupling matrix of the charged weak quark currents [1]. Yet in this theory the key to a deeper understanding of this symmetry violation is hidden behind the mystery of the flavour problem. On the other hand, CP-violating interactions are conceivable which have nothing to do with the fact that there are three generations of quarks and leptons with disparate mass spectra. Interactions of this kind naturally appear in popular and, so far, empirically acceptable extensions of the Standard Model. The question whether CP-symmetry breaking is due to a single "source" – which is most likely the Kobayashi-Maskawa phase [1] – or whether there are several CP-nonconserving interactions which will show up in different physical situations must be resolved experimentally. Another enigma of particle physics – often called "problem number one" – is the dynamics of electroweak symmetry breaking. Very probably these two dark corners are related: clarification of weak gauge symmetry breaking – which must also be achieved by experiments – would also shed light on the origin(s) of CP violation.

In these lectures I shall first review the salient features of CP violation in the Standard Model and in some of its extensions, notably multi-Higgs and supersymmetric extensions. The issue of explicit versus spontaneous CP breaking will also be discussed. Then a brief overview is given on the status and prospects of searches for CP non-conservation effects in weak decays of strange, charmed, and beauty hadrons, on the search for permanent electric dipole moments of particles, and on present and future high energy CP tests at colliders.

2 Models

The discussion in this section rests on the assumption that the Higgs mechanism – which requires at least one elementary scalar field multiplet with non-vanishing ground state expectation value – gives the correct description of electroweak gauge symmetry breaking. A priori, breaking this symmetry does not require elementary Higgs fields; it might have occurred "dynamically" by condensation of (new) fermion bilinears. These vacuum expectation values can have complex phases relative to each other, which induce observable CP violation. I shall not discuss this possibility of CP-breaking, which is not without problematic features, any further here (see, e.g., [2]).

Moreover, the following discussion remains within the context of four-dimensional gauge theories, where CP constitutes a discrete symmetry transformation. In higher dimensional theories, including string theories, CP can be a gauge symmetry which gets spontaneously broken [3].

In the framework of four-dimensional gauge theories with elementary Higgs fields one can also distinguish between two situations:
(a) CP invariance is violated explicitly at the Lagrangian level. That is, in the "Hamiltonian of the world", $H_{inv}+H'$, there is a term H' (which by a posteriori reasoning can usually be treated in perturbation theory) for which $[H', U_{CP}] \neq 0$. Here U_{CP} is the unitary operator which implements the CP transformation in the Hilbert space of the theory given by H_{inv}.
(b) One may have CP invariance of the full Hamiltonian, $[H, U_{CP}] = 0$, but this symmetry is spontaneously broken by the ground state, $U_{CP}|0> \neq e^{i\phi}|0>$. This scenario requires more than one Higgs multiplet. In the following we shall discuss both options.

2.1 The Kobayashi-Maskawa mechanism

CP violation in the three-generation Standard Model (SM) of electroweak interactions is related to the fact that Nature has chosen the option that, as far as quarks are concerned, the mass eigenstates are different from the weak interaction eigenstates. (This may also be the case in the lepton sector as recent experimental results on atmospheric neutrinos and their interpretation in terms of neutrino oscillations indicate.) In the weak basis the Yukawa interactions of the quarks with the SU(2) doublet Higgs field are described by two 3×3 coupling matrices. The requirement of hermiticity of the Hamiltonian does not preclude that these matrices are complex. After having transformed the quark fields from the weak basis to the mass basis, the various pieces of the SM Lagrangian $\mathcal{L}_{SM}$ are diagonal in generation space (in unitary gauge), except for the charged current interactions of quarks,

$$\mathcal{L}_{cc} = -\frac{g}{\sqrt{2}}\overline{U}_L\gamma^\mu V_{KM} D_L W_\mu^+ + \text{h.c.}, \tag{1}$$

which contains the Cabibbo-Kobayashi-Maskawa (KM) matrix [1], a 3×3 unitary matrix in generation space. Here $\bar{U} = (\bar{u}, \bar{c}, \bar{t})$ and $D = (d, s, b)^T$ are the quark fields in the mass basis. Five parameters of the CKM matrix elements can be eliminated by a change of phase of the quark fields

$$u_i \to e^{i\omega_i} u_i, \quad d_j \to e^{i\tilde{\omega}_j} d_j \quad \Rightarrow \quad V_{ij} \to e^{i(\omega_i - \tilde{\omega}_j)} V_{ij}, \tag{2}$$

where i,j = 1,2,3 are generation indices. Hence the matrix V_{KM} has four observable parameters, which may be chosen to be three Euler-type angles and a phase angle δ_{KM}. If $\delta_{KM} \neq 0, \pm\pi$ the charged current Hamiltonian $H_{cc} = -\int d^3x \mathcal{L}_{cc}(x)$ is non-invariant under a CP transformation.

In view of eq. (2), only functions of V_{ij} which are rephasing-invariant have a physical meaning. The simplest invariants are $|V_{ij}|$ and

$$Q_{ijkl} = V_{ij} V_{kl} V_{il}^* V_{kj}^*. \tag{3}$$

For three generations the unitarity of V_{KM} implies that the $|\text{Im}(Q_{ijkl})|$ are all equal [4], [5]. (In fact the various unitarity triangles, representing the orthogonality relations of the KM matrix elements, all have the same area which is equal to $|\text{Im}Q|/2$.) Therefore, for instance

$$\text{Im}Q = \text{Im}(V_{ud} V_{cb} V_{ub}^* V_{cd}^*) \tag{4}$$

is an invariant measure of CP violation à la KM. This expression immediately shows that the strength of KM CP violation is small even if the CP-violating phase angle were maximal. Insertion of the measured values of the moduli of the KM matrix elements into eq. (4) yields that $|\text{Im}Q|$ is smaller than 10^{-4}. (For a discussion of maximal CP violation in the SM, see [6].)

A deeper understanding of CP violation à la KM requires an answer to the "flavour problem", i.e., an answer to the question why there are three fermion generations and why is there such a hierarchy in the mass spectra of the u- and d-type quarks, respectively. If any two quarks with the same charge were mass-degenerate, the CP-phase in V_{KM} could be eliminated by a suitable unitary transformation of the quark fields. This feature of KM CP violation is exhibited by the well-known invariant [5], [7]

$$J_{CP} = \prod_{\substack{i>j \\ u,c,t}} (m_i^2 - m_j^2) \prod_{\substack{i>j \\ d,s,b}} (m_i^2 - m_j^2) \quad \text{Im}Q. \tag{5}$$

If neutrinos have non-degenerate masses then there can be KM-type CP violation in the lepton sector as well. For three generations, the leptonic analogue of the KM matrix, V_{lept}, which then parameterizes the relative strength of the leptonic charged current-induced transitions, can have 1 CP phase angle (3 CP phase angles) if the neutrinos are of the Dirac (Majorana) type.

KM CP violation is observable only in flavour-changing charged current reactions. As is obvious from eqs. (4) and (5), effects are in general small because of small mixing angles involved. In charm hadron and in top quark decays, which are Cabibbo-allowed, even a maximal CP phase in the KM matrix thus leads only to very small effects. K and B mesons, whose weak decays are at least singly Cabibbo-forbidden, are therefore *the* objects to test the KM mechanism.

A non-zero KM phase leads only to negligibly small effects in flavour-diagonal amplitudes. For instance it induces tiny electric dipole moments (EDM) of quarks [8] and even tinier ones for charged leptons (see section 4).

2.2 The strong CP problem

At this point it is appropriate to recall the strong CP problem, which is actually not a problem of Quantum Chromodynamics (QCD) in isolation, but of the theory of strong *and* weak interactions. In QCD topologically non-trivial quantum fluctuations (instantons) induce a parity- and time-reversal-violating term in the QCD quantum action of the form $S_\theta = (\theta g^2/32\pi^2) \int d^4x G^a_{\mu\nu} \tilde{G}^{a\mu\nu}$, where θ is the QCD vacuum angle. This term has observable consequences in flavour-diagonal amplitudes. Observables depend, however, on the parameter

$$\overline{\theta} = \theta - arg(det\mathcal{M}_q), \tag{6}$$

where $\mathcal{M}_q$ is the non-diagonal mass matrix of the u- and d-type quarks in the weak basis. The experimental upper bound on the neutron EDM implies [9] that $|\overline{\theta}| < 3 \times 10^{-10}$. We lack a deeper understanding why this parameter should be so small. Simply setting $\overline{\theta}$ equal to zero is unsatisfactory because, according to 't Hooft's naturalness condition [10], it does not increase the symmetries of the SM. After all, CP must not be a good symmetry of the SM if this theory is to explain the observed CP effect in K^0-$\overline{K^0}$ mixing. It requires δ_{KM} to be of order one and hence one would expect $arg(det\mathcal{M}_q)$ to be of the same order. So there is apparently severe fine tuning of θ required to bring $\overline{\theta}$ down to the level of 10^{-10}. For a more detailed discussion of this problem and of the possible ways out that have been proposed, see [11]. One may take a "just so" attitude and consider $\overline{\theta}$ to be just another one of the uncalculable parameters of the SM that happens to be (very close to) zero. However, many theorists believe that one cannot understand CP violation before one hasn't solved this problem.

2.3 Extensions of the Standard Model

There are a number of well-known arguments which motivate the belief in new physics beyond the Standard Model, to be discovered in particle physics experiments. Extensions of the SM, even if based on the gauge group $SU(3)_c \times SU(2)_L \times U(1)_Y$, almost invariably entail a larger non-gauge sector, that is to say, scalar self interactions and Yukawa interactions. In this way quite a number of "new" CP-violating (CPV) interactions for quarks *and* for leptons are conceivable in a natural way. (In the following, new CP-violating interactions refer to interactions that are not due to the KM phase). In particular CPV interactions with the following features may exist:

(a) Interactions that are unrelated to the mixing of quark generations and the

hierarchy of quark masses. Such interactions induce CP effects also in flavour-diagonal amplitudes.
(b) Higgs-type interactions whose strength increases with the mass of the fermion involved, leading to sizeable effects in the heavy flavour sector.

Explicit CP violation in multi-Higgs models The simplest, phenomenologically viable model with extra CPV besides the KM phase is, perhaps, the extension of the SM by an extra $SU(2)$ Higgs doublet. The two Higgs doublets Φ_1, Φ_2 are assumed to couple to quarks and leptons in such a way that there are no flavour-changing neutral couplings at the tree-level (see, e.g., [12]). This "natural flavour-conservation constraint" can be enforced by imposing a discrete symmetry. The different implementations of this symmetry define different models (see, for instance, [13]). Apart from complex Yukawa coupling matrices, which lead to the KM phase, the requirement of hermiticity, renormalizability, and $SU(2)_L \times U(1)_Y$ invariance of the Lagrangian does not preclude explicit CPV in the Higgs potential V_Φ. Requiring that the potential breaks the above-mentioned discrete symmetry only softly (that is, by terms with operator dimension less than four) one can have

$$V_\Phi = V_0(\Phi_1, \Phi_2) + [\kappa \Phi_1^\dagger \cdot \Phi_2 + h(\Phi_1^\dagger \cdot \Phi_2)^2 + \text{h.c.}]. \tag{7}$$

Here V_0 denotes the CP-invariant part of the potential. A CP transformation,

$$\Phi_{1,2}(\mathbf{x}, t) \quad \xrightarrow{CP} \quad e^{i\alpha_{1,2}} \Phi_{1,2}^*(-\mathbf{x}, t), \tag{8}$$

shows that the term in the square brackets of eq. (7) breaks CP if $\xi = \text{Im}(\kappa^2 h^*) \neq 0$. Note that it is unnatural to assume $\xi = 0$. Even if this were so at tree level, the non-trivial KM phase δ_{KM}, which is needed to explain the observed CPV, would induce a non-zero ξ through radiative corrections.

The spectrum of physical Higgs boson states in the two-doublet models consists of a charged Higgs boson and its antiparticle, $H^\pm$, and three neutral states. As far as CPV is concerned, $H^\pm$ carries the KM phase. It affects the CPV phenomenology of flavour-changing $|\Delta F| = 2$ neutral meson mixing and $|\Delta F| = 1$ weak decays of mesons and baryons. From experimental data on $b \to s + \gamma$ the lower bound $m_{H^+} > 210$ GeV on the mass of this particle was derived [14].
If ξ were zero, the set of neutral Higgs boson states would consist of two scalar (CP=1) and one pseudo-scalar (CP= −1) state. Explicit CPV in the Higgs potential has the consequence that these states mix [15] (note that the mixing occurs already at tree level), leading to three mass eigenstates $\varphi_{1,2,3}$ that no longer have a definite CP parity. That is, they couple both to scalar and to pseudo-scalar quark and lepton currents. The Yukawa interactions read (for ease of notation the same symbol is used for a field and its associated particle state)

$$\mathcal{L}_\varphi = -(\sqrt{2}G_F)^{1/2} \sum_f (a_f m_f \bar{f} f + \tilde{a}_f m_f \bar{f} i\gamma_5 f)\, \varphi. \tag{9}$$

The sum over the Higgs fields $i = 1, 2, 3$ is implicit. Here G_F is Fermi's constant, f denotes a quark or lepton field, m_f is the mass of the associated particle, and the dimensionless reduced Yukawa couplings $a_f, \tilde{a}_f$ depend on the parameters of the Higgs potential and on the type of model [16]. From LEP data one infers that the lightest of the three states φ_i should have a mass larger than about 50 GeV. (The precise lower bound depends on the parameters of the model.)

In terms of the parameters of eq. (9) CP violation in the neutral Higgs sector occurs if $a_f \cdot \tilde{a}_f \neq 0$. The following generic features arise: (a) The Yukawa interaction (9) leads to CPV in *flavour-diagonal* amplitudes for quarks and for leptons. (b) The induced CP effects are proportional to some power$(m_f)^p$, where one finds p=1,2,3 for reactions discussed in sections 4, 5 below. For example, neutral φ exchange at tree level induces an effective CPV interaction of the form $(\bar{f}f)(\bar{f}i\gamma_5 f)$ with a coupling strength proportional to m_f^2. Neutral φ exchange at one-loop in the γff, Zff, and Gqq amplitudes (G denotes a gluon) leads to T- respectively CP-violating electric, weak, and chromo-electric dipole moment form factors of the fermion involved which are proportional to m_f^3 [16]. Potentially large effects can be expected for top quarks.

In models with a more complicated scalar sector, for instance, in models with $n \geq 3$ Higgs doublets, there is more than one charged Higgs particle. The scalar potential can be such that these states mix in a CP-violating way which leads to a complex mass matrix for these bosons. Transforming all fields to their respective mass basis, the interaction of the quarks with the charged Higgs bosons then reads

$$\mathcal{L}_{ch} = -(2\sqrt{2}G_F)^{1/2} \sum_i (\alpha_i \bar{U}_L V_{KM} M_D D_R + \beta_i \bar{U}_R M_U V_{KM} D_L)\, H_i^+ + \text{h.c.}, \quad (10)$$

where $M_{U,D}$ denote the real, diagonal 3×3 mass matrices of the u- and d-type quarks, and, in general,

$$\text{Im}(\alpha_i \beta_i^*) \neq 0, \quad (11)$$

due to the complex phases in the mass matrix of the charged Higgs bosons. The interactions of $H_i^\pm$ with leptons are of the same structure. In models where the right-handed quarks q_R couple to several Higgs multiplets one can have additional CP phases such that the reduced Yukawa couplings in eq. (10) satisfy

$$\text{Im}(\alpha_i \alpha_j^*) \neq 0, \qquad \text{Im}(\beta_i \beta_j^*) \neq 0 \quad (i \neq j). \quad (12)$$

Charged Higgs exchange with couplings as in eq. (10) induces also CP violation in flavour-diagonal amplitudes. For instance, if eq. (11) holds, one- and two-loop contributions to electric dipole form factors of quarks and leptons are induced (see section 4). If eq. (12) holds, CPV chiral-invariant form factors in the $b\bar{b}ZG$ amplitude are generated at one-loop order [17].

In fact, even in two-Higgs doublet models charged Higgs boson exchange can provide CPV independent of the KM phase if one allows for general Yukawa interactions [18] (which imply flavour-changing neutral Higgs boson exchanges at tree-level).

Explicit CP violation in supersymmetric models In the minimal supersymmetric extension (MSSM) of the Standard Model [19], [20] CP-violating phases can arise, apart from the complex Yukawa interactions of the quarks yielding a non-trivial δ_{KM}, from the soft supersymmetry breaking terms. The requirement of gauge invariance and hermiticity of the Lagrangian allows for
i) complex Majorana masses M_i in the gaugino mass terms,

$$-\sum_i (M_i \lambda_i \lambda_i)/2 + \text{h.c.}, \tag{13}$$

ii) complex trilinear scalar couplings, that is, complex 3×3 matrices $A_{q,\ell}$ in generation space which contain the couplings of the scalar quarks and leptons to the Higgs doublets Φ_1, Φ_2. For instance[1]

$$\tilde{D}_R'^{\dagger} A_d \Phi_1^{\dagger} \cdot \tilde{Q}_L' + \text{h.c.}, \tag{14}$$

and analogous interactions with coupling matrices A_u and A_ℓ. The tilde and the prime denote scalar quark fields in the weak basis, capital letters denote as before vectors in generation space. The label L refers to $SU(2)_L$ doublets, $\tilde{Q}_L' = (\tilde{U}_L', \tilde{D}_L')^T$, and the label R in eq. (14) refers to $SU(2)_L$ singlets.
iii) a complex soft term in the Higgs potential

$$\kappa \Phi_1^{\dagger} \cdot \Phi_2 + \text{h.c.} \tag{15}$$

Motivated by supergravity models it is often assumed that the A_f's are proportional to the Yukawa coupling matrices

$$A_f = A Y_f, \qquad f = u, d, \ell, \tag{16}$$

where A is a complex parameter. The observable CP phases of the MSSM are readily counted [21]. Apart from the KM phase and the QCD $\bar{\theta}$ parameter, these are $arg(AM_i^*)$ and $arg(\kappa M_i^*)$, where M_i are the gaugino mass terms in eq. (13). If eq. (16) is not imposed then there are quite a number of independent phases.

After spontaneous symmetry breaking and after having transformed the various fields such that all mass matrices have become real and diagonal, the CP phases have been shifted into the fermion-sfermion-neutralino and -chargino interaction terms. Let us write down here only the gluino interaction, which involves the QCD coupling g_{QCD}. One arrives at the CP-violating quark-squark-gluino interaction Lagrangian in the mass basis, which reads for u-type quarks

$$\mathcal{L}_{\tilde{G}u\tilde{u}} = i\sqrt{2} g_{QCD} \sum_{j,l} (e^{-i\varphi_u} \bar{u}_{jL} \Gamma_{jl} \tilde{G}^a T^a \tilde{u}_l + e^{+i\varphi_u} \bar{u}_{jR} \Gamma'_{jl} \tilde{G}^a T^a \tilde{u}_l) + \text{h.c.}, \tag{17}$$

where j=1,2,3, and l=1,...,6, $\tilde{G}^a$ denote the gluino fields, T^a are the generators of $SU(3)_c$ in the fundamental representation, and Γ, Γ' are complex 3×6 matrices.

---

[1] In order to facilitate the comparison with the above models, the non-SUSY convention, i.e., the same hypercharge assignment for both $SU(2)$ Higgs doublets, $\Phi_i = (\phi_i^+, \phi_i^0)^T$, (i=1,2) is employed here.

(Recall that for each flavour there are two squark respectively slepton mass eigenstates, which are in general not mass-degenerate.) The $\tilde{d}d\tilde{G}$ interaction is of the same form. As already mentioned there are further CP-violating fermion-sfermion-neutralino and -chargino interactions (of similar structure as eq. (17)) with interesting phenomenological consequences.

If eq. (16) holds then the phase $\varphi_q = \varphi_{\tilde{G}} - \varphi_A$ is universal and Γ, Γ' depend, as far as CP phases are concerned, only on the KM phase. However, in the general case things are really more complex.
As the gluino interactions involve both flavour-diagonal and flavour-changing $\Delta Q = 0$ vertices, $\mathcal{L}_{\tilde{G}q\tilde{q}}$ induces CPV effects in neutral meson mixing, in flavour-changing decays of hadrons, and it leads to electric dipole moments (EDM), e.g. of the neutron, of considerable size. The latter constitutes a well-known conflict for the MSSM. The predictions for the neutron EDM come out too large as compared with the experimental upper bound if the CP phases of the soft terms i) ii) and iii) above were of order one and if the squark and gluino masses were about, say, 200 GeV. (See section 4). Therefore it is often assumed in the literature that the CP phases of the soft terms i) ii) and iii) are zero at a very high energy scale, which is usually taken to be a supposed grand unification scale or the Planck scale. Then CP violation at this scale is assumed to come only from the Yukawa couplings, i.e., the KM phase. When evolving the parameters of this constrained MSSM model down to lower energies, the KM phase induces, through renormalization, also small phases in the soft SUSY breaking terms [22], which are phenomenologically acceptable as far as EDMs are concerned. (For a discussion of the CP-violating phases and their phenomenological consequences in the supersymmetric grand unified $SO(10)$ model, see [23].)

What about Higgs sector CPV in supersymmetric extensions of the SM? In the MSSM the tree-level Higgs potential is, schematically, of the form

$$V_{tree} = V_0(\Phi_1, \Phi_2) + (\kappa \Phi_1^\dagger \cdot \Phi_2 + \text{h.c.}). \tag{18}$$

As is well-known (see, e.g. [13]) SUSY does not allow for independent quartic couplings in V_0. They are proportional to linear combinations of the $SU(2)$ and $U(1)$ gauge couplings squared. Moreover, a term of the form $(\Phi_1^\dagger \cdot \Phi_2)^2$ and two other quartic terms which are non-invariant under $\Phi_1 \to -\Phi_1$ are absent at tree-level. Hence by suitable adjustment of the phases α_i in eq. (8) a CP transformation on the scalar fields can be implemented such that $\int d^3x V_{tree}$ is CP-invariant. Thus there is no CPV mixing of the three physical neutral Higgs states at tree level. The CPV interactions in the MSSM discussed above generate a (small) complex coupling h (cf. eq. 7) in the effective potential at one-loop order, and the parameter ξ defined above can now become non-zero. The other quartic terms absent at tree-level will also be induced. Hence there can be CPV mixing at one-loop order of the two CP=1 and the CP= -1 neutral Higgs states in the MSSM. It is, however, expected to be small.

In next-to-minimal SUSY models with two $SU(2)$ Higgs doublet fields, Φ_1, Φ_2, and a gauge singlet field N the Higgs potential can explicitly violate CP at the tree level. For instance, this is the case for the potential

$$V_{tree} = V_{inv}(\Phi_1, \Phi_2, N) + (h_1 N^3 + h_2 \Phi_1^\dagger \cdot \Phi_2 N + h_3 \Phi_1^\dagger \cdot \Phi_2 N^2 + \text{h.c.}), \qquad (19)$$

where $h_{1,2,3}$ are arbitrary complex couplings and V_{inv} is the CP-invariant part of the potential. Appropriate field redefinitions show that there is one observable CPV phase in eq. (19). In this case the three CP=1 and the two CP= −1 physical neutral Higgs states mix at tree-level [24]. (There is, however, no mixing of the two scalar with the pseudo-scalar component of the two Higgs doublets.)

Spontaneous CP violation There is a potential cosmological problem when discrete symmetries are spontaneously broken [25]. When spontaneous CPV occurs in the early universe at some temperature, one expects that domains with different CP signatures (i.e., different signs of the CP-violating phase(s)) are formed. The energy density of the walls which separate these domains dissipate not rapidly enough when the universe expands further. Estimates yield that the energy density associated with these walls today exceeds the closure density of the universe by many orders of magnitude [25]. The problem is avoided if CP got broken before inflation took place. However, the connection to low energy phenomena then becomes very loose.

Ignoring this problem, it is nevertheless interesting to investigate multi-Higgs or supersymmetric extensions of the SM with spontaneous CPV at the weak scale. The simplest model in this respect is the original two-Higgs doublet model of T.D. Lee [26], respectively its more recent variants [27], [28], [18]. By construction the Lagrangians of these models, which have the same gauge group and the same particle content – apart from the Higgs sector – as the SM, are CP-invariant. Hence the Yukawa couplings can be taken to be real, without loss of generality. Gauge invariance, hermiticity, and renormalizability imply that the tree level Higgs potential has the general form [26]

$$\begin{aligned} V = {} & V_0(\Phi_1, \Phi_2) + (\kappa \Phi_1^\dagger \cdot \Phi_2 + \lambda_1 (\Phi_1^\dagger \cdot \Phi_2)^2 + \lambda_2 (\Phi_1^\dagger \cdot \Phi_2)(\Phi_1^\dagger \cdot \Phi_1) \\ & + \lambda_3 (\Phi_1^\dagger \cdot \Phi_2)(\Phi_2^\dagger \cdot \Phi_2) + \text{h.c.}), \end{aligned} \qquad (20)$$

where, contrary to eq. (7), the parameters in eq. (20) are real because V is required to be CP-invariant.
Minimisation of the potential yields the vacuum expectation values (VEV) of the two Higgs fields (the phase of Φ_1 has been adjusted such that the VEV of this field is real)

$$< 0|\phi_1^0|0 >= v_1/\sqrt{2}, \qquad < 0|\phi_2^0|0 >= v_2 e^{i\vartheta}/\sqrt{2}, \qquad (21)$$

where v_1, v_2 are real and positive parameters which have to satisfy the experimental constraint $\sqrt{v_1^2 + v_2^2} = 246$ GeV. There exists a range of parameters of V such that the absolute minimum is characterized by a non-trivial phase [26]

$$\vartheta = \arccos\left[\frac{2\kappa - \lambda_2 v_1^2 - \lambda_3 v_2^2}{4\lambda_1 v_1 v_2}\right]. \qquad (22)$$

The necessary condition for this to happen is

$$\lambda_1 > 0, \qquad \left|\frac{2\kappa - \lambda_2 v_1^2 - \lambda_3 v_2^2}{4\lambda_1 v_1 v_2}\right| < 1. \tag{23}$$

If the phase angle $\vartheta \neq \pm n\pi/2$, n = 0,1,2,.., then the ground state breaks CP spontaneously. It can be shown that there is then no unitary implementation of CP such that the Lagrangian and the vacuum remain invariant [29].

One consequence of spontaneous CPV is, again, CPV mixing of neutral Higgs states. Yet one must account for the observed CPV in K^0-$\bar{K}^0$ mixing, but the Yukawa couplings are real by construction. Therefore the construction principle of "natural flavour conservation" must be given up. The right-handed quark fields u_{iR}, d_{iR} are coupled both to Φ_1 and Φ_2, yielding the general Yukawa interactions

$$\mathcal{L}_Y = -\sum_{k=1}^{2}[Y_k^d \bar{Q}_L' \cdot \Phi_k D_R' + Y_k^u \bar{Q}_L' \cdot \tilde{\Phi}_k U_R'] + \text{h.c.}, \tag{24}$$

where primes denote the weak basis, $\tilde{\Phi} = i\sigma_2\Phi$, and Y_k^q denote four 3×3 real Yukawa matrices. Expanding around the ground state (21), one obtains the non-diagonal complex mass matrices

$$M_u = \frac{v_1}{\sqrt{2}}Y_1^u + \frac{v_2 e^{i\vartheta}}{\sqrt{2}}Y_2^u, \qquad M_d = \frac{v_1}{\sqrt{2}}Y_1^d + \frac{v_2 e^{i\vartheta}}{\sqrt{2}}Y_2^d. \tag{25}$$

It follows that $M_u M_u^\dagger$ and $M_d M_d^\dagger$ are arbitrary hermitian matrices. Diagonalization of these matrices leads to charged weak quark interactions of the usual form (1) with a complex mixing matrix V_{KM} whose CP phase depends on ϑ. In short, these models have only one CP parameter, namely the "vacuum phase angle" ϑ, but a rich CP phenomenology:
i) CPV in charged weak current reactions ($W^\pm$ and $H^\pm$ exchange) and
ii) CPV mediated by flavour-conserving and by flavour-changing neutral Higgs boson φ exchange.
Note that the observed CPV in $|\Delta S| = 2$ K^0-$\bar{K}^0$ mixing is dominantly generated by tree-level φ exchange, $s\bar{d} \leftrightarrow \bar{s}d$. In this respect, these models may be regarded as a realization of Wolfenstein's "superweak hypothesis". But in general there will be also other CPV $|\Delta F| = 2$ tree-level transitions. The problematic feature of these models is that fine tuning of the flavour-changing neutral current couplings (or the appeal to some approximate symmetry in flavour space) and/or rather large φ masses are required in order to avoid conflict with experiment (leaving aside the strong CP problem).

The simplest SM extension with spontaneous CPV *and* flavour conservation in neutral Higgs particle interactions at tree-level seems to be the three Higgs-doublet model of ref. [30]. CPV originates from two CPV vacuum phase angles ϑ_1, ϑ_2 which lead to CPV neutral Higgs mixing and to a complex mass matrix for the charged Higgs bosons. However, in this model V_{KM} remains real [31] and the observed CP violation in neutral kaon mixing must be accounted for by charged Higgs boson exchange (one-loop box diagrams). The model appears to be incompatible with experimental data: One has a hard time explaining why CPV charged Higgs boson exchange generates $\epsilon \approx 10^{-3}$ in the K meson system, but suppresses ϵ'/ϵ to the level of 10^{-3} or even below that number and the neutron electric dipole moment below $10^{-25}e$cm.

Is spontaneous CPV a viable concept for supersymmetric extensions of the SM? Let us first consider the MSSM and assume CP invariance of the Lagrangian $\mathcal{L}_{MSSM}$. It is clear from the discussion below eq. (20) that there is no spontaneous CPV at tree level, because the couplings $\lambda_1 = \lambda_2 = \lambda_3 = 0$ in the tree level potential (18). Chargino, neutralino, t, and $\tilde{t}$ contributions to the effective potential at one-loop [32] generate small, real couplings

$$\lambda_{1,2,3} \sim g^4/32\pi^2 \sim 5 \times 10^{-4}. \tag{26}$$

If the parameters are such that condition (23) is fulfilled then the model can have a stable ground state [32] which spontaneously breaks CP. It follows from eqs. (23) and (26) that this requires the parameter κ to be small, $\kappa = \mathcal{O}(\lambda_1 v^2)$. This implies, however, that the lightest of the three neutral Higgs bosons has a mass of about $m \approx \sqrt{(4\lambda_1 v^2 sin^2\vartheta)} \approx 5$ GeV, which is incompatible with experimental constraints. Hence this scenario is of no use for the MSSM. (The appearance of a light boson can be understood from the Georgi-Pais theorem [33].)

Spontaneous CPV in the next-to-minimal supersymmetric extension of the SM (see above) was investigated in [34], [35]. Radiative corrections to the tree-level scalar potential (19), with all couplings now being real because of CP invariance, are also essential in this model for the mechanism of spontaneous CPV to work [35]. (The masses of the scalar states are required to be positive.) Refs. [35] find that in this case the mass of the lightest neutral Higgs boson has an upper bound of about 36 GeV and the sum of the masses of the two lightest neutral Higgs bosons is not much above the mass of the Z boson. It is questionable whether this scenario is still compatible with data from LEP.

model	mechanism of CPV	required non-degeneracy in mass
SM	KM	u-type quarks, d-type quarks
massive neutrinos	KM-like	charged leptons, neutrinos
multi-Higgs models	neutral Higgs mixing	neutral φ_j bosons
	charged Higgs mixing	charged Higgs bosons $H_j^{\pm}$
MSSM	phases in SUSY breaking terms	scalar fermions of a given flavour $\tilde{f}_1, \tilde{f}_2$, $(\tilde{f} = \tilde{q}, \tilde{\ell})$

Miscellaneous issues As has been discussed in section 2.1, CP violation à la KM requires non-degeneracy of the masses of both u- and d-type quarks. In fact this is a generic feature of CP violation from the non-gauge sector (that is to say, ignoring the θ term of QCD). For the models discussed above this is exhibited in the table above. Invariants similar to J_{CP} of eq. (5) can be constructed also for the non-KM sources of CPV (cf., for instance, [36]).

So far the only hint for CPV beyond the KM phase comes from attempts to develop scenarios for explaining the baryon asymmetry of the universe on the basis of particle physics models and the big-bang model. It is well-known by now that within this framework baryogenesis requires [37] baryon number violation, C and CP violation, and thermal non-equilibrium, that is to say, an "arrow of time". Two types of scenarios have been intensely investigated in recent years:
a) Baryogenesis at the electroweak phase transition. Investigations of the nature of the phase transition lead to the conclusion that this scenario does not work within the SM. (For a recent review, see [38].) Even if non-SM interactions exist such that the electroweak phase transition was of first order, it is questionable whether KM CP violation did the job. (For reviews, see [39].) According to present knowledge it seems that, for instance, two-Higgs doublet extensions [39], [40] and the MSSM [41], with CPV as discussed in section 2, still provide viable alternatives.
b) Out-of-equilibrium decays of ultra-heavy Majorana neutrinos [42], with $(B\text{-}L)$ violation, at temperatures far above the electroweak phase transition.

It would be fascinating to relate the observed CP violation in the laboratory, which so far amounts to the ϵ parameter of neutral K meson mixing, to the fact that the visible universe contains matter, rather than anti-matter, with a baryon-to-photon ratio of $n_B/n_\gamma \sim 10^{-10}$. However, as suggested by the investigations of scenario b), it may be that the CP-violating interactions which were at work in the early universe are irrelevant for reactions that can be explored in, say, an atomic physics experiment, at a B meson factory, or even at the LHC. In any case, it is challenging enough to search for CPV phenomena in laboratories on the earth. In the following a number of those phenomena which are predicted by the KM mechanism and/or by some non-KM sources of CP violation are discussed.

3 Weak Decays

Observable CP violation à la KM requires quarks whose weak decays are Cabibbo suppressed. That is not the case for c and t quarks. Therefore CP searches involving these quarks will predominantly test for new interactions.

3.1 Kaons and Hyperons

CP violation in the kaon system manifests itself in the very existence of the decays $K_L \to 2\pi$ and in a non-zero CP asymmetry, δ, between the rates of the semi-leptonic decays $K_L \to \pi^\mp \ell^\pm \nu$. From these observations it can be inferred that CPV takes place in the $|\Delta S| = 2$ K^0-$\bar{K^0}$ mixing amplitude.

The strength of $|\Delta S| = 2$ CPV is characterized by the ϵ parameter. One has $\delta \approx 2\text{Re}(\epsilon) = 3.27(12)\times 10^{-3}$. The KM mechanism can naturally explain the order of magnitude of this number – given the moduli of the CKM matrix elements measured in other decays. Recall that CPV in the SM is small because of small inter-generation mixing angles involved (cf. eq. (4)). The parameter ϵ is determined in the SM by the ratio of the imaginary part to the real part of the well-known box diagram mixing amplitude. (To be precise, of its dispersive part). A simple counting of the CKM matrix elements involved shows that $\epsilon_{SM} \sim 10^{-3} \sin \delta_{KM}$.

The present experimental status of whether there is also "direct" $|\Delta S| = 1$ CP violation in the $K^0 \to 2\pi$ amplitudes is inconclusive [43], [44]. New experiments [45] aim at measuring the corresponding observable, namely $\text{Re}(\epsilon'/\epsilon)$, at the level of 10^{-4}. On the theoretical side considerable effort has been spent over the last years to calculate the next-to-leading order QCD corrections to the effective weak Hamiltonian within the SM, to pursue various approaches in determining weak matrix elements, and to get a handle on the various uncertainties involved in the prediction of ϵ'/ϵ. Recent detailed reviews [46], [47] of the current status estimate this quantity within the SM $\sim$ *a few* $\times 10^{-4}$. There are considerable uncertainties of this estimate due to (a) cancellations of the QCD penguin and electroweak penguin diagram contributions to ϵ', (b) uncertainties in the knowledge of some SM parameters, notably the mass of the strange quark, and (c) methodic uncertainties in calculating the non-leptonic weak decay matrix elements.

The present high statistics kaon decay experiments [45] can also search for and investigate several rare K decays. For instance, in the case of the decay $K_L \to \pi\pi e^+e^-$, there is a CP asymmetry in the distribution $d\Gamma/d\phi$ (ϕ is the angle between the normal vectors of the e^+e^- and $\pi\pi$ planes) generated by the observed CP violation in $K^0 - \bar{K}^0$ mixing. The asymmetry is predicted to be rather large, about 14 percent [48].

Hyperon decays also offer a possibility to search for CP violation in $\Delta S = 1$ decays – although detectable effects require, very probably, non SM CP-violating interactions. Consider for instance the decay of polarized $\Lambda \to p\pi^-$ and $\bar{\Lambda} \to \bar{p}\pi^+$. The differential decay distributions are proportional to $(1 + \alpha_\Lambda \boldsymbol{\omega}_\Lambda \cdot \hat{p}_p)$ and $(1 + \alpha_{\bar{\Lambda}} \boldsymbol{\omega}_{\bar{\Lambda}} \cdot \hat{p}_{\bar{p}})$, respectively, where $\boldsymbol{\omega}$ is the hyperon polarization vector and $\hat{p}$ is the (anti) proton direction of flight in the hyperon rest frame. The spin analyser quality factor α, which is parity-violating, is generated by the interference of S and P wave amplitudes. CP invariance requires that $\alpha_\Lambda = -\alpha_{\bar{\Lambda}}$. Hence a CP observable is

$$A_\Lambda = \frac{\alpha_\Lambda + \alpha_{\bar{\Lambda}}}{\alpha_\Lambda - \alpha_{\bar{\Lambda}}}. \tag{27}$$

Note that A_Λ is CP-odd but T-even, i.e., even under the reversal of momenta and spins. Hence a non-zero asymmetry (27) requires, apart from CP phases, also absorptive parts in the amplitudes. Neglecting isospin $I = 3/2$ contributions, an approximate expression for A_Λ is given by (see, for instance ref. [49])

$$A_\Lambda \simeq -\tan(\delta^P_{1/2} - \delta^S_{1/2}) \sin(\varphi^P_{1/2} - \varphi^S_{1/2}), \tag{28}$$

where $\delta_{1/2}^{S,P}$ and $\varphi_{1/2}^{S,P}$ are the S,P wave final state phase shifts and weak CP phases for the isospin $I = 1/2$ amplitudes, respectively.

In the Standard Model CP violation in $\Delta S = 1$ hyperon decays is induced by penguin amplitudes. Extensions of the SM may add charged Higgs penguin, gluino penguin contributions, etc. Predictions for hyperon CP observables like A_Λ are usually obtained [50], [51], [52] as follows: within a given model of CP violation one computes first the effective weak $\Delta S = 1$ Hamiltonian at the quark level. (In the SM its next-to-leading order QCD corrections are known [46].) The strong phase shifts $\delta_{1/2}^{S,P}$ are extracted from experimental data. The usual strategy in determining the weak phases $\varphi_{1/2}^{S,P}$ is to take the real parts of the matrix elements $< \pi p|H_{eff}|\Lambda >_{I=1/2}^{S,P}$ from experiment, whereas the CPV part is computed using various models for hadronic matrix elements. Although the theoretical uncertainties are quite large one may conclude [51], [52] from these calculations that within the SM the asymmetry A_Λ is about 4×10^{-5}. Contributions from non SM sources of CP violation can yield larger effects, but are constrained by the ϵ' and ϵ parameters from K decays. He and Valencia conclude that $|A_\Lambda^{non-SM}|$ cannot exceed $a\ few \times 10^{-4}$.

Data from a high statistics hyperon experiment [53] (E871) at Fermilab are presently being analysed. The decay chain $\Xi^- \to \Lambda\pi^- \to p\pi^+\pi^-$ and the corresponding decay chain for $\bar{\Xi}^+$ will be used. They Ξ's will be produced *unpolarized*. Then the Λ polarization is given by $\boldsymbol{\omega}_\Lambda = \alpha_\Xi \hat{\mathbf{p}}_\Lambda$, where $\hat{\mathbf{p}}_\Lambda$ is the Λ direction of flight in the Ξ rest frame. E871 measures the asymmetry

$$A = \frac{\alpha_\Lambda \alpha_\Xi - \alpha_{\bar{\Lambda}} \alpha_{\bar{\Xi}}}{\alpha_\Lambda \alpha_\Xi + \alpha_{\bar{\Lambda}} \alpha_{\bar{\Xi}}} \simeq A_\Lambda + A_\Xi . \tag{29}$$

A_Ξ is estimated to be smaller than A_Λ because of smaller phase shifts. E871 expect to produce about 10^9 events. They aim at a sensitivity $\delta A \simeq 10^{-4}$. If an effect will be observed at this level it will be, in view of the above, most probably of non SM origin.

3.2 Charm

$D^0 - \bar{D}^0$ mixing and associated CP violation in the $\Delta C = 2$ mixing amplitude, and direct CP violation in the $\Delta C = 1$ charm decay amplitudes are predicted to be very small in the SM.

In the SM direct CPV may be significant only for singly Cabibbo suppressed decays. In this case one has at the quark level two contributions to the decay amplitude, namely the usual "tree" amplitude and the penguin amplitude, that have different weak phases. At the hadron level the decay amplitude is of the form $Ae^{i\delta_A} + Be^{i\delta_B}$, where $\delta_{A,B}$ are strong interaction phase shifts. This leads to a CP asymmetry

$$A_D = \frac{\Gamma(D \to f) - \Gamma(\bar{D} \to \bar{f})}{\Gamma(D \to f) + \Gamma(\bar{D} \to \bar{f})} \propto \mathrm{Im}(AB^*) \sin(\delta_B - \delta_A) . \tag{30}$$

Buccella et al. [55] have investigated A_D within the SM for a number of Cabibbo suppressed channels. They calculated the strong phase shifts for the respective channels by assuming dominance of the nearest resonance. For some modes, for instance $D^+ \to \bar{K}^{*0} K^+$ and $D^+ \to \rho^+ \pi^0$ they find $A_D \sim 10^{-3}$. In some extensions of the SM like non-minimal supersymmetry [56] or left-right-symmetric models [57] A_D can be larger by about one order of magnitude. Moreover, asymmetries of the same order could also be generated in these models for Cabibbo allowed and doubly Cabibbo suppressed channels.

$D^0 - \bar{D}^0$ mixing is very small in the SM, $x = \Delta m_D / \Gamma_D << 10^{-2}$. However, quite a number of extensions of the SM, for instance multi-Higgs or supersymmetric extensions, can lead to $x \sim 10^{-2}$. In these models it is quite natural that there is (new) CP violation associated with $\Delta C = 2$ mixing. It is mostly these expectations [58] from SM extensions that nourish the hope of observable mixing and observable indirect and direct CP violation in proposed high statistics charm experiments [54] with 10^8 to 10^9 events.

3.3 Beauty

High statistics experiments with the aim of measuring CPV rate asymmetries in B decays will provide, in the years to come, the decisive tests of the KM mechanism [47], [59]. These asymmetries are characterized by the angles – conventionally called α, β, and γ – of a well-known CKM unitarity triangle, which visualises the following orthogonality relation of the CKM matrix elements:

$$V_{ud} V_{ub}^* + V_{cd} V_{cb}^* + V_{td} V_{tb}^* = 0. \tag{31}$$

Several fits (see, e.g., [60], [61] and for a more recent analysis [47]), using as input the value of ϵ parameter of the K system, $B_d^0 - \bar{B}_d^0$ mixing, etc., have been performed to constrain these angles. These fits yield in particular $0.3 \le \sin(2\beta) \le 0.8$, supporting the expectation that CP violation outside the K system will first be observed through an asymmetry between the rates of B_d^0 and $\bar{B}_d^0 \to J/\Psi + K_S$. The integrated rate asymmetry, which can be calculated in a clean way (that is to say, almost without uncertainties due to hadronic final state interaction phases), is proportional to $\sin(2\beta)$.

Similarly the time integrated rate asymmetry of $B_d^0, \bar{B}_d^0 \to \pi^+\pi^-$ is related to $\sin(2\alpha)$. However, apart from the fact that these modes have very small branching ratios, there is an uncertainty in the prediction of the CP asymmetry because of penguin diagrams contributing to the decay amplitudes. In principle this uncertainty can be eliminated by an isospin analysis [62]. (Recall that there is no QCD penguin contribution to the $I = 3/2$ component of the $B_d \to \pi\pi$ amplitude.) The method requires measuring $B_d^0 \to \pi^+\pi^-, \pi^0\pi^0$ and the conjugated channels, and $B^\pm \to \pi^\pm \pi^0$. It will be difficult to carry out.

The CP parameter $\sin(2\gamma)$ is for instance related to the time integrated asymmetry of the rates $B_s^0, \bar{B}_s^0 \to \rho K_S$. However, that is not a clean and feasible way of extracting $\sin(2\gamma)$: firstly because these modes have very small branching ratios and secondly because of theoretical complications in view of penguin contributions. One proposed alternative is as follows [63]: From the measured decay rates one has to determine the moduli of the decay amplitudes for $B^+ \to D^0 K^+, \bar{D}^0 K^+, D_{1,2}K^+$ and for the charge conjugated channels. ($D_{1,2}$ are the CP- even and odd eigenstates.) From the two triangle relations relating the three complex amplitudes for B^+ and for B^-, respectively, one can obtain $\sin^2\gamma$ up to an ambiguity which can in principle also be resolved. (For other proposals to measure the angles α and γ, see, e.g., the review [47].)

According to the KM mechanism for the three generation SM $\alpha+\beta+\gamma=\pi$. A deviation from this relation would provide evidence for new CP-violating interactions [64]. (If the sum of these angles turns out to be π, note that this does not necessarily imply absence of new CPV effects in the B system.) Of course, more specific searches for new CPV in the B system can be made, for instance by investigating CP observables that are predicted to be small in the SM, e.g., the asymmetry in the rate for $B_s^0 \to J/\Psi + \phi$ and its conjugated mode and, likewise, the rate asymmetry for $B^\pm \to J/\Psi + K^\pm$. (For investigations of non-KM CPV in B decays, see, e.g., [65].)

4 Electric Dipole Moments

The searches for permanent electric dipole moments (EDM), for instance, of the neutron or of an atom with non-degenerate ground state, are known to be very sensitive means to trace new CPV interactions. Recall that a non-zero EDM of a non-degenerate stationary state would signal P and T violation, that is, CP violation assuming CPT invariance. Consider the expectation value of the EDM operator $\mathbf{D} = \int d^3x\mathbf{x}\rho(\mathbf{x})$, where ρ the charge density operator, in a particle state $|j>$ of definite total angular momentum at rest. Rotational invariance tells us that

$$< j|\mathbf{D}|j >= d < j|\mathbf{J}|j >, \tag{32}$$

where $\mathbf{J}$ is the total angular momentum operator. With respect to parity and time-reversal transformations one has

$$\mathbf{D} \xrightarrow{P} -\mathbf{D}, \quad \mathbf{D} \xrightarrow{T} \mathbf{D}, \tag{33}$$

$$\mathbf{J} \xrightarrow{P} \mathbf{J}, \quad \mathbf{J} \xrightarrow{T} -\mathbf{J}. \tag{34}$$

Hence $d \neq 0$ signals P and T violation. This argument applies not only to elementary particles but to atoms and molecules as well, as long as the stationary state under consideration has no energy degeneracies besides those due to rotational invariance. (For an elaborate discussion, see [66].) The experimental signature for an EDM is a linear Stark effect in an external electric field.

A non-zero atomic EDM d_A could be due to a non-zero electron EDM d_e, non-zero nucleon EDMs, P- and T-violating nucleon-nucleon, and/or electron-nucleon interactions. Schematically,

$$d_A = R_A d_e + C_A^{eN} + C_A^N. \tag{35}$$

It has been shown long ago [67] that paramagnetic atoms can have large enhancement factors R_A. (See also [68] for a recent review.) More recent atomic physics calculations [69] obtained for instance for Thallium the factor $R_{Tl} \simeq -585$ with an estimated error of about 10%. For Thallium one has to good approximation $d_{Tl} \simeq d_e R_{Tl} + C_{Tl}^{eN}$. The nuclear contributions can be neglected for the following reasons: The nuclear ground state of ^{205}Tl has spin 1/2 and therefore cannot have a nuclear quadrupole moment. A potential (small) contribution of a Schiff moment of the Thallium nucleus is irrelevant at the present level of experimental sensitivity. From the experimental upper bound [70] on d_{Tl} and with R_{Tl} the upper bound $|d_e| < 4 \cdot 10^{-27} e$ cm was derived [70].
Very precise experimental upper bounds were obtained on the EDMs of certain diamagnetic atoms, in particular for mercury [71]. The mercury EDM, like that of other diamagnetic atoms, is not sensitive to d_e but to the Schiff moment of the ^{199}Hg nucleus which at the quark-parton level would be due to non-zero (chromo) EDMs of quarks and/or P- and T-violating quark-quark or gluonic effective interactions. As the transition from the level of partons to the level of a nucleus involves large uncertainties the experimental limits on the EDMs of diamagnetic atoms are difficult to interpret in terms of microscopic models of CP violation [72].
Experimental searches for a non-zero EDM of the neutron at Grenoble [73] and at Gatchina [74] have lead to the upper limit $|d_n| < 9 \cdot 10^{-26} e$ cm.

Theoretical predictions of the EDM of the electron – or of other leptons – usually constitutes a straightforward problem of perturbation theory because models of CPV are weak coupling theories a posteriori. However, a firm numerical prediction within a given extension of the SM would require knowledge of parameters like masses and couplings of new particles, apart from CP phases. The calculation of d_n and of T-violating nucleon-nucleon interactions, etc. involves in addition methodological uncertainties. For a given model of CPV one can usually construct with reasonable precision the relevant effective P- and T-violating low energy Hamiltonian at the quark gluon level which contains (chromo) EDM operators of quarks, the $G\tilde{G}$ and $GG\tilde{G}$ operators, etc. The transition to the nucleon/nuclear level, that is, the computation of T-violating hadronic matrix elements involves large uncertainties. In computing/estimating the neutron EDM naive dimensional estimates, the quark and the MIT bag model [75], sum rule techniques [76], [77], [78], and experimental constraints on the quark contribution to the nucleon spin [79] have been used.

As was discussed in section 2.1, the KM phase induces only tiny CP-violating effects in flavour-diagonal amplitudes. Hence the SM predicts tiny particle EDMs (barring the strong CP problem of QCD; i.e., assuming $\Theta_{QCD} = 0$). A typical estimate [75] for the neutron is $|(d_n)^{KM}| < 10^{-30} e$ cm. In the SM with massless neutrinos CPV in the lepton sector occurs only as a spill-over from the quark sector: estimates [80] yield $|(d_e)^{KM}| < 10^{-37} e$ cm.

Quite a number of other CPV interactions are conceivable that lead to neutron and electron EDMs of the same order of magnitude as the present experimental upper bounds. (For reviews, see [75], [80], [81].) Multi-Higgs extensions of the SM can contain neutral Higgs particles with indefinite CP parity (cf. section 2.3). Exchange of these bosons induces quark and lepton EDMs already at one loop. For light quarks and leptons the dominant effect occurs at two loops [82]. In two-Higgs doublet extensions [83], [84] of the SM with maximal CPV in the neutral Higgs sector and a light neutral Higgs particle with mass of order 100 GeV neutron and electron EDMs as large as $10^{-25}e$ cm and *a few* $\times 10^{-27}e$ cm, respectively, can be induced. Contributions from charged Higgs boson exchanges can have a similar order of magnitude [85].

In the minimal supersymmetric extension of the SM (MSSM) there are in general, apart from the KM phase, extra CP phases due to complex soft SUSY breaking terms (cf. section 2.3). These phases are not bound to be small a priori. They generate quark and lepton EDMs and chromo EDMS of quarks at one-loop order [86], [79], [89] which can be quite large. (Unless the gaugino, squark or slepton masses are close [87] to 1 TeV which causes, however, other problems.) In particular, the prediction for the electron, which is not clouded by hadronic uncertainties, is $d_e \simeq 10^{-25} \sin\varphi_e$ (e cm) for neutralino and $\tilde{e}$ masses of the order of 100 GeV. That means the leptonic SUSY phase φ_e must be quite small, $\varphi_e \sim 0.01$, which seems unnatural in the generic MSSM case. The constrained versions of the MSSM, mentioned in section 2.3, lead to substantially smaller predictions for the neutron and electron EDMs [22], [88].

In supersymmetric grand unified theories the small phase problem eases by construction, too. In the SO(10) model considered in refs. [23], [89] the phases in the soft terms are assumed to be zero at the Planck scale. Unification of the quarks and leptons of a generation into a single multiplet leads, apart from the KM phase, to extra CKM phases entering the fermion-sfermion gaugino (higgsino) interactions at the weak scale. GIM cancellations lead to a smaller d_n and d_e than in the generic MSSM – but d_e can be close to its experimental upper bound.

Clearly, the present experimental EDM bounds have an impact on the parameter spaces of popular extensions of the SM. In particular the bound on d_e is important in view of the "theoretically clean" predictions. Further improvement of experimental sensitivity is highly desirable. As to future low-energy T violation experiments: A number of proposals [68], [90], [91] have been made to improve the experimental sensitivity to d_e and to the EDMs of certain atoms by factors of 10 to 100. The Berkeley Tl experiment [70] will improve its sensitivity to d_e significantly. An experiment is underway [90] with the paramagnetic molecule YbF, which is very challenging but has the incentive of having a high sensitivity to d_e. There is also a new idea [92] to measure the neutron EDM with substantially improved sensitivity.

The present experimental sensitivity to EDMs of quarks and leptons from the second and third fermion generation is typically of the order 10^{-16} to $10^{-18}e$ cm (see below and [68]). Although this is orders of magnitude larger than the present limit on d_e it constitutes nevertheless interesting information. Some CP-violating interactions, for instance CPV Higgs boson or leptoquark exchange, lead to EDMs in the heavy flavour sector that are much larger than d_e or d_n.

5 High Energy Searches

Many proposals and studies for CP symmetry tests in high energetic e^+e^-, $p\bar{p}$, and pp collisions have been made (see [93], [94], [95], [98], [99] for early studies). In particular the production and decay of τ leptons, b, and t quarks are suitable for this purpose, as it allows for searches of new CPV interactions that become stronger in the heavy flavour sector. Contributions from the KM phase to the phenomena discussed below are negligibly small. Typically one pursues statistical tests with suitable asymmetries or correlations. Consider a reaction where the initial and the final states are eigenstates of CP. This means that the various contributions to the scattering amplitude $\mathcal{T}$, and the observables associated with this reaction, can be classified as being even or odd under a CP transformation. CP tests are to be made with CP-odd observables $\mathcal{O}_{\mathcal{CP}}$ which change sign under a CP transformation. If the scattering amplitude of the reaction is affected by CPV interactions in a significant way, $\mathcal{T} = \mathcal{T}_{inv} + \mathcal{T}_{CPV}$, then the interference of the CP-invariant and the CPV part generates a non-zero expectation value

$$< \mathcal{O}_{CP} > \;=\; \frac{\int d\sigma \mathcal{O}_{CP}}{\int d\sigma} \;\neq\; 0. \tag{36}$$

Because an unpolarized $f\bar{f}$ state is a CP eigenstate in its c.m. frame it can be shown [96] that unpolarized (and transversely polarized) e^+e^- and $p\bar{p}$ collisions allowfor "theoretically clean" CP symmetry tests: in these cases $< \mathcal{O}_{CP} >$ cannot be faked by CP-invariant interactions as long as the phase space cuts are CP-blind. The "self conjugate" situation discussed above can be realized in these cases by comparing data from the reaction $i \to f$ with those of the CP-conjugated one $i \to \bar{f}$. In the case of pp collisions potential contributions from CP-invariant interactions to an observable being used for a CP symmetry test (for instance, T-odd[2] observables will in general receive contributions from QCD absorptive parts) must be carefully discussed.

In order to maximize the sensitivity to CPV couplings it is often useful to consider so-called optimal observables [97] that maximize the signal-to-noise ratio. For a given reaction and a given model of CPV – or a model independent description of CPV using effective Lagrangians or form factors – with only one or a few small parameters these observables can be constructed in a straightforward fashion.

5.1 $e^+e^- \to \tau^+\tau^-$

CPV effects in tau lepton production with e^+e^- collisions and in τ decay were discussed in [98], [99], [100], [101], [102], [103], [104], [105], [106], [107]. CPV in $e^+e^- \to \tau^+\tau^-$ can be traced back to non-zero EDM and weak dipole moment (WDM) form factors [99], [100] $d_\tau^\gamma(s)$ and $d_\tau^Z(s)$, respectively, where $s = E_{c.m.}^2$.

[2] Recall that "T-odd" refers to being odd under the reversal of momenta and spins. The initial and final state are not interchanged.

These form factors induce a number of CP-odd tau polarization asymmetries and spin-spin correlations, for instance a non-zero $d_\tau^Z(s)$ (more precisely, the real part of that form factor) leads to a difference in the polarizations of τ^+ and τ^- orthogonal to the scattering plane. Because the taus auto-analyse their spins through their parity-violating weak decays the tau polarization asymmetries and spin-spin correlations transcribe to a number of CP-odd angular correlations $< \mathcal{O}_{CP} >$ among the final states from $\tau^+\tau^-$ decay.

In their pioneering work the OPAL and ALEPH collaborations [108], [109], [110], [111] at LEP have demonstrated that CP tests in high energy e^+e^- collisions can be performed with an accuracy at the few per mill level. In the meantime the four LEP experiments measured a number of CP-odd correlations in $e^+e^- \to \tau^+\tau^-$. They turned out to be consistent with zero. From these results upper limits on the real and imaginary parts of the WDM form factors were derived. The combined upper limit on the real part is [112] $|\text{Re}d_\tau^Z(s = m_Z^2)| < 3.6 \cdot 10^{-18}e$ cm (95% CL).

As already mentioned above the tau EDM and WDM form factors can be much larger than the electron EDM. There are a number of SM extensions where the dominant contributions to these form factors are one-loop effects, being not suppressed by small fermion masses. In these models one has $d_\tau = e\,\delta/m_Z$ with δ of order α/π. For multi Higgs models one finds [105] that d_τ can reach $10^{-20}e$ cm, whereas CPV scalar leptoquark exchange [105] can lead to d_τ as large as $3 \cdot 10^{-19}e$ cm. In [113] the EDM and WDM form factors were computed in the minimal supersymmetric extension of the SM. These authors obtained d_τ^Z of order $10^{-21}e$ cm.

5.2 $e^+e^- \to b\,\bar{b}$ *gluon(s)*

CP violation in this neutral current reaction would signal new interactions. At the parton level these interactions would affect correlations among parton momenta/energies and parton spins. While the partonic momentum directions can be reconstructed from the jet directions of flight the spin-polarization of the b quark cannot, in general, be determined with reliable precision due to fragmentation. This implies that useful CP observables are primarily those which originate from partonic momentum correlations [94]. With these correlations only chirality-conserving effective couplings can be probed with reasonable sensitivity. Several correlations were proposed and studied [94], [114], [115], [116]. This situation is in contrast to $\tau^+\tau^-$ and $t\bar{t}$ production (see below) where the fermion polarizations can be traced in the decays. That is why in these cases searches for CPV dipole form factors, which are chirality-flipping, can be made with good precision.

In the framework of $\text{SU}(2)_L$-invariant effective Lagrangians it can be shown that chiral invariant CPV effective $Zb\bar{b}G$ interactions of dimension $d = 6$ (after spontaneous symmetry breaking) exist [94], [96]. In multi-Higgs extensions of the SM these interactions can be induced to one-loop order [17]. They remain non-zero in the limit of vanishing b quark mass. Note that these CPV effective interactions are chiral-invariant *and* flavour-diagonal which is a remarkable feature. A dimensionless coupling $\hat{h}_b$ associated with these interactions [115] turns out to be of the order of a typical one-loop radiative correction, i.e., a few percent if CP phases are maximal. This coupling could be larger in models with excited quarks.

At the Z resonance the above reaction provides an excellent possibility to probe for this type of interactions. The ALEPH collaboration [117] has made a CP study with their sample of $Z \to b\bar{b}G$ events. They obtained a limit of $|\hat{h}_b| < 0.59$ at 95% CL.

5.3 Top Quarks and Higgs Bosons

Because of their extremely short lifetime top quarks decay on average before they can hadronize. This means that the spin properties of t quarks can be inferred with good accuracy from their weak decays. (The SM predicts that $t \to W\, b$ is the main decay mode.) Like in the case of the tau lepton a number of t spin-polarization and spin-spin correlation effects may be used to search for non-SM physics. Because of their heavy mass, top quarks – once they are available in sufficiently large numbers – will be a good probe of the electroweak symmetry breaking sector through their Yukawa couplings. In particular they will be a good probe of Higgs sector CP violation. Many CP tests involving top quarks have been proposed. These proposals include $t\bar{t}$ production in high energy e^+e^- collisions [118], [16], [119], [120], [121], [122], [123], [124], [125] and in $p\bar{p}$ and pp collisions [126], [127], [128], [129], [130], [131], [132], [133] at Tevatron and LHC energies, respectively. (As already mentioned, in the latter case no genuine CP tests in the way described above can be made. One must carefully discuss and compute potential fake effects.) Useful channels for these tests are the final states from semi-leptonic decay of both t and $\bar{t}$ and those from semi-leptonic (non-leptonic) $t(\bar{t})$ decay plus the charge conjugated channels. (The charged lepton from semi-leptonic t decay is known to be the most efficient t spin analyzer. Non-leptonic t decays, on the other hand, allow for reconstruction of the top momentum.) Observables $\mathcal{O}_{CP}$ include triple correlations, energy asymmetries, etc. and their optimized versions. Computations of $< \mathcal{O}_{CP} >$ have been made in a model-independent way using effective Lagrangians, form factor parameterizations of the t production and decay vertices, and within several extensions of the SM, notably two-Higgs doublet and supersymmetric extensions. At the upgraded Tevatron one can reach an interesting sensitivity to the chromo EDM form factor of the top of about [126], [130], [132] $\delta d_t^{chromo} \simeq 10^{-18}e$ cm. Multi Higgs extensions of the SM can induce top EDM, WDM, and chromo EDM form factors of this order of magnitude [16], [134]. The minimal SUSY extension of the SM leads to smaller predictions for these form factors [120], [135]. EDM and WDM form factors could be searched for most efficiently in $e^+e^- \to t\bar{t}$ not far above threshold [118], [120], [123]. It was shown [120] that, within two-Higgs doublet extensions of the SM, neutral Higgs sector CP violation induces effects at the percent level in this reaction.

A possibility to check for CPV Yukawa couplings of the t quark would be associated $t\bar{t}$ Higgs boson production. CP effects can be large [136], but the cross sections are quite small.

If neutral Higgs boson(s) φ will be discovered and at least one of them can be produced in reasonably large numbers then the CP properties of the scalar sector could be determined directly by checking whether φ has $J^{PC} = 0^{++}$, 0^{-+}, or whether it has undefined CP parity as predicted by multi-Higgs extensions of the SM with Higgs sector CPV. A number of suggestions and theoretical studies in this respect were made [137], [138], [139], [140], [141], [142], [143], [144], [145], [146], [147]. (Some of them follow the text book descriptions of how to determine the CP parity of π^0.) In the fermion-antifermion decay of a neutral Higgs particle with undefined CP parity CP violation occurs at *tree level* and manifests itself in spin-spin correlations [137]. One of them is CP-odd and can be as large as 0.5. These correlations could be traced in $\varphi \to \tau^+\tau^-$ and, for heavy φ, in $\varphi \to t\bar{t}$; for instance, when φ is produced in high-energetic electron positron collisions. In the case of LHC production, $pp \to \varphi + X \to t\bar{t} + X$, interference with the non-resonant $t\bar{t}$ background diminishes the effect [128], [137].

A "Compton collider" realized by backscattering laser photons off high energy e^- or e^+ beams would be an excellent tool to study Higgs bosons [148] by tuning the beams to resonantly produce φ. The CP properties of φ could be checked by appropriate asymmetries and correlations [141], [145], [146].

6 Summary

The gauge theory paradigm, which describes physics so well up to the highest energy scales presently attainable, suggests that, if there is physics beyond the Standard Theory, there can be a number of different types of CP-violating interactions which manifest themselves in different physical situations. Hence searches for CP violation effects should be made in as many particle reactions as possible. Present experimental investigations of K decays and of hyperon decays search for "direct" CP violation in $|\Delta S| = 1$ weak transitions at the level of 10^{-4}. While an effect of this order of magnitude can be induced by the KM phase in K decays, it would point towards a new source of CPV in the case of hyperon decays. However, in order to be eventually able to discriminate better between different models of CPV improved calculations of hadronic matrix elements both for $K \to 2\pi$ and for non-leptonic hyperon decays are needed. The decisive tests of the KM mechanism will hopefully be provided by the B meson factories in the years to come. The searches for a neutron EDM, atomic EDMs, or other T-violation effects in atoms or molecules remain a unique low energy window to physics beyond the SM. Searches of non-SM CP violation can also be made at present and future high energy colliders. Experiments at LEP have already demonstrated that high-energy CP tests can attain sensitivities at the sub-percent level. Specifically, if Higgs sector CPV exists, effects of up to a few percent are possible in the top quark system. Moreover, when Higgs boson(s) will be discovered and eventually produced in large numbers it is also conceivable to study their CP properties directly.

While at present CP non-conservation may still be considered, from an agnostic point of view, as a curious and small effect of mysterious origin in the neutral kaon system, one can be optimistic that, in view of the activities outlined above, we will have a clearer understanding of the cause of this symmetry violation in the not too distant future.

Acknowledgements

I am indebted to E. Commins, E.A. Hinds, S.K. Lamoreaux, and K.B. Luk for information about their present and planned experiments. Moreover, I wish to thank the organizers of the 1998 Schladming Winter School, W. Plessas and his colleagues, for inviting me to this pleasant meeting.

References

[1] M. Kobayashi and T. Maskawa, *Prog. Theor. Phys.* 49 (1973) 652.
[2] T. Inagaki, *Nucl. Phys. (Proc. Suppl.)* 37A (1994) 197.
[3] M. Dine, R. G. Leigh and D. A. MacIntire, *Phys. Rev. Lett.* 69 (1992) 2030; K. Choi, D. B. Kaplan and A. D. Nelson, *Nucl. Phys.* B391 (1993) 515.
[4] L. L. Chau and W. Y. Keung, *Phys. Rev. Lett.* 53 (1984) 1802; G. C. Branco and L. Lavoura, *Phys. Lett.* B208 (1988) 123; J. D. Bjorken, *Phys. Rev.* D39 (1989) 1396.
[5] C. Jarlskog, *Phys. Rev. Lett.* 55 (1985) 1039.
[6] H. Fritzsch, lectures given at this school.
[7] J. Bernabéu, G. C. Branco and M. Gronau, *Phys. Lett.* 169B (1986) 243.
[8] A. Czarnecki and B. Krause, *Phys. Rev. Lett.* 78 (1997) 4339.
[9] V. Baluni, *Phys. Rev.* D19 (1979) 2227; R. Crewther et al., *Phys. Lett.* 88B (1979) 123; *ibid.* 91B (1980) 487 (E).
[10] G. 't Hooft, in: *Recent Developments in Gauge Theories*, Proccedings of the 1979 Cargèse Summer Institute, G. 't Hooft et al. (Eds.), Plenum Press (1980).
[11] R. Peccei, preprint hep-ph/9606475 (1996), and lectures given at this school.
[12] S. Weinberg, *Phys. Rev.* D42 (1990) 860.
[13] J. Gunion, H. Haber, G. Kane and S. Dawson, *The Higgs Hunter's Guide*, Addison-Wesley, New York (1990).
[14] F. M. Borzumati and C. Greub, preprint hep-ph/9802391 (1998).
[15] N. G. Deshpande and E. Ma, *Phys. Rev.* D16 (1977) 1583.
[16] W. Bernreuther, T. Schröder and T. N. Pham, *Phys. Lett.* B279 (1992) 389.
[17] W. Bernreuther, A. Brandenburg, P. Haberl, and O. Nachtmann, *Phys. Lett.* B387 (1996) 155.
[18] Y. L. Wu and L. Wolfenstein, *Phys. Rev. Lett.* 73 (1994) 1762.
[19] H. P. Nilles, *Phys. Rep.* 110C (1984) 1; and lectures given at this school.
[20] H. E. Haber and G. L. Kane, *Phys. Rep.* 117C (1985) 75.
[21] M. Dugan, M. Grinstein and L. J. Hall, *Nucl. Phys.* B255 (1985) 413.
[22] S. Bertolini and F. Vissani, *Phys. Lett.* B324 (1994) 164.

[23] S. Dimopoulos and L. J. Hall, *Phys. Lett.* B344 (1995) 185.
[24] M. Matsuda and M. Tanimoto, *Phys. Rev.* D52 (1995) 3100;
N. Haba, *Prog. Theor. Phys.* 97 (1997) 301.
[25] Y. Zeldovich, I. Kobzarev and L. Okun, *Sov. Phys.* JETP 40 (1974) 1.
[26] T. D. Lee, *Phys. Rev.* D8 (1973) 1226; *Phys. Rep.* C9 (1974) 143.
[27] G. C. Branco and M. N. Rebelo, *Phys. Lett.* B160 (1985) 117.
[28] J. Liu and L. Wolfenstein, *Nucl. Phys.* B289 (1987) 1.
[29] G. C. Branco, J. M. Gérard and W. Grimus, *Phys. Lett.* 136B (1984) 383.
[30] S. Weinberg, *Phys. Rev. Lett.* 37 (1976) 657.
[31] G. C. Branco, *Phys. Rev. Lett.* 44 (1980) 504.
[32] N. Maekawa, *Phys. Lett.* B282 (1992) 387;
A. Pomarol, *Phys. Lett.* B287 (1992) 331;
N. Haba, *Phys. Lett.* B398 (1997) 305;
O. C. W. Kong and F. L. Lin, *Phys. Lett.* B419 (1998) 217.
[33] H. Georgi and A. Pais, *Phys. Rev.* D10 (1974) 1246.
[34] J. Romao, *Phys. Lett.* B173 (1886) 309;
A. Pomarol, *Phys. Rev.* D47 (1993) 273.
[35] K. S. Babu and S. M. Barr, *Phys. Rev.* D49 (1994) R2156;
N. Haba, in: KEK proceedings 97-12 (1997).
[36] A. Mendez and A. Pomarol, *Phys. Lett.* B272 (1991) 313;
L. Lavoura and J. P. Silva, *Phys. Rev.* D50 (1994) 4619.
[37] A. Sakharov, *JETP Lett.* 5 (1967) 24.
[38] Z. Fodor, KEK preprint KEK-TH-552 (1997).
[39] A. G. Cohen, D.B. Kaplan and A. E. Nelson, *Annu. Rev. Nucl. Part. Sci.* 43 (1993) 27;
V. A. Rubakov and M. E. Shaposhnikov, *Usp. Fiz. Nauk* 166 (1996) 493;
A. D. Dolgov, hep-ph/9707419 (1997).
[40] J. M. Cline and P. A. Lemieux, *Phys. Rev.* D56 (1997) 3873.
[41] B. de Carlos and J. R. Espinosa, *Nucl. Phys.* B503 (1997) 24;
J. M. Cline and G. D. Moore, preprint hep-ph/9806354 (1998).
[42] M. Fukugita and T. Yanagida, *Phys. Lett.* B174 (1986) 45;
M. A. Luty, *Phys. Rev.* D45 (1992) 455;
M. Flanz, E. Paschos and U. Sarkar, *Phys. Lett.* B345 (1995) 248;
W. Buchmüller and M. Plümacher, *Phys. Lett.* B389 (1996) 73;
M. Plümacher, *Z. Phys.* C74 (1997) 549.
[43] G. D. Barr *et al.*, *Phys. Lett.* B317 (1993) 233.
[44] L. K. Gibbons *et al.*, *Phys. Rev. Lett.* 70 (1993) 1203.
[45] H. Wahl, lectures given at this school.
[46] A. J. Buras, M. Jamin and M. E. Lautenbacher, *Phys. Lett.* B389 (1996) 749;
G. Buchalla, A. J. Buras and M. E. Lautenbacher, *Rev. Mod. Phys.* 68 (1996) 1125;
A. J. Buras, preprint hep-ph/9806471 (1998).
[47] A.J. Buras and R. Fleischer, preprint hep-ph/9704376 (1997).
[48] L. M. Sehgal and M. Wanninger, *Phys. Rev.* D46 (1992) 1035; *ibid.* D46 (1992) 5209 (E);
L. M. Sehgal and P. Heiliger, *Phys. Rev.* D48 (1993) 4146.
[49] W. Grimus, *Fortschr. Phys.* 36 (1988) 201.
[50] J. F. Donoghue, X. G. He and S. Pakvasa, *Phys. Rev.* D34 (1986) 833;
J. F. Donoghue, B. R. Holstein and G. Valencia, *Phys. Lett.* B178 1986 319;
E. A. Paschos and Y. L. Wu, *Mod. Phys. Lett.* A6 (1991) 93.

[51] X. G. He, H. Steger and G. Valencia, *Phys. Lett.* B272 (1991) 411.
[52] X. G. He and G. Valencia, *Phys. Rev.* D52 (1995) 5257;
G. Valencia, preprint hep-ph/9511242 (1995).
[53] J. Antos *et al.*, Fermilab Proposal P-871 (1994);
K. B. Luk, LBL Berkeley preprint LBNL-41269 (1998).
[54] D. Kaplan, *Nucl. Phys. (Proc. Suppl.)* B50 (1996) 260.
[55] F. Buccella *et al.*, *Phys. Rev.* D51 (1995) 3478.
[56] I. Bigi, preprint hep-ph/9412227 (1994).
[57] A. Le Yaouanc et al. preprint hep-ph/9504270 (1995).
[58] G. Blaylock, A. Seiden and Y. Nir, *Phys. Lett.* B335 (1995) 555;
L. Wolfenstein, *Phys. Rev. Lett.* 75 (1995) 2460.
[59] I. Bigi et al., in: *CP Violation*, C. Jarlskog (Ed.) World Scientific, Singapore (1988), p. 175;
Y. Nir and H. Quinn, in: *B Decays* S. Stone (Ed.) World Scientific, Singapore (1992), p. 362;
I. Dunietz, *ibid.* p. 393.
[60] A. Ali and D. London, preprint hep-ph/9508272 (1995).
[61] A. Pich and J. Prades, *Phys. Lett.* B346 (1995) 342.
[62] M. Gronau and D. London, *Phys. Rev. Lett.* 65 (1990) 3381.
[63] M. Gronau and D. Wyler, *Phys. Lett.* B265 (1991) 172.
[64] I. Bigi and F. Gabbiani, *Nucl. Phys.* B352 (1991) 309.
[65] M. P. Worah, preprint hep-ph/9711265 (1997);
A. L. Kagan and M. Neubert, preprint hep-ph/9803368 (1998).
[66] H. Rupertsberger, *Found. Phys.* 20 (1990) 1079.
[67] P. G. H. Sandars, *Phys. Lett.* 14 (1965) 194.
[68] E. D. Commins, *Electric Dipole Moments of Leptons*, Berkeley preprint (1997), to be published in *Advances in Atomic and Molecular Physics* (1998).
[69] Z. W. Liu and H. P. Kelly, *Phys. Rev.* A45 (1992) R4210.
[70] E. D. Commins, S. B. Ross, D. DeMille and B. C. Regan, *Phys. Rev.* A50 (1994) 2960.
[71] J. P. Jacobs *et al.*, *Phys. Rev.* A52 (1995) 3521.
[72] V. M. Khatsymovsky, I. B. Khriplovich and A. S. Yelkhovsky, *Ann. Phys.* 186 (1988) 1.
[73] K. F. Smith et al., *Phys. Lett.* B234 (1990) 191.
[74] I. S. Altarev et al., *Phys. Lett.* B276 (1992) 242.
[75] X. G. He, B. H. J. McKellar and S. Pakvasa, *Int. J. Mod. Phys.* A4 (1989) 5011; *ibid.* A6 (1991) 1063 (E).
[76] M. Chemtob, *Phys. Rev.* D45 (1992) 1649.
[77] I. B. Khriplovich, *Phys. Lett.* B382 (1996) 145.
[78] I. B. Khriplovich and K. N. Zyablyuk, *Phys. Lett.* B383 (1996) 429.
[79] J. Ellis and R. Flores, *Phys. Lett.* B377 (1996) 83.
[80] W. Bernreuther and M. Suzuki, *Rev. Mod. Phys.* 63 (1991) 313.
[81] S. M. Barr, *Int. J. Mod. Phys.* A8 (1993) 209.
[82] S. M. Barr and A. Zee, *Phys. Rev. Lett.* 65 (1990) 21.
[83] R. G. Leigh, S. Paban and R. M. Xu, *Nucl. Phys.* B352 (1991) 45;
J. F. Gunion and R. Vega, *Phys. Lett.* B251 (1990) 157;
D. Chang, W. Y. Keung and T. C. Yuan, *Phys. Rev.* D43 (1991) 14.
[84] T. Hayashi et al., *Phys. Lett.* B348 (1995) 489.
[85] D. B. Chao, D. Chang and W. Y. Keung, *Phys. Rev. Lett.* 79 (1997) 1988.

[86] J. Ellis, S. Ferrara and D. V. Nanopoulos, *Phys. Lett.* B114 (1982) 231;
W. Buchmüller and D. Wyler, *Phys. Lett.* B121 (1983) 321;
J. Polchinski and M. B. Wise, *Phys. Lett.* B125 (1983) 393;
F. del Aguila et al., *Phys. Lett.* B126 (1983) 71;
J. M. Frère and M. B. Gavela, *Phys. Lett.* B132 (1983) 107;
F. del Aguila et al., *Phys. Lett.* B126 (1983) 71;
E. Franco and M. Mangano, *Phys. Lett.* B135 (1984) 445;
J. M. Gérard et al., *Phys. Lett.* B140 (1984) 349;
P. Nath, *Phys. Rev. Lett.* 66 (1991) 2565.
[87] Y. Kizukuri and N. Oshimo, *Phys. Rev.* D46 (1992) 3025.
[88] T. Inui et al., *Nucl. Phys.* B449 (1995) 49;
S. A. Abel, W. N. Cottingham and I. B. Whittingham, *Phys. Lett.* B370 (1996) 106.
[89] R. Barbieri, A. Romanino and A. Strumia, *Phys. Lett.* B369 (1996) 283.
[90] E. A. Hinds, *Physica Scripta* T70 (1997) 34;
B. E. Sauer, J. Wang, and E. A. Hinds, *J. Chem. Phys.* 105 (1996) 7412.
[91] N. F. Ramsey and A. Weis, *Phys. Bl.* 52 (1996) 859.
[92] R. Golub and S. K. Lamoreaux, *Phys. Rep.* C237 (1994) 2.
[93] J. F. Donoghue and G. Valencia, *Phys. Rev. Lett.* 58 (1987) 451;
ibid. 60 (1988) 243 (E).
[94] W. Bernreuther, U. Löw, J. P. Ma and O. Nachtmann, *Z. Phys.* C43 (1989) 117.
[95] M. B. Gavela, F. Iddir, A. Le Yaouanc, L. Oliver, O. Pène and J. C. Raynal, *Phys. Rev.* D39 (1989) 1870.
[96] W. Bernreuther and O. Nachtmann, *Phys. Lett.* B268 (1991) 424.
[97] D. Atwood and A. Soni, *Phys. Rev.* D45 (1992) 2405;
M. Davier, L. Duflot, F. Le Diberder and A. Rougé, *Phys. Lett.* B306 (1993) 411;
P. Overmann, Dortmund preprint DO-TH 93-24 (1993);
M. Diehl and O. Nachtmann, *Z. Phys.* C62 (1994) 397.
[98] F. Hoogeveen and L. Stodolsky, *Phys. Lett.* B212 (1988) 505.
[99] W. Bernreuther and O. Nachtmann, *Phys. Rev. Lett.* 63 (1989) 2787.
[100] W. Bernreuther, G. W. Botz, O. Nachtmann and P. Overmann, *Z. Phys.* C52 (1991) 567.
[101] S. Goozovat and C. A. Nelson, *Phys. Lett.* B267 (1991) 128.
[102] C. A. Nelson, *Phys. Lett.* B355 (1995) 561. C. A. Nelson et al., *Phys. Rev.* D50 (1994) 4544.
[103] J. Bernabéu, G. A. Gonzáles-Sprinberg and J. Vidal, *Phys. Lett.* B326 (1994) 168.
[104] B. Ananthanarayan and A. S. Rindani, *Phys. Rev.* D51 (1995) 5996.
[105] W. Bernreuther, A. Brandenburg, and P. Overmann, *Phys. Lett.* B391 (1997) 413; ibid. B412 (1997) 425 (E).
[106] S. Y. Choi, K. Hagiwara, and M. Tanabashi, *Phys. Rev.* D52 (1995) 1614.
[107] J. H. Kühn and E. Mirkes, *Phys. Lett.* B398 (1997) 407.
[108] P. D. Acton et al. (OPAL collab.), *Phys. Lett.* B281 (1992) 405.
[109] R. Akers et al. (OPAL collab.), *Z. Phys.* C66 (1995) 31.
[110] D. Buskulic et al. (ALEPH collab.), *Phys. Lett.* B297 (1992) 459.
[111] D. Buskulic et al. (ALEPH collab.), *Phys. Lett.* B346 (1995) 371.
[112] N. Wermes, *Nucl. Phys. (Proc. Suppl.)* 55C (1997) 313.
[113] W. Hollik et al., *Phys. Lett.* B425 (1998) 322.
[114] J. Körner, J. P. Ma, R. Münch, O. Nachtmann and R. Schöpf, *Z. Phys.* C49 (1991) 447.

[115] W. Bernreuther, G. W. Botz, D. Bruß, P. Haberl and O. Nachtmann, *Z. Phys.* C68 (1995) 73.
[116] K. J. Abraham and B. Lampe, *Phys. Lett.* B326 (1994) 175.
[117] D. Buskulic et al. (ALEPH collab.), *Phys. Lett.* B384 (1996) 365.
[118] W. Bernreuther, O. Nachtmann, P. Overmann and T. Schröder, *Nucl. Phys.* B388 (1992) 53; *ibid.* B406 (1993) 516 (E).
[119] G.L. Kane, G. A. Ladinsky and C.-P. Yuan: *Phys. Rev.* D45 (1991) 124.
[120] W. Bernreuther and P. Overmann, *Z. Phys.* C61 (1994) 599;
W. Bernreuther and P. Overmann, *Z. Phys.* C72 (1996) 461.
[121] T. Arens and L.M. Sehgal, *Phys. Rev.* D50 (1994) 4372.
[122] B. Grzadkowski, *Phys. Lett.* B305 (1993) 384.
[123] F. Cuypers and S.D. Rindani, *Phys. Lett.* B343 (1995) 333;
S. Poulose and S.D. Rindani, *Phys. Lett.* B349 (1995) 379.
[124] A. Pilaftsis and M. Nowakowski, *Int. Journ. of Mod. Phys.* A9 (1994) 1097;
D. Atwood and A. Soni, preprint hep-ph/9607481 (1996).
[125] A. Bartl, E. Christova and W. Majerotto, *Nucl. Phys.* B460 (1996) 235; *ibid.* B465 (1996) 365 (E)
A. Bartl, E. Christova, T. Gajdosik, and W. Majerotto, preprint hep-ph/9802352.
[126] J.P. Ma and A. Brandenburg, *Z. Phys.* C56 (1992) 97;
A. Brandenburg and J.P. Ma, *Phys. Lett.* B298 (1993) 211.
[127] C.R. Schmidt and M.E. Peskin, *Phys. Rev. Lett.* 69 (1992) 410.
[128] W. Bernreuther and A. Brandenburg, *Phys. Rev.* D49 (1994) 4481.
[129] C. R. Schmidt, *Phys. Lett.* B293 (1992) 111.
[130] P. Haberl, O. Nachtmann and A. Wilch, *Phys. Rev.* D53 (1996) 4875.
[131] D. Atwood et al., *Phys. Rev.* D54 (1996) 5412.
[132] B. Grzadkowski, B. Lampe and K. J. Abraham, *Phys. Lett.* B415 (1997) 193.
[133] H. Y. Zhou, preprint hep-ph/9805538 (1998).
[134] A. Soni and R. M. Xu, *Phys. Rev. Lett.* 69 (1992) 33.
[135] A. Bartl, E. Christova, T. Gajdosik, and W. Majerotto, *Nucl. Phys.* B507 (1997) 35.
[136] S. Bar-Shalom et al., *Phys. Rev.* D53 (1996) 1162;
J.F. Gunion, B. Grzadkowski and X. G. He, preprint UCD-96-14 (1996).
[137] W. Bernreuther and A. Brandenburg, *Phys. Lett.* B314 (1993) 104;
W. Bernreuther, A. Brandenburg and M. Flesch, *Phys. Rev.* D56 (1997) 90.
[138] A. Skjold and P. Osland, *Phys. Lett.* B311 (1993) 261;
A. Skjold and P. Osland, *Phys. Lett.* B329 (1994) 305.
[139] D. Chang, W. Y. Keung and I. Phillips, *Phys. Rev.* D48 (1993) 3225.
[140] X. G. He, J. P. Ma and B. H. J. McKellar, *Phys. Rev.* D49 (1994) 4548.
[141] M. Krämer, J. Kühn, M. Stong and P. Zerwas, *Z. Phys.* C64 (1994) 21.
[142] V. Barger, K. Cheung, A. Djouadi, B. Kniehl and P. Zerwas, *Phys. Rev.* D49 (1994) 79.
[143] T. Arens, U. Gieseler and L. M. Sehgal, *Phys. Lett.* B339 (1994) 127;
T. Arens and L. M. Sehgal, *Z. Phys.* C66 (1995) 89.
[144] W. N. Cottingham and I. B. Whittingham, *Phys. Rev.* D52 (1995) 539.
[145] B. Grzadkowski and J.F. Gunion, *Phys. Lett.* B294 (1992) 361.
[146] H. Anlauf, W. Bernreuther and A. Brandenburg, *Phys. Rev.* D52 (1995) 3803; *ibid.* D53 (1996) 1725 (E).
[147] A. Pilaftsis, *Phys. Rev. Lett.* 77 (1996) 4996.

[148] Proceedings of the European Linear Collider Workshop 1995: e^+e^- *Collisions at TeV Energies*, P. Zerwas (Ed.), DESY orange report DESY 96-123D (1996).

CP Violation: Experimental Status and Prospects

H. Wahl

CERN, CH-1211 Geneva 23, Switzerland

Abstract. Since the discovery of CP violation through the observation of decays of longlived neutral kaons into two pions a number of experiments have been performed to reveal its origin. Precision measurements in the neutral kaon system are discussed and compared to the effects expected in the Standard Model. The basic parameters in the phenomenology of CP violation have been determined from the decay amplitudes into two pions and from the semileptonic charge asymmetry as follows:

$$|\eta_{+-}| = (2.28 \pm 0.02) \times 10^{-3}$$

$$|\eta_{00}/\eta_{+-}|^2 = 0.991 \pm 0.003$$

$$\Phi_{+-} = 43.7^\circ \pm 0.6^\circ$$

$$\Phi_{00} = 43.5^\circ \pm 1.0^\circ$$

$$\mathrm{Re}\,\varepsilon = (1.65 \pm 0.06) \times 10^{-3}$$

They can be used to determine the complex phase of the CKM-matrix and to perform stringent tests of the CPT theorem. The parameter ε implies a mixing between CP even and odd states in short- and longlived physical states at the level of 2×10^{-3} in amplitude. State mixing is the dominant source of CP violation. The observed small difference between the ratio of amplitudes for decays into charged and neutral pions indicates also a directly CP violating contribution at the level of 10^{-6} to the decay of the CP odd state K_2 into two pions.

CP violation is well established in the $K^0 - \overline{K}^0$ system. All experimental results are consistent with the Standard Model of electroweak interactions. CP violation has not been found anywhere else but is expected to be significant also in the B^0 system. A definite confirmation of direct CP violation remains an important issue. Ongoing experimental activities are reviewed as well as prospects to study CP violating effects in K- and B-meson decays in fixed target experiments, with dedicated electron-positron colliders and in hadronic heavy-quark production at high energies.

Spontaneously Broken Symmetries

H. Narnhofer and W. Thirring

Institut für Theoretische Physik, Universität Wien
Boltzmanngasse 5, A-1090 Wien, Austria

Abstract. Symmetries are automorphisms of the observable algebra $\mathcal{A}$. There are two notions of spontaneous symmetry breaking. The stronger refers to a state ω over $\mathcal{A}$ such that the symmetry does not extend to the strong closure $\pi_\omega(\mathcal{A})''$ where π_ω is the GNS-representation associated to ω. The standard example for that are the Schwinger terms $[j(x), j(x')] = i\delta'(x - x')$, $x \in \mathbf{R}$ which are not invariant under the parity $j(x) \to j(-x)$ though the j's are constructed from fermion fields $\psi(x)$ as strong limits and $\psi(x) \to \psi(-x)$ is a symmetry of the fermion algebra. The weaker notion which can also be realized in elementary quantum mechanics refers to a time evolution $\tau_t \in \text{Aut}\,\mathcal{A}$. A symmetry σ is said to be spontaneously broken if there is a state ω with $\omega \circ \tau_t = \omega$, $\omega \circ \sigma \neq \omega$ though $[\sigma, \tau_t] = 0$.

We are interested in the latter case, i.e. in states that are invariant under time evolution but not under a gauge symmetry that commutes with time evolution*. The possibility, if such states exist, can be decided on the level of the gauge invariant subalgebra by using the crossed product construction and there reduces to an eigenvalue problem. The method is applied to the Fermi algebra and spin systems. In particular, it is found that a translation invariant state cannot break the symmetry $\psi \to -\psi$ where ψ is a fermionic field operator.

*) To appear in the "Annales de l'Institut Henri Poincaré, Physique Théorique".

Chiral Symmetry

Gerhard Ecker

Institut für Theoretische Physik, Universität Wien
Boltzmanngasse 5, A-1090 Wien, Austria

Abstract. Broken chiral symmetry has become the basis for a unified treatment of hadronic interactions at low energies. After reviewing mechanisms for spontaneous chiral symmetry breaking, I outline the construction of the low-energy effective field theory of the Standard Model called chiral perturbation theory. The loop expansion and the renormalization procedure for this nonrenormalizable quantum field theory are developed. Evidence for the standard scenario with a large quark condensate is presented, in particular from high-statistics lattice calculations of the meson mass spectrum. Elastic pion–pion scattering is discussed as an example of a complete calculation to $O(p^6)$ in the low-energy expansion. The meson–baryon system is the subject of the last lecture. After a short summary of heavy baryon chiral perturbation theory, a recent analysis of pion–nucleon scattering to $O(p^3)$ is reviewed. Finally, I describe some very recent progress in the chiral approach to the nucleon–nucleon interaction.

1 The Standard Model at Low Energies

My first Schladming Winter School took place exactly 30 years ago. Recalling the program of the 1968 School (Urban 1968), many of the topics discussed at the time are still with us today. In particular, chiral symmetry was very well represented in 1968, with lectures by S. Glashow, F. Gursey and H. Leutwyler. In those pre–QCD days, chiral Lagrangians were already investigated in much detail but the prevailing understanding was that due to their nonrenormalizability such Lagrangians could not be taken seriously beyond tree level. The advent of renormalizable gauge theories at about the same time seemed to close the chapter on chiral Lagrangians.

More than ten years later, after an influential paper of Weinberg (1979) and especially through the systematic analysis of Gasser and Leutwyler (1984, 1985), effective chiral Lagrangians were taken up again when it was realized that in spite of their nonrenormalizability they formed the basis of a consistent quantum field theory. Although QCD was already well established by that time the chiral approach was shown to provide a systematic low–energy approximation to the Standard Model in a regime where QCD perturbation theory was obviously not applicable.

Over the years, different approaches have been pursued to investigate the Standard Model in the low–energy domain. Most of them fall into the following three classes:

i. QCD–inspired models
   There is a large variety of such models with more or less inspiration from

QCD. Most prominent among them are different versions of the Nambu–Jona-Lasinio model (Nambu and Jona-Lasinio 1961; Bijnens 1996 and references therein) and chiral quark models (Manohar and Georgi 1984; Bijnens et al. 1993). Those models have provided a lot of insight into low–energy dynamics but in the end it is difficult if not impossible to disentangle the model dependent results from genuine QCD predictions.

ii. Lattice QCD

iii. Chiral perturbation theory (CHPT)
The underlying theory with quarks and gluons is replaced by an effective field theory at the hadronic level. Since confinement makes a perturbative matching impossible, the traditional approach (Weinberg 1979; Gasser and Leutwyler 1984, 1985; Leutwyler 1994) relies only on the symmetries of QCD to construct the effective field theory. The main ingredient of this construction is the spontaneously (and explicitly) broken chiral symmetry of QCD.

The purpose of these lectures is to introduce chiral symmetry as a leit–motiv for low–energy hadron physics. The first lecture starts with a review of spontaneous chiral symmetry breaking. In particular, I discuss a recent classification of possible scenarios of chiral symmetry breaking by Stern (1998) and a connection between the quark condensate and the V, A spectral functions in the large–N_c limit (Knecht and de Rafael 1997). The ingredients for constructing the effective chiral Lagrangian of the Standard Model are put together. This Lagrangian can be organized in two different ways depending on the chiral counting of quark masses: standard vs. generalized CHPT. To emphasize the importance of renormalizing a nonrenormalizable quantum field theory like CHPT, the loop expansion and the renormalization procedure for the mesonic sector are described in some detail. After a brief review of quark mass ratios from CHPT, I discuss the evidence from lattice QCD in favour of a large quark condensate. The observed linearity of the meson masses squared as functions of the quark masses is consistent with the standard chiral expansion to $O(p^4)$. Moreover, it excludes small values of the quark condensate favoured by generalized CHPT. Elastic pion–pion scattering is considered as an example of a complete calculation to $O(p^6)$ in the low–energy expansion. Comparison with forthcoming experimental data will allow for precision tests of QCD in the confinement regime. Once again, the quark condensate enters in a crucial way. In the meson–baryon sector, the general procedure of heavy baryon CHPT is explained for calculating relativistic amplitudes from frame dependent amplitudes. As an application, I review the analysis of Mojžiš (1998) for elastic πN scattering to $O(p^3)$. Finally, some promising new developments in the chiral treatment of the nucleon–nucleon interaction are discussed.

1.1 Broken Chiral Symmetry

The starting point is an idealized world where $N_f = 2$ or 3 of the quarks are massless (u, d and possibly s). In this chiral limit, the QCD Lagrangian

$$\begin{aligned}\mathcal{L}^0_{\mathrm{QCD}} &= \overline{q} i\gamma^\mu \left(\partial_\mu + i g_s \frac{\lambda_\alpha}{2} G^\alpha_\mu\right) q - \frac{1}{4} G^\alpha_{\mu\nu} G^{\alpha\mu\nu} + \mathcal{L}_{\text{heavy quarks}} \\ &= \overline{q_L} i \not{D} q_L + \overline{q_R} i \not{D} q_R - \frac{1}{4} G^\alpha_{\mu\nu} G^{\alpha\mu\nu} + \mathcal{L}_{\text{heavy quarks}} \\ q_{R,L} &= \frac{1}{2}(1 \pm \gamma_5) q \qquad\qquad q = \begin{pmatrix} u \\ d \\ [s] \end{pmatrix}\end{aligned} \tag{1}$$

exhibits a global symmetry

$$\underbrace{SU(N_f)_L \times SU(N_f)_R}_{\text{chiral group } G} \times U(1)_V \times U(1)_A \,.$$

At the effective hadronic level, the quark number symmetry $U(1)_V$ is realized as baryon number. The axial $U(1)_A$ is not a symmetry at the quantum level due to the Abelian anomaly ('t Hooft 1976; Callan et al. 1976; Crewther 1977) that leads for instance to $M_{\eta'} \neq 0$ even in the chiral limit.

A classical symmetry can be realized in quantum field theory in two different ways depending on how the vacuum responds to a symmetry transformation. With a charge $Q = \int d^3x J^0(x)$ associated to the Noether current $J^\mu(x)$ of an internal symmetry and for a translation invariant vacuum state $|0\rangle$, the two realizations are distinguished by the

Goldstone alternative

$Q\vert 0\rangle = 0$	$\vert\vert Q\vert 0\rangle\vert\vert = \infty$
Wigner–Weyl	Nambu–Goldstone
linear representation	nonlinear realization
degenerate multiplets	massless Goldstone bosons
exact symmetry	spontaneously broken symmetry

There is compelling evidence both from phenomenology and from theory that the chiral group G is indeed spontaneously broken :

i Absence of parity doublets in the hadron spectrum.
ii. The $N_f^2 - 1$ pseudoscalar mesons are by far the lightest hadrons.

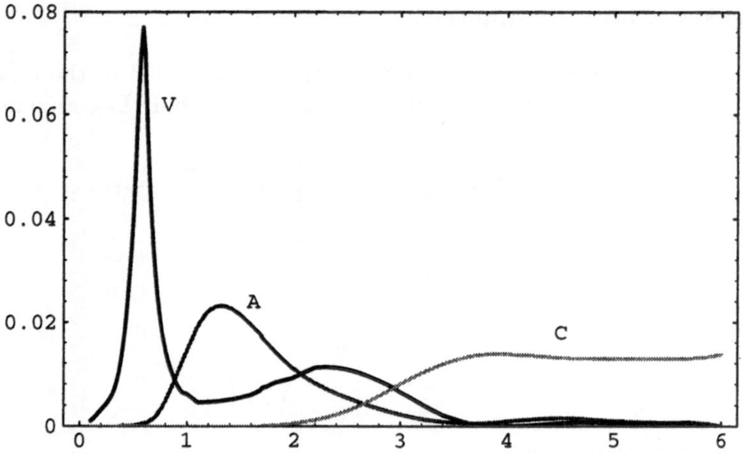

Fig. 1. Vector and axial–vector spectral functions in the $I = 1$ channel as functions of s (in GeV^2) from Donoghue and Perez (1997). V, A stand for the isovector resonance contributions and C denotes the (common) continuum contribution.

iii. The vector and axial–vector spectral functions are quite different as shown in Fig. 1.
iv. The anomaly matching conditions ('t Hooft 1980; Frishman et al. 1981; Coleman and Grossman 1982) together with confinement require the spontaneous breaking of G for $N_f \geq 3$.
v. In vector–like gauge theories like QCD (with the vacuum angle $\theta_{QCD} = 0$), vector symmetries like the diagonal subgroup of G, $SU(N_f)_V$, remain unbroken (Vafa and Witten 1984).
vi. There is by now overwhelming evidence from lattice gauge theories (see below) for a nonvanishing quark condensate.

All these arguments together suggest very strongly that the chiral symmetry G is spontaneously broken to the vectorial subgroup $SU(N_f)_V$ (isospin for $N_f = 2$, flavour $SU(3)$ for $N_f = 3$):

$$G \longrightarrow H = SU(N_f)_V \ . \tag{2}$$

To investigate the underlying mechanism further, let me recall one of the standard proofs of the Goldstone theorem (Goldstone 1961): starting with the charge operator in a finite volume V, $Q^V = \int_V d^3x J^0(x)$, one assumes the existence of a (local) operator A such that

$$\lim_{V\to\infty} \langle 0|[Q^V(x^0), A]|0\rangle \neq 0 \ , \tag{3}$$

which is of course only possible if

$$Q|0\rangle \neq 0 \ . \tag{4}$$

Then the Goldstone theorem tells us that there exists a massless state $|G\rangle$ with

$$\langle 0|J^0(0)|G\rangle\langle G|A|0\rangle \neq 0 \ . \tag{5}$$

The left–hand side of Eq. (3) is called an order parameter of the spontaneous symmetry breaking. The relation (5) contains two nonvanishing matrix elements. The first one involves only the symmetry current and it is therefore independent of the specific order parameter:

$$\langle 0|J^0(0)|G\rangle \neq 0 \tag{6}$$

is a necessary and sufficient condition for spontaneous breaking. The second matrix element in (5), on the other hand, does depend on the order parameter considered. Together with (6), its nonvanishing is sufficient but of course not necessary for the Nambu–Goldstone mechanism.

In QCD, the charges in question are the axial charges

$$Q^i_A = Q^i_R - Q^i_L \qquad (i = 1, \ldots, N_f^2 - 1) \ . \tag{7}$$

Which is (are) the order parameter(s) of spontaneous chiral symmetry breaking in QCD? From the discussion above, we infer that the operator A in (3) must be a colour–singlet, pseudoscalar quark–gluon operator. The unique choice for a local operator in QCD with lowest operator dimension three is[1]

$$A_i = \overline{q}\gamma_5\lambda_i q \tag{8}$$

with

$$[Q^i_A, A_j] = -\frac{1}{2}\overline{q}\{\lambda_i, \lambda_j\}q \ . \tag{9}$$

If the vacuum is invariant under $SU(N_f)_V$,

$$\langle 0|\overline{u}u|0\rangle = \langle 0|\overline{d}d|0\rangle \, [= \langle 0|\overline{s}s|0\rangle] \ . \tag{10}$$

Thus, a nonvanishing quark condensate

$$\langle 0|\overline{q}q|0\rangle \neq 0 \tag{11}$$

is sufficient for spontaneous chiral symmetry breaking. As already emphasized, (11) is certainly not a necessary condition. Increasing the operator dimension, the next candidate is the so–called mixed condensate of dimension five,

$$\langle 0|\overline{q}\sigma_{\mu\nu}\lambda_\alpha q G^{\alpha\mu\nu}|0\rangle \neq 0 \ , \tag{12}$$

and there are many more possibilities for operator dimensions ≥ 6. All order parameters are in principle equally good for triggering the Goldstone mechanism. As we will see later on, the quark condensate enjoys nevertheless a special status. Although the following statement will have to be made more precise, we are going

[1] Here, the λ_i are the generators of $SU(N_f)_V$ in the fundamental representation.

to investigate whether the quark condensate is the dominant order parameter of spontaneous chiral symmetry breaking in QCD.

To analyse the possible scenarios, it is useful to consider QCD in a Euclidean box of finite volume $V = L^4$. The Lagrangian for a massive quark in a given gluonic background is

$$\mathcal{L} = \overline{q}(\not{D} + m)q \tag{13}$$

with hermitian $i\not{D}$. In a finite volume, the Dirac operator has a discrete spectrum :

$$i\not{D}u_n = \lambda_n u_n \tag{14}$$

with real eigenvalues λ_n and orthonormal spinorial eigenfunctions u_n. Spontaneous chiral symmetry breaking is related to the infrared structure of this spectrum in the limit $V \to \infty$ (Banks and Casher 1980; Vafa and Witten 1984; Leutwyler and Smilga 1992; ...; Stern 1998).

The main reason for working in Euclidean space is the following. Because of

$$i\not{D}u_n = \lambda_n u_n \longrightarrow i\not{D}\gamma_5 u_n = -\lambda_n \gamma_5 u_n \ , \tag{15}$$

the nonzero eigenvalues come in pairs $\pm\lambda_n$. Therefore, the fermion determinant in a given gluon background is real and positive (for $\theta_{QCD} = 0$) :

$$\det(\not{D} + m) = m^\nu \prod_{\lambda_n \neq 0} (m - i\lambda_n) = m^\nu \prod_{\lambda_n > 0} (m^2 + \lambda_n^2) > 0 \ , \tag{16}$$

where ν is the multiplicity of the zero modes. The fermion integration yields a real, positive measure for the gluonic functional integral. Thus, many statements for correlation functions in a given gluon background will survive the functional average over the gluon fields.

The quark two–point function for coinciding arguments can be written as (the subscript G denotes the gluon background)

$$\langle \overline{q}(x) q(x) \rangle_G = -\sum_n \frac{u_n^\dagger(x) u_n(x)}{m - i\lambda_n} \tag{17}$$

implying[2]

$$\frac{1}{V}\int d^4x \langle \overline{q}(x) q(x) \rangle_G = -\frac{1}{V}\sum_n \frac{1}{m - i\lambda_n} = -\frac{2m}{V}\sum_{\lambda_n > 0} \frac{1}{m^2 + \lambda_n^2} \ . \tag{18}$$

This relation demonstrates that the chiral and the infinite–volume limits do not commute. Taking the chiral limit $m \to 0$ for fixed volume yields $\langle \overline{q} q \rangle_G = 0$, in accordance with the fact that there is no spontaneous symmetry breaking in a finite volume. The limit of interest is therefore first $V \to \infty$ for fixed m and then $m \to 0$.

[2] The zero modes will not be relevant in the infinite–volume limit.

For $V \to \infty$, the eigenvalues λ_n become dense and we must replace the sum over eigenvalues by an integral over a density $\rho(\lambda)$:

$$\frac{1}{V}\sum_n \stackrel{V\to\infty}{\longrightarrow} \int d\lambda \rho(\lambda) \ .$$

Averaging the relation (18) over gluon fields and taking the infinite–volume limit, one gets

$$\langle 0|\overline{q}q|0\rangle = -\int_{-\infty}^{\infty} \frac{d\lambda\rho(\lambda)}{m-i\lambda} = -2m\int_0^{\infty}\frac{d\lambda\rho(\lambda)}{m^2+\lambda^2} \ . \tag{19}$$

In the chiral limit, we obtain the relation of Banks and Casher (1980) :

$$\lim_{m\to 0}\langle 0|\overline{q}q|0\rangle = -\pi\rho(0) \ . \tag{20}$$

For free fields, $\rho(\lambda) \sim \lambda^3$ near $\lambda = 0$. Thus, the eigenvalues must accumulate near zero to produce a nonvanishing quark condensate. Although the Banks–Casher relation does not tell us which gauge field configurations could be responsible for $\rho(0) \neq 0$, many suggestions are on the market (instantons, monopoles, ...).

This is a good place to recall the gist of the Vafa–Witten argument for the conservation of vector symmetries (Vafa and Witten 1984):

$$\begin{aligned}\langle 0|\overline{u}u-\overline{d}d|0\rangle &= -\int_{-\infty}^{\infty} d\lambda\rho(\lambda)\left(\frac{1}{m_u-i\lambda}-\frac{1}{m_d-i\lambda}\right)\\ &= (m_u-m_d)\int_{-\infty}^{\infty}\frac{d\lambda\rho(\lambda)}{(m_u-i\lambda)(m_d-i\lambda)}\\ &\stackrel{m_u\to m_d}{\longrightarrow} 0 \ .\end{aligned} \tag{21}$$

Unlike in the chiral limit, the integrand in (21) does not become singular in the equal–mass limit and the vacuum remains $SU(N_f)_V$ invariant.

The previous discussion concentrated on one specific order parameter for spontaneous chiral symmetry breaking, the quark condensate. Stern (1998) has recently performed a similar analysis for a quantity that is directly related to the Goldstone matrix element (6). Consider the correlation function

$$\Pi_{LR}^{\mu\nu}(q)\delta_{ij} = 4i\int d^4x e^{iqx}\langle 0|TL_i^\mu(x)R_j^\nu(0)|0\rangle \tag{22}$$

$$L_i^\mu = \overline{q_L}\gamma^\mu\frac{\lambda_i}{2}q_L \ , \qquad R_i^\mu = \overline{q_R}\gamma^\mu\frac{\lambda_i}{2}q_R \ .$$

In the chiral limit, the correlator vanishes for any q unless the vacuum is asymmetric. In particular, one finds in the chiral limit

$$\lim_{m_q\to 0}\Pi_{LR}^{\mu\nu}(0) = -F^2 g^{\mu\nu} \tag{23}$$

where the constant F (the pion decay constant in the chiral limit) characterizes the Goldstone matrix element (6):

$$\langle 0|\bar{q}\gamma^\mu\gamma_5\frac{\lambda_i}{2}q|\varphi_j(p)\rangle = i\delta_{ij}F\left[1+O(m_q)\right]p^\mu e^{-ipx} \; . \tag{24}$$

Thus, $\Pi^{\mu\nu}_{LR}(0) \neq 0$ is a necessary and sufficient condition for spontaneous chiral symmetry breaking.

Introducing the average (over all gluon configurations) number of states $N(\varepsilon, L)$ with $|\lambda| \leq \varepsilon$, Stern (1998) defines a mean eigenvalue density $\widehat{\rho}$ in finite volume as

$$\widehat{\rho}(\varepsilon, L) = \frac{N(\varepsilon, L)}{2\varepsilon V} \; . \tag{25}$$

Of course,

$$\rho(0) = \lim_{\varepsilon\to 0}\lim_{L\to\infty}\widehat{\rho}(\varepsilon, L) \tag{26}$$

with the previously introduced density ρ. With similar techniques as before (again in Euclidean space), Stern (1998) has derived a relation for the decay constant F :

$$F^2 = \pi^2 \lim_{\varepsilon\to 0}\lim_{L\to\infty} L^4 J(\varepsilon, L)\widehat{\rho}(\varepsilon, L)^2 \tag{27}$$

in terms of an average transition probability between states with $|\lambda| \leq \varepsilon$:

$$J(\varepsilon, L) = \frac{1}{N(\varepsilon, L)^2} << \sum_{kl}^{\varepsilon} J_{kl} >> \, , \quad J_{kl} = \frac{1}{4}\sum_\mu \left| \int d^4x u_k^\dagger(x)\gamma_\mu u_l(x)\right|^2 \tag{28}$$

where $<< \ldots >>$ denotes an average over gluon configurations. The formula (27) closely resembles the Greenwood–Kubo formula for electric conductivity (see Stern 1998).

As already emphasized, the eigenvalues λ_n must accumulate near zero to trigger spontaneous chiral symmetry breaking. A crucial parameter is the critical exponent κ defined as (Stern 1998)

$$<< \lambda_n >> \sim L^{-\kappa} \tag{29}$$

for λ_n near zero and $L \to \infty$. Up to higher powers in ε, the average number of states and the mean eigenvalue density depend on κ as

$$N(\varepsilon, L) = \left(\frac{2\varepsilon}{\mu}\right)^{\frac{4}{\kappa}} (\mu L)^4 + \ldots \tag{30}$$

$$\widehat{\rho}(\varepsilon, L) = \left(\frac{2\varepsilon}{\mu}\right)^{\frac{4}{\kappa}-1} \mu^3 + \ldots \tag{31}$$

in terms of some energy scale μ. As is obvious from the definition (29) and from the expressions (30),(31), the eigenvalues with maximal κ are the relevant ones.

The completeness sum rule $\sum_l J_{kl} = 1$ for the transition probabilities yields an upper bound for F^2 (Stern 1998) :

$$F^2 \leq \pi^2 \mu^2 \lim_{\varepsilon \to 0} \left(\frac{2\varepsilon}{\mu} \right)^{\frac{4}{\kappa} - 2} . \tag{32}$$

Therefore, while $\kappa = 1$ for free fields, spontaneous chiral symmetry breaking requires $\kappa \geq 2$. With the same notation, we also have

$$\langle 0|\bar{q}q|0\rangle = -\pi\mu^3 \lim_{\varepsilon \to 0} \left(\frac{2\varepsilon}{\mu} \right)^{\frac{4}{\kappa} - 1} \tag{33}$$

leading to $\kappa = 4$ for a nonvanishing quark condensate (Leutwyler and Smilga 1992). On rather general grounds, the critical index is bounded by

$$1 \leq \kappa \leq 4 . \tag{34}$$

Stern (1998) has argued that the existence of an effective chiral Lagrangian analytic in the quark masses suggests that the exponent $4/\kappa$ is actually an integer[3]. In this case, only $\kappa = 1$ or $\kappa = 2, 4$ would be allowed, the latter two cases being compatible with spontaneous chiral symmetry breaking.

There are then two preferred scenarios for spontaneous chiral symmetry breaking (Stern 1998):

i. $\kappa = 2$:
The density of states near $\varepsilon = 0$ is too small to generate a nonvanishing quark condensate, but the high "quark mobility" J induces $F \neq 0$.

ii. $\kappa = 4$:
Here, the density of states is sufficiently large for $\rho(0) \neq 0$. This option is strongly supported by lattice data (see below) favouring a nonvanishing quark condensate. With hindsight, the scenario most likely realized in nature is at least consistent with the previous analyticity hypothesis.

Are there other indications for a large quark condensate? Knecht and de Rafael (1997) have recently found an interesting relation between chiral order parameters and the vector and axial-vector spectral functions in the limit of large N_c. They consider again the correlation function (22). In the chiral limit, it can be expressed in terms of a single scalar function $\Pi_{LR}(Q^2)$:

$$\Pi^{\mu\nu}_{LR}(q) = (q^\mu q^\nu - g^{\mu\nu} q^2)\Pi_{LR}(Q^2) , \qquad Q^2 = -q^2 . \tag{35}$$

[3] There are explicit counterexamples to this analyticity assumption in less than four dimensions (L. Alvarez-Gaumé, H. Grosse and J. Stern, private communications).

Because it is a (nonlocal) order parameter, $\Pi_{LR}(Q^2)$ vanishes in all orders of QCD perturbation theory for a symmetric vacuum. The asymptotic behaviour for large and small Q^2 ($Q^2 \geq 0$) is

$$\Pi_{LR}(Q^2) = -\frac{4\pi}{Q^6}[\alpha_s + O(\alpha_s^2)]\langle\overline{u}u\rangle^2 + O(\frac{1}{Q^8}) \tag{36}$$

$$-Q^2\Pi_{LR}(Q^2) = F^2 + O(Q^2) \; . \tag{37}$$

For the large-Q^2 behaviour (36) (Shifman et al. 1979), $N_c \to \infty$ has already been assumed to factorize the four-quark condensate into the square of the (two-)quark condensate. In the same limit, the correlation function $\Pi_{LR}(Q^2)$ is determined by an infinite number of stable vector and axial-vector states:

$$-Q^2\Pi_{LR}(Q^2) = F^2 + \sum_A \frac{F_A^2 Q^2}{M_A^2 + Q^2} - \sum_V \frac{F_V^2 Q^2}{M_V^2 + Q^2} \; , \tag{38}$$

where $M_I, F_I (I = V, A)$ are the masses and the coupling strengths of the spin-1 mesons to the respective currents. Comparison with the asymptotic behaviour (36) yields the two Weinberg sum rules (Weinberg 1967)

$$\sum_V F_V^2 - \sum_A F_A^2 = F^2 \tag{39}$$

$$\sum_V F_V^2 M_V^2 - \sum_A F_A^2 M_A^2 = 0 \tag{40}$$

and allows (38) to be rewritten as

$$-Q^2\Pi_{LR}(Q^2) = \sum_A \frac{F_A^2 M_A^4}{Q^2(M_A^2 + Q^2)} - \sum_V \frac{F_V^2 M_V^4}{Q^2(M_V^2 + Q^2)} \; . \tag{41}$$

This expression can now be matched once more to the asymptotic behaviour (36). Referring to Knecht and de Rafael (1997) for a general discussion, I concentrate here on the simplest possibility assuming that the V, A spectral functions can be described by single resonance states plus a continuum. The experimental situation for the $I = 1$ channel shown in Fig. 1 is clearly not very far from this simplest case. In addition to the inequality $M_V < M_A$ following from the Weinberg sum rules (39), (40), the matching condition requires

$$4\pi[\alpha_s + O(\alpha_s^2)]\langle\overline{u}u\rangle^2 = F^2 M_V^2 M_A^2 \tag{42}$$

or approximately

$$4\pi\alpha_s\langle\overline{u}u\rangle^2 \simeq F_\pi^2 M_\rho^2 M_{A_1}^2 \; . \tag{43}$$

From the last relation, Knecht and de Rafael (1997) extract a quark condensate

$$\langle\overline{u}u\rangle(\nu = 1 \text{ GeV}) \simeq -(303 \text{ MeV})^3 \tag{44}$$

with ν the QCD renormalization scale in the $\overline{\mathrm{MS}}$ scheme. In view of the assumptions made, especially the large–N_c limit, this value is quite compatible with

$$\langle\overline{u}u\rangle(\nu = 1 \text{ GeV}) = -\left[(229 \pm 9) \text{ MeV}\right]^3 \tag{45}$$

from a recent compilation of sum rule estimates (Dosch and Narison 1998).

The conclusion is that the V, A spectrum is fully consistent with both sum rule and lattice estimates for the quark condensate. We come back to this issue in the discussion of light quark masses.

1.2 Effective Field Theory

The pseudoscalar mesons are not only the lightest hadrons but they also have a special status as (pseudo–) Goldstone bosons. In the chiral limit, the interactions of Goldstone bosons vanish as their energies tend to zero. In other words, the interactions of Goldstone bosons become arbitrarily weak for decreasing energy no matter how strong the underlying interaction is. This is the basis for a systematic low–energy expansion with an effective chiral Lagrangian that is organized in a derivative expansion.

There is a standard procedure for implementing a symmetry transformation on Goldstone fields (Coleman et al. 1969; Callan et al. 1969). Geometrically, the Goldstone fields $\varphi = [\pi, K, \eta_8]$ can be viewed as coordinates of the coset space G/H. They are assembled in a matrix field $u(\varphi) \in G/H$, the basic building block of chiral Lagrangians. Different forms of this matrix field (e.g., the exponential representation) correspond to different parametrizations of coset space. Since the chiral Lagrangian is generically nonrenormalizable, there is no distinguished choice of field variables as for renormalizable quantum field theories.

An element g of the symmetry group G induces in a natural way a transformation of $u(\varphi)$ by left translation:

$$u(\varphi) \stackrel{g\in G}{\longrightarrow} gu(\varphi) = u(\varphi')h(g,\varphi) \ . \tag{46}$$

The so–called compensator field $h(g,\varphi)$ is an element of the conserved subgroup H and it accounts for the fact that a coset element is only defined up to an H transformation. For $g \in H$, the symmetry is realized in the usual linear way (Wigner–Weyl) and $h(g)$ does not depend on the Goldstone fields φ. On the other hand, for $g \in G$ corresponding to a spontaneously broken symmetry ($g \notin H$), the symmetry is realized nonlinearly (Nambu–Goldstone) and $h(g,\varphi)$ does depend on φ.

For the special case of chiral symmetry $G = SU(N_f)_L \times SU(N_f)_R$, parity relates left– and right–chiral transformations. With a standard choice of coset representatives, the general transformation (46) takes the special form

$$u(\varphi') = g_R u(\varphi)h(g,\varphi)^{-1} = h(g,\varphi)u(\varphi)g_L^{-1} \tag{47}$$

$$g = (g_L, g_R) \in G \ .$$

For practical purposes, one never needs to know the explicit form of $h(g,\varphi)$, but only the transformation property (47). In the mesonic sector, it is often more convenient to work with the square of $u(\varphi)$. Because of (47), the matrix field $U(\varphi) = u(\varphi)^2$ has a simpler linear transformation behaviour:

$$U(\varphi) \stackrel{G}{\to} g_R U(\varphi) g_L^{-1} \,. \tag{48}$$

It is therefore frequently used as basic building block for mesonic chiral Lagrangians.

When non–Goldstone degrees of freedom like baryons or meson resonances are included in the effective Lagrangians, the nonlinear picture with $u(\varphi)$ and $h(g,\varphi)$ is more appropriate. If a generic hadron field Ψ (with $M_\Psi \neq 0$ in the chiral limit) transforms under H as

$$\Psi \stackrel{h\in H}{\to} \Psi' = h_\Psi(h)\Psi \tag{49}$$

according to a given representation h_Ψ of H, the compensator field in this representation furnishes immediately a realization of all of G:

$$\Psi \stackrel{g\in G}{\to} \Psi' = h_\Psi(g,\varphi)\Psi \,. \tag{50}$$

This transformation is not only nonlinear in φ but also space–time dependent requiring the introduction of a chirally covariant derivative. We will come back to this case in the last lecture on baryons and mesons.

Before embarking on the construction of an effective field theory for QCD, we pause for a moment to realize that there is in fact no chiral symmetry in nature. In addition to the spontaneous breaking discussed so far, chiral symmetry is broken explicitly both by nonvanishing quark masses and by the electroweak interactions of hadrons.

The main assumption of CHPT is that it makes sense to expand around the chiral limit. In full generality, chiral Lagrangians are therefore constructed by means of a two–fold expansion in both

- derivatives ($\sim$ momenta) and
- quark masses :

$$\mathcal{L}_{\text{eff}} = \sum_{i,j} \mathcal{L}_{ij} \,, \qquad \mathcal{L}_{ij} = O(p^i m_q^j) \,. \tag{51}$$

The two expansions become related by expressing the pseudoscalar meson masses in terms of the quark masses m_q. If the quark condensate is nonvanishing in the chiral limit, the squares of the meson masses start out linear in m_q (see below). The constant of proportionality is a quantity B with

$$B = -\frac{\langle \overline{u}u \rangle}{F^2} \tag{52}$$

in the chiral limit. Assuming the linear terms to provide the dominant contributions to the meson masses corresponds to a scale (the product Bm_q is scale invariant)

$$B(\nu = 1 \text{ GeV}) \simeq 1.4 \text{ GeV} . \tag{53}$$

This standard scenario of CHPT (Weinberg 1979; Gasser and Leutwyler 1984, 1985; Leutwyler 1994) is compatible with a large quark condensate as given for instance in (45). The standard chiral counting

$$m_q = O(M^2) = O(p^2) \tag{54}$$

reduces the two-fold expansion (51) to

$$\mathcal{L}_{\text{eff}} = \sum_n \mathcal{L}_n , \qquad \mathcal{L}_n = \sum_{i+2j=n} \mathcal{L}_{ij} . \tag{55}$$

For mesons, the chiral expansion proceeds in steps of two (n = 2,4,6,...) because the index i is even.

Despite the evidence in favour of the standard scenario, the alternative of a much smaller or even vanishing quark condensate (e.g., for $\kappa = 2$ in the previous classification of chiral symmetry breaking) is actively being pursued (Fuchs et al. 1991; Stern et al. 1993; Knecht et al. 1993, 1995; Stern 1997 and references therein). This option is characterized by

$$B(\nu = 1 \text{ GeV}) \sim O(F_\pi) \tag{56}$$

with the pion decay constant $F_\pi = 92.4$ MeV. The so-called generalized CHPT amounts to a reordering of the effective chiral Lagrangian (55) on the basis of a modified chiral counting with $m_q = O(p)$. We will come back to generalized CHPT in several instances, in particular during the discussion of quark masses, but for most of these lectures I will stay with the mainstream of standard CHPT.

Both conceptually and for practical purposes, the best way to keep track of the explicit breaking is through the introduction of external matrix fields (Gasser and Leutwyler 1984, 1985) v_μ, a_μ, s, p. The QCD Lagrangian (1) with N_f massless quarks is extended to

$$\mathcal{L} = \mathcal{L}^0_{\text{QCD}} + \overline{q}\gamma^\mu(v_\mu + a_\mu\gamma_5)q - \overline{q}(s - ip\gamma_5)q \tag{57}$$

to include electroweak interactions of quarks with external gauge fields v_μ, a_μ and to allow for nonzero quark masses by setting the scalar matrix field $s(x)$ equal to the diagonal quark mass matrix. The big advantage is that one can perform all calculations with a (locally) G invariant effective Lagrangian in a manifestly chiral invariant manner. Only at the very end, one inserts the appropriate external fields to extract the Green functions of quark currents or matrix elements of interest. The explicit breaking of chiral symmetry is automatically taken care of by this spurion technique. In addition, electromagnetic gauge invariance is manifest.

Table 1. The effective chiral Lagrangian of the Standard Model

$\mathcal{L}_{\text{chiral dimension}}$ (# of LECs)	loop order
$\mathcal{L}_2(2) + \mathcal{L}_4^{\text{odd}}(0) + \mathcal{L}_2^{\Delta S=1}(2) + \mathcal{L}_0^{\gamma}(1)$ $+ \mathcal{L}_1^{\pi N}(1) + \mathcal{L}_2^{\pi N}(7) + \ldots$	$L=0$
$+ \underline{\mathcal{L}_4^{\text{even}}(10)} + \underline{\mathcal{L}_6^{\text{odd}}(32)} + \underline{\mathcal{L}_4^{\Delta S=1}(22, \text{octet})} + \underline{\mathcal{L}_2^{\gamma}(14)}$ $+ \underline{\mathcal{L}_3^{\pi N}(23)} + \mathcal{L}_4^{\pi N}(?) + \ldots$	$L=1$
$+ \underline{\mathcal{L}_6^{\text{even}}(112 \text{ for } SU(N_f))} + \ldots$	$L=2$

Although this procedure produces all Green functions for electromagnetic and weak currents, the method must be extended in order to include virtual photons (electromagnetic corrections) or virtual W bosons (nonleptonic weak interactions). The present status of the effective chiral Lagrangian of the Standard Model is summarized in Table 1. The purely mesonic Lagrangian is denoted as $\mathcal{L}_2+\mathcal{L}_4+\mathcal{L}_6$ and will be discussed at length in the following lecture. Even (odd) refers to terms in the meson Lagrangian without (with) an ε tensor. The pion–nucleon Lagrangian $\sum_n \mathcal{L}_n^{\pi N}$ will be the subject of the last lecture. The chiral Lagrangians for virtual photons (superscript γ) and for nonleptonic weak interactions (superscript $\Delta S = 1$) will not be treated in these lectures. The numbers in brackets denote the number of independent coupling constants or low–energy constants (LECs) for the given Lagrangian. They apply in general for $N_f = 3$ except for the πN Lagrangian ($N_f = 2$) and for the mesonic Lagrangian of $O(p^6)$ (general N_f). The different Lagrangians are grouped together according to the chiral order that corresponds to the indicated loop order. The underlined parts denote completely renormalized Lagrangians.

A striking feature of Table 1 is the rapidly growing number of LECs with increasing chiral order. Those constants describe the influence of all states that are not represented by explicit fields in the effective chiral Lagrangians. Although the general strategy of CHPT has been to fix those constants from experiment and then make predictions for other observables there is obviously a natural limit for such a program. This is the inescapable consequence of a nonrenormalizable effective Lagrangian that is constructed solely on the basis of symmetry considerations. Nevertheless, I will try to convince you that even with 112 coupling constants one can make reliable predictions for low–energy observables.

2 Chiral Perturbation Theory with Mesons

The effective chiral Lagrangian for the strong interactions of mesons is constructed in terms of the basic building blocks $U(\varphi)$ and the external fields v_μ, a_μ, s and p. With the standard chiral counting described previously, the chiral Lagrangian starts at $O(p^2)$ with

$$\mathcal{L}_2 = \frac{F^2}{4}\langle D_\mu U D^\mu U^\dagger + \chi U^\dagger + \chi^\dagger U\rangle \tag{58}$$

$$\chi = 2B(s+ip) \qquad D_\mu U = \partial_\mu U - i(v_\mu + a_\mu)U + iU(v_\mu - a_\mu)$$

where $\langle\ldots\rangle$ stands for the N_f–dimensional trace. We have already encountered both LECs of $O(p^2)$. They are related to the pion decay constant and to the quark condensate:

$$F_\pi = F[1+O(m_q)] = 92.4\,\text{MeV} \tag{59}$$
$$\langle 0|\bar{u}u|0\rangle = -F^2B[1+O(m_q)]\ .$$

Expanding the Lagrangian (58) to second order in the meson fields and setting the external scalar field equal to the quark mass matrix, one can immediately read off the pseudoscalar meson masses to leading order in m_q, e.g.,

$$M^2_{\pi^+} = (m_u + m_d)B\ . \tag{60}$$

As expected, for $B \neq 0$ the squares of the meson masses are linear in the quark masses to leading order. The full set of equations ($N_f = 3$) for the masses of the pseudoscalar octet gives rise to several well–known relations:

$$F_\pi^2 M_\pi^2 = -(m_u+m_d)\langle 0|\bar{u}u|0\rangle \quad \text{(Gell-Mann et al. 1968)} \tag{61}$$

$$\frac{M_\pi^2}{m_u+m_d} = \frac{M^2_{K^+}}{m_s+m_u} = \frac{M^2_{K^0}}{m_s+m_d} \quad \text{(Weinberg 1977)} \tag{62}$$

$$3M^2_{\eta_8} = 4M_K^2 - M_\pi^2 \quad \text{(Gell-Mann 1957; Okubo 1962)} \tag{63}$$

Having determined the two LECs of $O(p^2)$, we may now calculate from the Lagrangian (58) any Green function or S–matrix amplitude without free parameters. The resulting tree–level amplitudes are the leading expressions in the low–energy expansion of the Standard Model. They are given in terms of F_π and meson masses and they correspond to the current algebra amplitudes of the sixties if we adopt the standard chiral counting.

The situation becomes more involved once we go to next–to–leading order, $O(p^4)$. Before presenting the general procedure, we observe that no matter how many higher–order Lagrangians we include, tree amplitudes will always be real. On the other hand, unitarity and analyticity require complex amplitudes in general. A good example is elastic pion–pion scattering where the partial–wave amplitudes $t_l^I(s)$ satisfy the unitarity constraint

$$\Im m\, t_l^I(s) \geq (1 - \frac{4M_\pi^2}{s})^{\frac{1}{2}} |t_l^I(s)|^2 . \tag{64}$$

Since $t_l^I(s)$ starts out at $O(p^2)$ (for $l < 2$), the partial–wave amplitudes are complex from $O(p^4)$ on.

This example illustrates the general requirement that a systematic low–energy expansion entails a loop expansion. Since loop amplitudes are in general divergent, regularization and renormalization are essential ingredients of CHPT. Any regularization is in principle equally acceptable, but dimensional regularization is the most popular method for well–known reasons.

Although the need for regularization is beyond debate, the situation is more subtle concerning renormalization. Here are two recurrent questions in this connection:

- Why bother renormalizing a quantum field theory that is after all based on a nonrenormalizable Lagrangian?
- Why not use a "physical" cutoff instead?

The answer to both questions is that we are interested in predictions of the Standard Model itself rather than of some cutoff version no matter how "physical" that cutoff may be. Renormalization ensures that the final results are independent of the chosen regularization method. As we will now discuss in some detail, renormalization amounts to absorbing the divergences in the LECs of higher–order chiral Lagrangians. The renormalized LECs are then measurable, although in general scale dependent quantities. In any physical amplitude, this scale dependence always cancels the scale dependence of loop amplitudes.

2.1 Loop Expansion and Renormalization

This part of the lectures is on a more technical level than the rest. Its purpose is to demonstrate that we are taking the quantum field theory aspects of chiral Lagrangians seriously.

The strong interactions of mesons are described by the generating functional of Green functions (of quark currents)

$$e^{iZ[j]} = < 0\ \mathrm{out}|0\ \mathrm{in} >_j = \int [d\varphi] e^{iS_{\mathrm{eff}}[\varphi, j]} \tag{65}$$

where

$$j \sim v, a, s, p$$

denotes collectively the external fields.

The chiral expansion of the action

$$S_{\text{eff}}[\varphi, j] = S_2[\varphi, j] + S_4[\varphi, j] + S_6[\varphi, j] + \ldots \tag{66}$$
$$S_n[\varphi, j] = \int d^4x \mathcal{L}_n(x)$$

is accompanied by a corresponding expansion of the generating functional :

$$Z[j] = Z_2[j] + Z_4[j] + Z_6[j] + \ldots \tag{67}$$

Functional integration of the quantum fluctuations around the classical solution gives rise to the loop expansion. The classical solution is defined as

$$\left. \frac{\delta S_2[\varphi, j]}{\delta \varphi_i} \right|_{\varphi=\varphi_{\text{cl}}} = 0 \qquad \Rightarrow \qquad \varphi_{\text{cl}}[j] \tag{68}$$

and it can be constructed iteratively as a functional of the external fields j. Note that we define $\varphi_{\text{cl}}[j]$ through the lowest-order Lagrangian $\mathcal{L}_2(\varphi, j)$ at any order in the chiral expansion. In this case, $\varphi_{\text{cl}}[j]$ carries precisely the tree structure of $O(p^2)$ allowing for a straightforward chiral counting. This would not be true any more if we had included higher-order chiral Lagrangians in the definition of the classical solution.

With a mass-independent regularization method like dimensional regularization, it is straightforward to compute the degree of homogeneity of a generic Feynman amplitude as a function of external momenta and meson masses. This number is called the chiral dimension D of the amplitude and it characterizes the order of the low-energy expansion. For a connected amplitude with L loops and with N_n vertices of $O(p^n)$ ($n =$ 2,4,6,...), it is given by (Weinberg 1979)

$$D = 2L + 2 + \sum_n (n-2) N_n \ , \qquad n = 4, 6, \ldots \tag{69}$$

For a given amplitude, the chiral dimension obviously increases with L. In order to reproduce the (fixed) physical dimension of the amplitude, each loop produces a factor $1/F^2$. Together with the geometric loop factor $(4\pi)^{-2}$, the loop expansion suggests

$$4\pi F_\pi = 1.2 \text{ GeV} \tag{70}$$

as natural scale of the chiral expansion (Manohar and Georgi 1984). Restricting the domain of applicability of CHPT to momenta $|p| \lesssim O(M_K)$, the natural expansion parameter of chiral amplitudes is therefore expected to be of the order

$$\frac{M_K^2}{16\pi^2 F_\pi^2} = 0.18 \ . \tag{71}$$

As we will see soon, these terms often appear multiplied with chiral logarithms. Substantial higher-order corrections in the chiral expansion are therefore to be expected for chiral $SU(3)$. On the other hand, for $N_f = 2$ and for momenta $|p| \lesssim O(M_\pi)$ the chiral expansion is expected to converge considerably faster.

The formula (69) implies that $D = 2$ is only possible for $L = 0$: the tree–level amplitudes from the Lagrangian $\mathcal{L}_2$ are then polynomials of degree 2 in the external momenta and masses. The corresponding generating functional is given by the classical action:

$$Z_2[j] = \int d^4x \mathcal{L}_2(\varphi_{\rm cl}[j], j) \ . \tag{72}$$

Already at next–to–leading order, the amplitudes are not just polynomials of degree $D = 4$, but they are by definition of the chiral dimension always homogeneous functions of degree D in external momenta and masses. For $D = 4$, we have two types of contributions: either $L = 0$ with $N_4 = 1$, i.e., exactly one vertex of $O(p^4)$, or $L = 1$ and only vertices of $O(p^2)$ (which, as formula (69) demonstrates, do not modify the chiral dimension). Explicitly, the complete generating functional of $O(p^4)$ consists of

$L=0$	$\int d^4x \mathcal{L}_4(\varphi_{\rm cl}[j], j)$	chiral action of $O(p^4)$
	$Z_{\rm WZW}[\varphi_{\rm cl}[j], v, a]$	chiral anomaly
$L=1$	$Z_4^{(L=1)}[j]$	one–loop functional

In addition to the Wess–Zumino–Witten functional $Z_{\rm WZW}$ (Wess and Zumino 1971; Witten 1983) accounting for the chiral anomaly, the $L = 0$ part involves the general chiral Lagrangian $\mathcal{L}_4$ with 10 LECs (Gasser and Leutwyler 1985):

$$\begin{aligned}
\mathcal{L}_4 = {} & L_1\langle D_\mu U^\dagger D^\mu U\rangle^2 + L_2\langle D_\mu U^\dagger D_\nu U\rangle\langle D^\mu U^\dagger D^\nu U\rangle \\
& + L_3\langle D_\mu U^\dagger D^\mu U D_\nu U^\dagger D^\nu U\rangle + L_4\langle D_\mu U^\dagger D^\mu U\rangle\langle \chi^\dagger U + \chi U^\dagger\rangle \\
& + L_5\langle D_\mu U^\dagger D^\mu U(\chi^\dagger U + U^\dagger \chi)\rangle + L_6\langle \chi^\dagger U + \chi U^\dagger\rangle^2 + L_7\langle \chi^\dagger U - \chi U^\dagger\rangle^2 \\
& + L_8\langle \chi^\dagger U \chi^\dagger U + \chi U^\dagger \chi U^\dagger\rangle - iL_9\langle F_R^{\mu\nu} D_\mu U D_\nu U^\dagger + F_L^{\mu\nu} D_\mu U^\dagger D_\nu U\rangle \\
& + L_{10}\langle U^\dagger F_R^{\mu\nu} U F_{L\mu\nu}\rangle + \text{ 2 contact terms} = \sum_i L_i P_i
\end{aligned} \tag{73}$$

where $F_R^{\mu\nu}$, $F_L^{\mu\nu}$ are field strength tensors associated with the external gauge fields. This is the most general Lorentz invariant Lagrangian of $O(p^4)$ with (local) chiral symmetry, parity and charge conjugation.

The one–loop functional can be written in closed form as

$$Z_4^{(L=1)}[j] = \frac{i}{2} \ln \det D_2 = \frac{i}{2} \text{ Tr } \ln D_2 \tag{74}$$

in terms of the determinant of a differential operator associated with the Lagrangian $\mathcal{L}_2$. In accordance with general theorems of renormalization theory (e.g., Collins 1984), its divergent part takes the form of a local action with all the symmetries of $\mathcal{L}_2$ and thus of QCD. Since the chiral dimension of this divergence action is 4, it must be of the form (73) with divergent coefficients:

$$\mathcal{L}_{4,\rm div}^{(L=1)} = -\Lambda(\mu)\sum_i \Gamma_i P_i \tag{75}$$

$$\Lambda(\mu) = \frac{\mu^{d-4}}{(4\pi)^2}\left\{\frac{1}{d-4} - \frac{1}{2}[\ln 4\pi + 1 + \Gamma'(1)]\right\}$$

Table 2. Phenomenological values of the renormalized LECs $L_i^r(M_\rho)$, taken from Bijnens et al. (1995), and β functions Γ_i for these coupling constants.

i	$L_i^r(M_\rho) \times 10^3$	source	Γ_i
1	0.4 ± 0.3	$K_{e4}, \pi\pi \to \pi\pi$	3/32
2	1.35 ± 0.3	$K_{e4}, \pi\pi \to \pi\pi$	3/16
3	-3.5 ± 1.1	$K_{e4}, \pi\pi \to \pi\pi$	0
4	-0.3 ± 0.5	Zweig rule	1/8
5	1.4 ± 0.5	$F_K : F_\pi$	3/8
6	-0.2 ± 0.3	Zweig rule	11/144
7	-0.4 ± 0.2	Gell-Mann–Okubo, L_5, L_8	0
8	0.9 ± 0.3	$M_{K^0} - M_{K^+}, L_5,$ $(2m_s - m_u - m_d) : (m_d - m_u)$	5/48
9	6.9 ± 0.7	$\langle r^2 \rangle_V^\pi$	1/4
10	-5.5 ± 0.7	$\pi \to e\nu\gamma$	$-1/4$

with the conventions of Gasser and Leutwyler (1985) for $\overline{\text{MS}}$. The coefficients Γ_i are listed in Table 2.

Renormalization to $O(p^4)$ proceeds by decomposing

$$L_i = L_i^r(\mu) + \Gamma_i \Lambda(\mu) \tag{76}$$

such that

$$Z_4 - Z_{\text{WZW}} = Z_4^{(L=1)} + \int d^4x \mathcal{L}_4(L_i) = Z_{4,\text{fin}}^{(L=1)}(\mu) + \int d^4x \mathcal{L}_4(L_i^r(\mu)) \tag{77}$$

is finite and independent of the arbitrary scale μ. The generating functional and therefore the amplitudes depend on scale dependent LECs that obey the renormalization group equations

$$L_i^r(\mu_2) = L_i^r(\mu_1) + \frac{\Gamma_i}{(4\pi)^2} \ln \frac{\mu_1}{\mu_2} \,. \tag{78}$$

The current values of these constants come mainly from phenomenology to $O(p^4)$ and are listed in Table 2.

Many recent investigations in CHPT have included effects of $O(p^6)$ (see below for a discussion of elastic $\pi\pi$ scattering). The following contributions are also shown pictorially in Fig. 2:

$$\begin{aligned} D = 6: \quad & L = 0, \quad N_6 = 1 \\ & L = 0, \quad N_4 = 2 \\ & L = 1, \quad N_4 = 1 \\ & L = 2 \end{aligned}$$

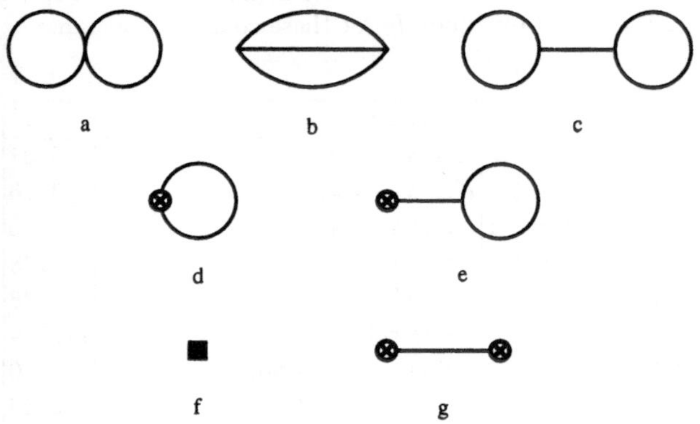

Fig. 2. Skeleton diagrams of $O(p^6)$. Normal vertices are from $\mathcal{L}_2$, crossed circles denote vertices from $\mathcal{L}_4$ and the square in diagram f stands for a vertex from $\mathcal{L}_6$. The propagators and vertices carry the full tree structure associated with the lowest-order Lagrangian $\mathcal{L}_2$.

Unlike for the one-loop functional (74), no simple closed form for the two-loop functional $Z_6^{(L=2)}[j]$ is known. General theorems of renormalization theory guarantee that

- the sum of the irreducible loop diagrams a, b, d in Fig. 2 is free of subdivergences, and that
- the sum of the one-particle-reducible diagrams c, e, g is finite and scale independent (at least for the form of $\mathcal{L}_4$ given in (73)).

As a consequence, $Z_{6,\mathrm{div}}^{(L=2)}$ is again a local action with all the symmetries of $\mathcal{L}_2$ and the corresponding divergence Lagrangian is of the general form $\mathcal{L}_6$ with divergent coefficients. For general N_f, this Lagrangian has 115 terms (Bijnens et al. 1998b), 112 measurable LECs and three contact terms. For $N_f = 3$, this Lagrangian was first written down by Fearing and Scherer (1996) but some of their terms are redundant.

How does renormalization at $O(p^6)$ work in practice? To simplify the discussion, we consider chiral $SU(2)$ with a single mass scale M (the pion mass at lowest order). The LECs in chiral $SU(2)$ and their associated β functions are usually denoted l_i, γ_i (Gasser and Leutwyler 1984). Since the divergences occur only in polynomials in the external momenta and masses, we consider a generic dimensionless coefficient Q of such a polynomial, e.g., m_6, f_6 in the chiral expansions of the pion mass and decay constant, respectively (Bürgi 1996; Bijnens et al. 1997):

$$M_\pi^2 = M^2 \left\{ 1 + m_4 \frac{M^2}{F^2} + m_6 \frac{M^4}{F^4} + O(F^{-6}) \right\} \tag{79}$$

$$F_\pi = F \left\{ 1 + f_4 \frac{M^2}{F^2} + f_6 \frac{M^4}{F^4} + O(F^{-6}) \right\} . \tag{80}$$

Working from now on in d dimensions, we obtain from the (irreducible) diagrams a,b,d and f

$$Q = Q_{\text{loop}} + Q_{\text{tree}} \tag{81}$$

with

$$Q_{\text{loop}}(d) = \underbrace{J(0)^2 x(d)}_{\text{diagrams a,b}} + \underbrace{J(0) \sum_i l_i y_i(d)}_{\text{diagram d}} \tag{82}$$

$$J(0) = \frac{1}{i} \int \frac{d^d k}{(2\pi)^d} \frac{1}{(k^2 - M^2)^2} .$$

The coefficients $x(d), y_i(d)$ are expanded to $O(\omega^2)$ in $\omega = \frac{1}{2}(d-4)$:

$$x(d) = x_0 + x_1\omega + x_2\omega^2 + O(\omega^3) \tag{83}$$
$$y_i(d) = y_{i0} + y_{i1}\omega + y_{i2}\omega^2 + O(\omega^3) .$$

Likewise, for $J(0)$ and the (unrenormalized) l_i we perform a Laurent expansion in ω:

$$J(0) = \frac{M^{2\omega}\Gamma(-\omega)}{(4\pi)^{2+\omega}} = \frac{(c\mu)^{2\omega}}{(4\pi)^2} \left(\frac{M}{c\mu}\right)^{2\omega} \frac{\Gamma(-\omega)}{(4\pi)^\omega} \tag{84}$$
$$= \frac{(c\mu)^{2\omega}}{(4\pi)^2} \left[-\frac{1}{\omega} + b(M/\mu) + a(M/\mu)\omega + O(\omega^2) \right]$$
$$l_i = \frac{(c\mu)^{2\omega}}{(4\pi)^2} \left[\frac{\gamma_i}{2\omega} + \beta_i(\mu) + \alpha_i(\mu)\omega + O(\omega^2) \right] . \tag{85}$$

In the $\overline{MS}$ scheme with

$$2 \ln c = -1 - \ln 4\pi - \Gamma'(1) \tag{86}$$

one gets

$$b(M/\mu) = -2 \ln \frac{M}{\mu} - 1 \tag{87}$$
$$\beta_i(\mu) = (4\pi)^2 l_i^r(\mu)$$

where the $l_i^r(\mu)$ are the standard renormalized LECs of Gasser and Leutwyler (1984).

An important consistency check is due to the absence of nonlocal divergences of the type

$$\frac{\ln M/\mu}{\omega}$$

implying (Weinberg 1979)

$$4x_0 = \sum_i \gamma_i y_{i0} \; . \tag{88}$$

For $SU(N_f)$, there are 115 such relations between two–loop and one–loop quantities due to the 115 independent monomials in the chiral Lagrangian of $O(p^6)$. We have recently verified these conditions by explicit calculation (Bijnens et al. 1998b).

With the summation convention for i implied, the complete loop contribution

$$\begin{aligned} Q_{\text{loop}} = \frac{\mu^{4\omega}}{(4\pi)^4} \Bigg\{ & -\frac{x_0}{\omega^2} + \frac{[x_1 - \beta_i(\mu) y_{i0} - \frac{1}{2}\gamma_i y_{i1}]}{\omega} \\ & + x_0 b(M/\mu)^2 + \left[-2x_1 + \beta_i(\mu) y_{i0} + \frac{1}{2}\gamma_i y_{i1}\right] b(M/\mu) \\ & + x_2 - \beta_i(\mu) y_{i1} - \frac{1}{2}\gamma_i y_{i2} - \alpha_i(\mu) y_{i0} + O(\omega) \Bigg\} \end{aligned} \tag{89}$$

is renormalized by the tree–level contribution from $\mathcal{L}_6$:

$$\begin{aligned} Q_{\text{tree}}(d) &= z(d) \\ &= \frac{\mu^{4\omega}}{(4\pi)^4} \left\{ \frac{x_0}{\omega^2} - \frac{[x_1 - \beta_i(\mu) y_{i0} - \frac{1}{2}\gamma_i y_{i1}]}{\omega} + (4\pi)^4 z^r(\mu) + O(\omega) \right\} \end{aligned} \tag{90}$$

where z is the appropriate combination of (unrenormalized) LECs of $O(p^6)$. The total contribution from diagrams a,b,d,f is now finite and scale independent:

$$\begin{aligned} Q &= \lim_{d \to 4} [Q_{\text{loop}}(d) + Q_{\text{tree}}(d)] \\ &= \frac{1}{(4\pi)^4} \Bigg\{ x_0 \left[1 + 2\ln\frac{M}{\mu}\right]^2 + \left[2x_1 - \frac{1}{2}\gamma_i y_{i1} - (4\pi)^2 l_i^r(\mu) y_{i0}\right] \left(1 + 2\ln\frac{M}{\mu}\right) \\ &\quad + x_2 - \frac{1}{2}\gamma_i y_{i2} - (4\pi)^2 l_i^r(\mu) y_{i1} + (4\pi)^4 \widehat{z}^r(\mu) \Bigg\} \end{aligned} \tag{91}$$

in terms of a redefined[4] combination $\widehat{z}^r(\mu)$ of LECs,

$$\widehat{z}^r(\mu) = z^r(\mu) - \frac{\alpha_i(\mu) y_{i0}}{(4\pi)^4} \tag{92}$$

that obeys the renormalization group equation

$$\mu \frac{d\widehat{z}^r(\mu)}{d\mu} = \frac{2}{(4\pi)^4} [2x_1 - (4\pi)^2 l_i^r(\mu) y_{i0} - \gamma_i y_{i1}] \; . \tag{93}$$

[4] This process independent (Bijnens et al. 1998b) redefinition absorbs the redundant expansion coefficients $\alpha_i(\mu)$.

Table 3. Complete calculations to $O(p^6)$ in standard CHPT.

$\gamma\gamma \to \pi^0\pi^0$	Bellucci et al. (1994)
$\gamma\gamma \to \pi^+\pi^-$	Bürgi (1996)
$\pi \to l\nu_l\gamma$	Bijnens and Talavera (1997)
$\pi\pi \to \pi\pi$	Bijnens et al. (1996, 1997)
π form factors	Bijnens et al. (1998a)
VV, AA	Golowich and Kambor (1995, 1997)
form factors	Post and Schilcher (1997)

Remarks:

i. Weinberg's relation (88) implies that the coefficient of the leading chiral log $\ln^2 M/\mu$ can be extracted from a one–loop calculation (cf. Kazakov 1988).
ii. There are in general additional finite contributions (including chiral logs) from the reducible diagrams c,e,g of Fig. 2.

In Table 3, I list the complete two–loop calculations that have been performed up to now. The first five entries are for chiral $SU(2)$, the last two for $N_f = 3$.

2.2 Light Quark Masses

In the framework of standard CHPT, the (current) quark masses m_q always appear in the combination $m_q B$ in chiral amplitudes. Without additional information on B through the quark condensate [cf. Eq. (59)], one can only extract ratios of quark masses from CHPT amplitudes.

The lowest–order mass formulas (62) together with Dashen's theorem on the lowest–order electromagnetic contributions to the meson masses (Dashen 1969) lead to the ratios (Weinberg 1977)

$$\frac{m_u}{m_d} = 0.55 , \qquad \frac{m_s}{m_d} = 20.1 . \tag{94}$$

Generalized CHPT, on the other hand, does not fix these ratios even at lowest order but only yields bounds (Fuchs et al. 1990), e.g.,

$$6 \le r := \frac{m_s}{\hat{m}} \le r_2 := \frac{2M_K^2}{M_\pi^2} - 1 \simeq 26 \tag{95}$$

with $2\hat{m} := m_u + m_d$. The ratios (94) receive higher–order corrections. The most important ones are corrections of $O(p^4) = O(m_q^2)$ and $O(e^2 m_s)$. Gasser and Leutwyler (1985) found that to $O(p^4)$ the ratios

$$\frac{M_K^2}{M_\pi^2} = \frac{m_s + \hat{m}}{m_u + m_d}[1 + \Delta_M + O(m_s^2)] \tag{96}$$

$$\frac{(M_{K^0}^2 - M_{K^+}^2)_{\rm QCD}}{M_K^2 - M_\pi^2} = \frac{m_d - m_u}{m_s - \hat{m}}[1 + \Delta_M + O(m_s^2)] \tag{97}$$

depend on the same correction Δ_M of $O(m_s)$. The ratio of these two ratios is therefore independent of Δ_M and it determines the quantity

$$Q^2 := \frac{m_s^2 - \hat{m}^2}{m_d^2 - m_u^2} \; . \tag{98}$$

Without higher–order electromagnetic corrections for the meson masses,

$$Q = Q_D = 24.2 \; ,$$

but those corrections reduce Q by up to 10% (Donoghue et al. 1993; Bijnens 1993; Duncan et al. 1996; Kambor et al. 1996; Anisovich and Leutwyler 1996; Leutwyler 1996a; Baur and Urech 1996; Bijnens and Prades 1997; Moussallam 1997). Plotting m_s/m_d versus m_u/m_d leads to an ellipse (Leutwyler 1990). In Fig. 3, the relevant quadrant of the ellipse is shown for $Q = 24$ (upper curve) and $Q = 21.5$ (lower curve).

Kaplan and Manohar (1986) pointed out that due to an accidental symmetry of $\mathcal{L}_2 + \mathcal{L}_4$ the separate mass ratios m_u/m_d and m_s/m_d cannot be calculated to $O(p^4)$ from S–matrix elements or V, A Green functions only. Some additional input is needed like resonance saturation (for (pseudo-)scalar Green functions), large–N_c expansion, baryon mass splittings, etc. Some of those constraints are also shown in Fig. 3. A careful analysis of all available information on the mass ratios was performed by Leutwyler (1996b, 1996c), with the main conclusion that the quark mass ratios change rather little from $O(p^2)$ to $O(p^4)$. In Table 4, I compare the so–called current algebra mass ratios of $O(p^2)$ with the ratios including $O(p^4)$ corrections, taken from Leutwyler (1996b, 1996c). The errors are Leutwyler's estimates of the theoretical uncertainties as of 1996. Although theoretical errors are always open to debate, the overall stability of the quark mass ratios is evident.

Let me now turn to the absolute values of the light quark masses. Until recently, the results from QCD sum rules (de Rafael 1998 and references therein) tended to be systematically higher than the quark masses from lattice QCD. Some lattice determinations were actually in conflict with rigorous lower bounds on the quark masses (Lellouch et al. 1997). Recent progress in lattice QCD (e.g., Lüscher 1997) has led to a general increase of the (quenched) lattice values. Table 5 contains the most recent determinations of both $\hat{m}$ and m_s that I am aware of. Judging only on the basis of the entries in Table 5, sum rule and lattice values for the quark masses now seem to be compatible with each other. The values are given at the $\overline{MS}$ scale $\nu = 2$ GeV as is customary in lattice QCD.

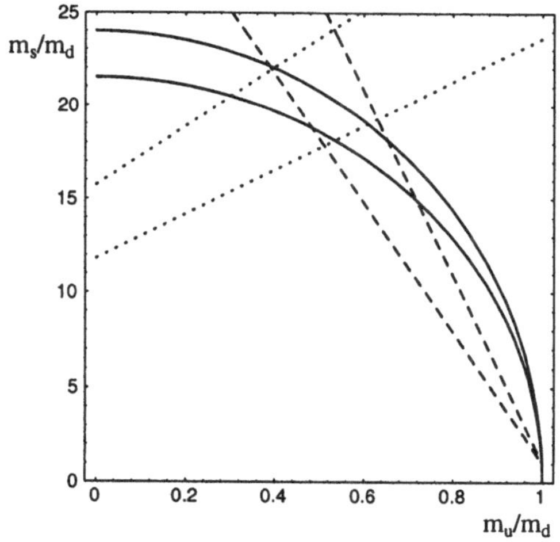

Fig. 3. First quadrant of Leutwyler's ellipse for $Q = 24$ (upper curve) and $Q = 21.5$ (lower curve). The dotted lines correspond to $\Theta_{\eta\eta'} = -15^0$ (upper line) and -25^0 (lower line) for the $\eta - \eta'$ mixing angle. The bounds defined by the two dashed lines come from baryon mass splittings, $\rho - \omega$ mixing and $\Gamma(\psi' \to \psi\pi^0)/\Gamma(\psi' \to \psi\eta)$ (Leutwyler 1996b, 1996c) for the ratio $R = (m_s - \hat{m})/(m_d - m_u)$ $(35 \leq R \leq 50)$.

Table 4. Quark mass ratios at $O(p^2)$ (Weinberg 1977) and to $O(p^4)$ (Leutwyler 1996b, 1996c).

	m_u/m_d	m_s/m_d	$m_s/\hat{m}$
$O(p^2)$	0.55	20.1	25.9
$O(p^4)$	0.55 ± 0.04	18.9 ± 0.8	24.4 ± 1.5

Except for chiral logs, the squares of the meson masses are polynomials in m_q. It is remarkable if not puzzling that many years of lattice studies have not seen any indications for terms higher than linear in the quark masses. An impressive example from the high-statistics spectrum calculation of the CP-PACS Collaboration (Aoki et al. 1998) is shown in Fig. 4. The ratio $M^2/(m_1 + m_2)$ appears to be flat over the whole range of quark masses accessible in the simulations. The different values of β stand for different lattice spacings but for each β the ratio is constant to better than 5%. Since lattice calculations have found evidence for nonlinear quark mass corrections to baryon masses (e.g., Aoki

Table 5. Light quark masses in MeV at the $\overline{MS}$ scale $\nu = 2$ GeV. The most recent values from QCD sum rules and (quenched) lattice calculations are listed.

$\hat{m}$		m_s
4.9 ± 0.9	sum rules	125 ± 25
Prades (1998)		Jamin (1998)
$5.7 \pm 0.1 \pm 0.8$	lattice	$130 \pm 2 \pm 18$
	Giménez et al. (1998)	

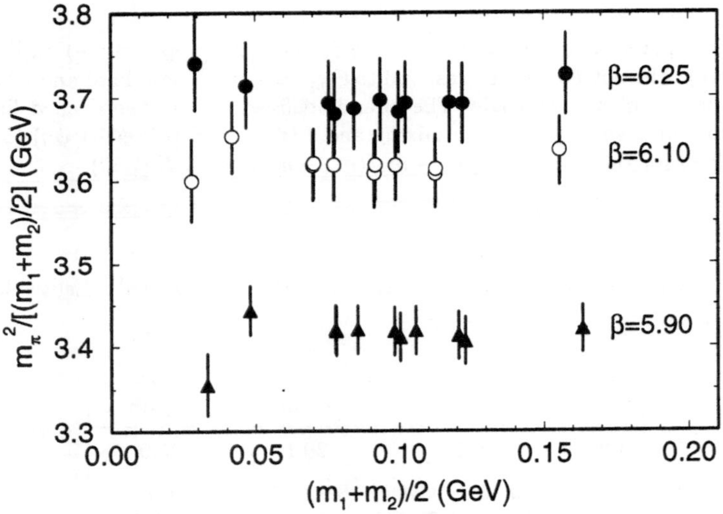

Fig. 4. $2M_\pi^2/(m_1 + m_2)$ as a function of $(m_1 + m_2)/2$ (from Aoki et al. 1998).

et al. 1998), it is difficult to blame this conspicuous linearity[5] between M^2 and m_q on the limitations of present–day lattice methods only.

In order to see whether the lattice findings are consistent with CHPT, I take the $O(p^4)$ result (Gasser and Leutwyler 1985) for M_K^2 and vary $m_1 = \hat{m}$, $m_2 = m_s$. Since the actual quark masses on the lattice are still substantially bigger than $\hat{m}$, the $SU(2)$ result for M_π^2 cannot be used for this comparison.

[5] Quenching effects are estimated to be $\sim 5\%$ at the lightest m_q presently available on the lattice (Sharpe 1997; Golterman 1997).

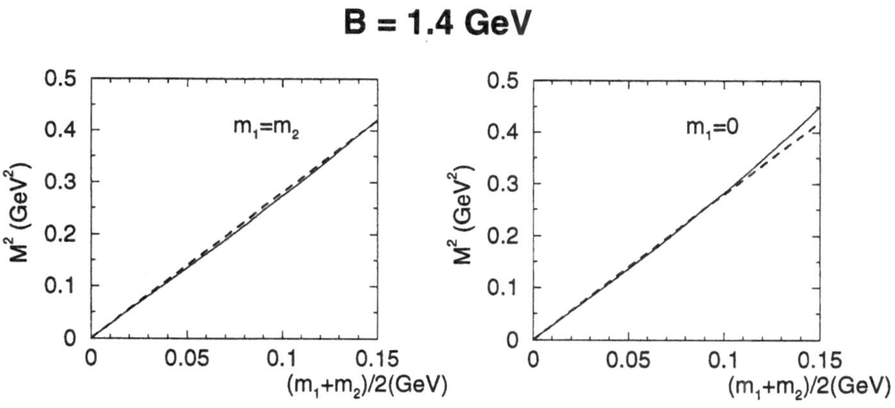

Fig. 5. $M^2(\text{GeV}^2)$ as function of the average quark mass (in GeV) in standard CHPT. The dashed lines are the lowest-order predictions; the full lines correspond to the results of $O(p^4)$ in Eq. (99). The two graphs are for $m_1 = m_2$ and $m_1 = 0$, respectively.

Writing M^2 instead of M_K^2 for general m_1, m_2, one finds

$$M^2 = (m_1 + m_2)B\left\{1 + \frac{(m_1 + 2m_2)B}{72\pi^2 F^2}\ln\frac{2(m_1 + 2m_2)B}{3\mu^2}\right. \tag{99}$$

$$\left. + \frac{8(m_1 + m_2)B}{F^2}(2L_8^r(\mu) - L_5^r(\mu)) + \frac{16(2m_1 + m_2)B}{F^2}(2L_6^r(\mu) - L_4^r(\mu))\right\}$$

with the scale dependent LECs given in Table 2. As can easily be checked with the help of Eq. (78), M^2 in (99) is independent of the arbitrary scale μ as it should be.

Since the L_i are by definition independent of quark masses, it is legitimate to use the values in Table 2 also when varying m_1, m_2. Let me first consider the standard scenario with $B(\nu = 1 \text{ GeV}) = 1.4 \text{ GeV}$[6] together with the mean values of the $L_i^r(M_\rho)$ in Table 2. In Fig. 5, M^2 is plotted as a function of the average quark mass $(m_1 + m_2)/2$ for two extreme cases: $m_1 = m_2$ or $m_1 = 0$. The second case with a massless quark can of course not be implemented on the lattice. As the figure demonstrates, there is little deviation from linearity at least up to $M \simeq 600$ MeV although this deviation is in general bigger than suggested by Fig. 4 (for the range of LECs in Table 2).

In order to demonstrate that the near-linearity is specific for standard CHPT, we now lower the value of B as suggested by the proponents of generalized CHPT. Remember that $B = O(F_\pi)$ is considered to be a reasonable value in

[6] Note that the quark masses in Fig. 4 correspond to $\nu = 2$ GeV, however.

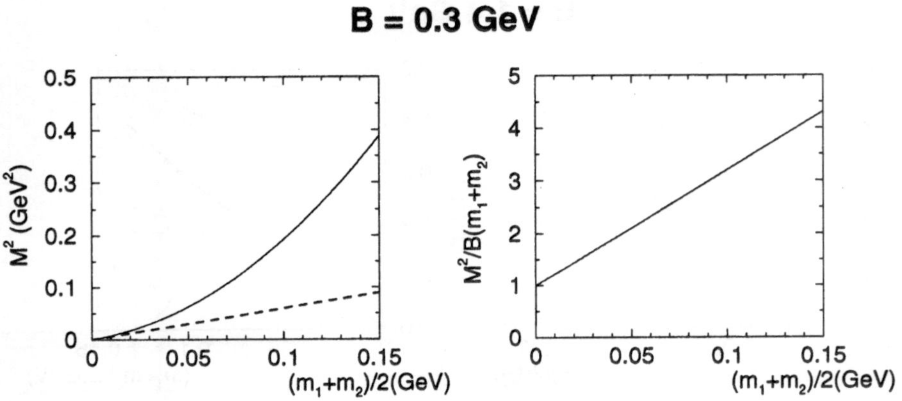

Fig. 6. $M^2(\mathrm{GeV}^2)$ as function of the average quark mass (in GeV) for B=0.3 GeV. The notation in the first graph is as in the previous figure. The second graph shows the ratio $M^2/B(m_1+m_2)$ from (99).

that scenario. To show the dramatic changes required by a small B, I choose an intermediate value $B = 0.3$ GeV. Of course, in order to obtain the observed meson masses, at least some of the LECs have to be scaled up. Leaving the signs of the LECs unchanged, Eq. (99) requires to scale up L_8^r to obtain realistic meson masses for a similar range of quark masses as before. But this is precisely the suggestion of generalized CHPT that the LECs associated with mass terms in $\mathcal{L}_4$ may have been underestimated (Stern 1997) by standard CHPT. For the following plot, I therefore take $L_8^r(M_\rho) = 20{\cdot}10^{-3}$. The two cases considered before ($m_1 = m_2$ or $m_1 = 0$) are now practically indistinguishable and they lead to a strong deviation from linearity as exhibited in the first graph of Fig. 6. The second graph can be compared with the lattice results in Fig. 4. Please make sure to compare the scales of the ordinates: whereas the lattice ratios vary by at most 5%, this ratio would now have to change by more than a factor of four (!) over the same range of quark masses.

The conclusion of this exercise is straightforward: lattice QCD is incompatible with a small quark condensate. Unless lattice simulations for the meson mass spectrum are completely unreliable, the observed linearity of M^2 in the quark masses favours standard CHPT and excludes values of B substantially smaller than the standard value.

2.3 Pion–Pion Scattering

There are several good reasons for studying elastic pion–pion scattering:

i. The elastic scattering of the lightest hadrons is a fundamental process for testing CHPT: the only particles involved are $SU(2)$ pseudo–Goldstone bosons. One may rightfully expect good convergence of the low–energy expansion near threshold.
ii. The behaviour of the scattering amplitude near threshold is sensitive to the mechanism of spontaneous chiral symmetry breaking (Stern et al. 1993), or more precisely, to the size of the quark condensate.
iii. After a long period without much experimental activity, there are now good prospects for significant improvements in the near future. K_{e4} experiments to extract pion–pion phase shifts due to the final–state interactions of the pions are already in the analysis stage at Brookhaven (Lowe 1997) or will start this year at the Φ factory DAΦNE in Frascati (Baillargeon and Franzini 1995; Lee-Franzini 1997). In addition, the ambitious DIRAC experiment (Adeva et al. 1994; Schacher 1997) is being set up at CERN to measure a combination of S–wave scattering lengths through a study of $\pi^+\pi^-$ bound states.

In the isospin limit $m_u = m_d$, the scattering amplitude is determined by one scalar function $A(s,t,u)$ of the Mandelstam variables. In terms of this function, one can construct amplitudes with definite isospin ($I = 0,1,2$) in the s–channel. A partial–wave expansion gives rise to partial–wave amplitudes $t_l^I(s)$ that are described by real phase shifts $\delta_l^I(s)$ in the elastic region $4M_\pi^2 \leq s \leq 16M_\pi^2$ in the usual way:

$$t_l^I(s) = (1 - \frac{4M_\pi^2}{s})^{-1/2} \exp i\delta_l^I(s) \sin \delta_l^I(s) \ . \tag{100}$$

The behaviour of the partial waves near threshold is of the form

$$\Re e \, t_l^I(s) = q^{2l}\{a_l^I + q^2 b_l^I + O(q^4)\} \ , \tag{101}$$

with q the center–of–mass momentum. The quantities a_l^I and b_l^I are referred to as scattering lengths and slope parameters, respectively.

The low–energy expansion for $\pi\pi$ scattering has been carried through to $O(p^6)$ where two–loop diagrams must be included. Before describing the more recent work, let me recall the results at lower orders.

$O(p^2)$ $(L = 0)$

As discussed previously in this lecture, only tree diagrams from the lowest–order Lagrangian $\mathcal{L}_2$ contribute at $O(p^2)$. The scattering amplitude was first written down by Weinberg (1966):

$$A_2(s,t,u) = \frac{s - M_\pi^2}{F_\pi^2} \ . \tag{102}$$

At the same order in the standard scheme, the quark mass ratios are fixed in terms of meson mass ratios, e.g., $r = r_2$ in the notation of Eq. (95).

In generalized CHPT, some of the terms in $\mathcal{L}_4$ in the standard counting appear already at lowest order. Because there are now more free parameters, the relation $r = r_2$ is replaced by the bounds (95). The $\pi\pi$ scattering amplitude of lowest order in generalized CHPT is (Stern et al. 1993)

$$A_2(s,t,u) = \frac{s - \frac{4}{3}M_\pi^2}{F_\pi^2} + \alpha \frac{M_\pi^2}{3F_\pi^2} \tag{103}$$

$$\alpha = 1 + \frac{6(r_2 - r)}{r^2 - 1}\,, \qquad \alpha \geq 1\,.$$

The amplitude is correlated with the quark mass ratio r. Especially the S–wave is very sensitive to α: the standard value of $a_0^0 = 0.16$ for $\alpha = 1$ $(r = r_2)$ moves to $a_0^0 = 0.26$ for a typical value of $\alpha \simeq 2$ $(r \simeq 10)$ in the generalized scenario. As announced before, the S–wave amplitude is indeed a sensitive measure of the quark mass ratios and thus of the quark condensate. To settle the issue, the lowest–order amplitude is of course not sufficient.

$\underline{O(p^4)\ (L \leq 1)}$

To next–to–leading order, the scattering amplitude was calculated by Gasser and Leutwyler (1983):

$$\begin{aligned} F_\pi^4 A_4(s,t,u) &= c_1 M_\pi^4 + c_2 M_\pi^2 s + c_3 s^2 + c_4 (t-u)^2 \\ &\quad + F_1(s) + G_1(s,t) + G_1(s,u)\,. \end{aligned} \tag{104}$$

F_1, G_1 are standard one–loop functions and the constants c_i are linear combinations of the LECs $l_i^r(\mu)$ and of the chiral log $\ln(M_\pi^2/\mu^2)$. It turns out that many observables are dominated by the chiral logs. This applies for instance to the $I = 0$ S–wave scattering length that increases from 0.16 to 0.20. This relatively big increase of 25% makes it necessary to go still one step further in the chiral expansion.

$\underline{O(p^6)\ (L \leq 2)}$

Two different approaches have been used. In the dispersive treatment (Knecht et al. 1995), $A(s,t,u)$ was calculated explicitly up to a crossing symmetric subtraction polynomial

$$[b_1 M_\pi^4 + b_2 M_\pi^2 s + b_3 s^2 + b_4 (t-u)^2]/F_\pi^4 + [b_5 s^3 + b_6 s (t-u)^2]/F_\pi^6 \tag{105}$$

with six dimensionless subtraction constants b_i. Including experimental information from $\pi\pi$ scattering at higher energies, Knecht et al. (1996) evaluated four of those constants $(b_3, \ldots, b_6)$ from sum rules. The amplitude is given in a form compatible with generalized CHPT.

The field theoretic calculation involving Feynman diagrams with $L = 0, 1, 2$ was performed in the standard scheme (Bijnens et al. 1996, 1997). Of course, the diagrammatic calculation reproduces the analytically nontrivial part of the dispersive approach. To arrive at the final renormalized amplitude, one needs in addition the following quantities to $O(p^6)$: the pion wave function renormalization constant (Bürgi 1996), the pion mass (Bürgi 1996) and the pion decay constant (Bijnens et al. 1996, 1997). Moreover, in the field theoretic approach the previous subtraction constants are obtained as functions

$$b_i(M_\pi/F_\pi, M_\pi/\mu; l_i^r(\mu), k_i^r(\mu)) , \tag{106}$$

where the k_i^r are six combinations of LECs of the $SU(2)$ Lagrangian of $O(p^6)$.

Compared to the dispersive approach, the diagrammatic method offers the following advantages:

i. The full infrared structure is exhibited to $O(p^6)$. In particular, the b_i contain chiral logs of the form $(\ln M_\pi/\mu)^n$ $(n \leq 2)$ that are known to be numerically important, especially for the infrared–dominated parameters b_1 and b_2.

ii. The explicit dependence on LECs makes phenomenological determinations of these constants and comparison with other processes possible. This is especially relevant for determining l_1^r, l_2^r to $O(p^6)$ accuracy (Colangelo et al. 1998).

iii. The fully known dependence on the pion mass allows one to evaluate the amplitude even at unphysical values of the quark mass (remember that we assume $m_u = m_d$). One possible application is to confront the CHPT amplitude with lattice calculations of pion–pion scattering (Colangelo 1997).

In the standard picture, the $\pi\pi$ amplitude depends on four LECs of $O(p^4)$ and on six combinations of $O(p^6)$ couplings. The latter have been estimated with meson resonance exchange that is known to account for the dominant features of the $O(p^4)$ constants (Ecker et al. 1989). It turns out (Bijnens et al. 1997) that the inherent uncertainties of this approximation induce small (somewhat bigger) uncertainties for the low (higher) partial waves. The main reason is that the higher partial waves are more sensitive to the short–distance structure.

However, as the chiral counting suggests, the LECs of $O(p^4)$ are much more important. Eventually, the $\pi\pi$ amplitude of $O(p^6)$ will lead to a more precise determination of some of those constants (Colangelo et al. 1998) than presently available. For the time being, one can investigate the sensitivity of the amplitude to the l_i^r. In Table 6, some of the threshold parameters are listed for three sets of the l_i^r (Bijnens et al. 1997; Ecker 1997): set I is mainly based on phenomenology to $O(p^4)$ (Gasser and Leutwyler 1984; Bijnens et al. 1994), for set II the $\pi\pi$ D–wave scattering lengths to $O(p^6)$ are used as input to fix l_1^r, l_2^r, whereas for set III resonance saturation is assumed for the l_i^r renormalized at $\mu = M_\eta$. Although some of the entries in Table 6 are quite sensitive to the choice of the l_i^r, two points are worth emphasizing:

- The S–wave threshold parameters are very stable, especially the $I = 0$ scattering length, whereas the higher partial waves are more sensitive to the choice of LECs of $O(p^4)$ (and also of $O(p^6)$).
- The resonance dominance prediction (set III) is in perfect agreement with the data although the agreement becomes less impressive for $\mu > M_\eta$.

Table 6. Threshold parameters in units of M_{π^+} for three sets of LECs l_i^r (Bijnens et al. 1997; Ecker 1997). The values of $O(p^4)$ correspond to set I. The experimental values are from Dumbrajs et al. (1983).

	$O(p^2)$	$O(p^4)$	$O(p^6)$ set I	$O(p^6)$ set II	$O(p^6)$ set III	experiment
a_0^0	0.16	0.20	0.217	0.206	0.209	0.26 ± 0.05
b_0^0	0.18	0.25	0.275	0.249	0.261	0.25 ± 0.03
$2a_0^0 - 5a_0^2$	0.55	0.61	0.641	0.634	0.626	0.66 ± 0.05
$-10\ b_0^2$	0.91	0.73	0.72	0.80	0.75	0.82 ± 0.08
$10\ a_1^1$	0.30	0.37	0.40	0.38	0.37	0.38 ± 0.02
$10^2 a_2^0$	0	0.18	0.27	input	0.19	0.17 ± 0.03

In Fig. 7, the phase shift difference $\delta_0^0 - \delta_1^1$ is plotted as function of the center–of–mass energy and compared with the available low–energy data. The two–loop phase shifts describe the K_{e4} data (Rosselet et al. 1977) very well for both sets I and II, with a small preference for set I. The curve for set III is not shown in the figure, it lies between those of sets I and II.

To conclude this part on $\pi\pi$ scattering, let me stress the main features:

- The low–energy expansion converges reasonably well. The main uncertainties are not due to the corrections of $O(p^6)$, but they are related to the LECs of $O(p^4)$. This will in turn make a better determination of those constants possible (Colangelo et al. 1998).
- Many observables , especially the S–wave threshold parameters, are infrared dominated by the chiral logs. This is the reason why the $I = 0$ S–wave scattering length is rather insensitive to the LECs of $O(p^4)$. From the calculations in standard CHPT, a value

$$a_0^0 = 0.21 \div 0.22 \tag{107}$$

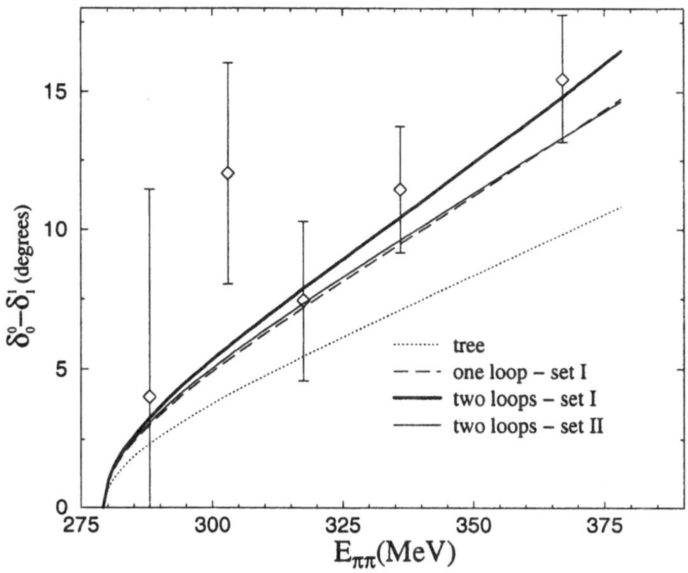

Fig. 7. Phase shift difference $\delta_0^0 - \delta_1^1$ at $O(p^2)$, $O(p^4)$ and $O(p^6)$ (set I and II) from Bijnens et al. (1997).

is well established. This will be a crucial test for the standard framework once the data become more precise. On the basis of available experimental information, there is at present no indication against the standard scenario of chiral symmetry breaking with a large quark condensate.

- Altogether, there is good agreement with the present low–energy data as both Table 6 and Fig. 7 demonstrate.
- Isospin violation and electromagnetic corrections have to be included. First results are already available (Meißner et al. 1997; Knecht and Urech 1997).

3 Baryons and Mesons

A lot of effort has been spent on the meson–baryon system in CHPT (e.g., Bernard et al. 1995; Walcher 1998). Nevertheless, the accuracy achieved is not comparable to the meson sector. Here are some of the reasons.

- The baryons are not Goldstone particles. Therefore, their interactions are less constrained by chiral symmetry than for pseudoscalar mesons.

- Due to the fermionic nature of baryons, there are terms of every positive order in the chiral expansion. In the meson case, only even orders can contribute.
- There are no "soft" baryons because the baryon masses stay finite in the chiral limit. Only baryonic three–momenta may be soft.
- In a manifestly relativistic framework (Gasser et al. 1988), the baryon mass destroys the correspondence between loop and chiral expansion that holds for mesons.

In this lecture, I will only consider chiral $SU(2)$, i.e., pions and nucleons only. Some of the problems mentioned have to do with the presence of the "big" nucleon mass that is in fact comparable to the scale $4\pi F_\pi$ of the chiral expansion. This comparison suggests a simultaneous expansion in

$$\frac{\mathbf{p}}{4\pi F} \qquad \text{and} \qquad \frac{\mathbf{p}}{m}$$

where $\mathbf{p}$ is a small three–momentum and m is the nucleon mass in the chiral limit. On the other hand, there is an essential difference between F and m: whereas F appears only in vertices, the nucleon mass enters via the nucleon propagator. To arrive at a simultaneous expansion, one therefore has to shift m from the propagator to the vertices of some effective Lagrangian. That is precisely the procedure of heavy baryon CHPT (Jenkins and Manohar 1991; Bernard et al. 1992), in close analogy to heavy quark effective theory.

3.1 Heavy Baryon Chiral Perturbation Theory

The main idea of heavy baryon CHPT is to decompose the nucleon field into "light" and "heavy" components. In fact, the light components will be massless in the chiral limit. The heavy components are then integrated out not unlike other heavy degrees of freedom. This decomposition is necessarily frame dependent but it does achieve the required goal: at the end, we have an effective chiral Lagrangian with only light degrees of freedom where the nucleon mass appears only in inverse powers in higher–order terms of this Lagrangian.

Since the derivation of the effective Lagrangian of heavy baryon CHPT is rather involved, I will exemplify the method only for the trivial case of a free nucleon with Lagrangian

$$\mathcal{L}_0 = \overline{\Psi}(i\partial\!\!\!/ - m)\Psi \ . \tag{108}$$

In terms of a time–like unit four–vector v (velocity), one introduces projectors $P_v^\pm = \frac{1}{2}(1 \pm v\!\!\!/)$. In the rest system with $v = (1,0,0,0)$, for instance, the $P_v^\pm$ project on upper and lower components of the Dirac field in the standard representation of γ matrices. With these projectors, one defines (Georgi 1990) velocity–dependent fields N_v, H_v:

$$\begin{aligned} N_v(x) &= \exp[imv\cdot x]P_v^+\Psi(x) \\ H_v(x) &= \exp[imv\cdot x]P_v^-\Psi(x) \ . \end{aligned} \tag{109}$$

The Dirac Lagrangian is now rewritten in terms of these fields:

$$\begin{aligned}\mathcal{L}_0 &= \overline{(N_v + H_v)} e^{imv\cdot x}(i\partial\!\!\!/ - m)e^{-imv\cdot x}(N_v + H_v) \\ &= \overline{N_v} iv\cdot\partial N_v - \overline{H_v}(iv\cdot\partial + 2m)H_v + \text{ mixed terms} .\end{aligned} \tag{110}$$

After integrating out the heavy components H_v in the functional integral with the fully relativistic pion–nucleon Lagrangian (Gasser et al. 1988), one arrives indeed at an effective chiral Lagrangian for the field N_v (and pions) only, with a massless propagator

$$\frac{iP_v^+}{v\cdot k + i\varepsilon} . \tag{111}$$

At every order except the leading one, $O(p)$, this Lagrangian consists of two pieces: the first one is the usual chiral Lagrangian of $O(p^n)$ with a priori unknown LECs. The second part comes from the expansion in $1/m$ and it is completely given in terms of LECs of lower than n-th order. Since the only nucleon field in this Lagrangian is N_v with a massless propagator, there is a straightforward analogue to chiral power counting in the meson sector given by formula (69). For a connected L–loop amplitude with E_B external baryon lines and N_{n,n_B} vertices of chiral dimension n (with n_B baryon lines at the vertex), the analogue of (69) is (Weinberg 1990, 1991)

$$D = 2L + 2 - \frac{E_B}{2} + \sum_{n,n_B}(n - 2 + \frac{n_B}{2})N_{n,n_B} . \tag{112}$$

However, as we will discuss later on in connection with nucleon–nucleon scattering, this formula is misleading for $E_B \geq 4$. On the other hand, no problems arise for the case of one incoming and one outgoing nucleon ($E_B = 2$) where

$$D = 2L + 1 + \sum_n [(n-2)N_{n,0} + (n-1)N_{n,2}] \geq 2L + 1 . \tag{113}$$

This formula is the basis for a systematic low–energy expansion for single–nucleon processes, i.e., for processes of the type $\pi N \to \pi \ldots \pi N$, $\gamma N \to \pi \ldots \pi N$, $l\,N \to l\,\pi \ldots \pi N$ (including nucleon form factors), $\nu_l N \to l\,\pi \ldots \pi N$. The corresponding effective chiral Lagrangian is completely known to $O(p^3)$ (Bernard et al. 1992; Ecker and Mojžiš 1996; Fettes et al. 1998) including the full renormalization at $O(p^3)$ (Ecker 1994):

$$\begin{aligned}\mathcal{L}_{\pi N} &= \mathcal{L}_{\pi N}^{(1)} + \mathcal{L}_{\pi N}^{(2)} + \mathcal{L}_{\pi N}^{(3)} + \ldots \\ \mathcal{L}_{\pi N}^{(1)} &= \overline{N_v}(iv\cdot\nabla + g_A S\cdot u)N_v\end{aligned} \tag{114}$$

$$u_\mu = i(u^\dagger\partial_\mu u - u\partial_\mu u^\dagger) + \text{ external gauge fields} , \qquad S^\mu = i\gamma_5\sigma^{\mu\nu}v_\nu/2$$

with a chiral and gauge covariant derivative ∇ and with g_A the axial–vector coupling constant in the chiral limit.

Two remarks are in order at this point.

Table 7. Relations between relativistic covariants and the corresponding quantities in the initial nucleon rest frame ($v = p_{\rm in}/m_N$, $q = p_{\rm out} - p_{\rm in}$, $t = q^2$) with $\overline{u}(p_{\rm out})\Gamma u(p_{\rm in}) = \overline{u}(p_{\rm out})P_v^+\widehat{\Gamma}P_v^+u(p_{\rm in})$.

Γ	$\widehat{\Gamma}$
1	1
γ_5	$\dfrac{q\cdot S}{m_N(1-t/4m_N^2)}$
γ^μ	$\left(1-t/4m_N^2\right)^{-1}\left(v^\mu+\dfrac{q^\mu}{2m_N}+\dfrac{i}{m_N}\varepsilon^{\mu\nu\rho\sigma}q_\nu v_\rho S_\sigma\right)$
$\gamma^\mu\gamma_5$	$2S^\mu-\dfrac{q\cdot S}{m_N(1-t/4m_N^2)}v^\mu$
$\sigma^{\mu\nu}$	$2\varepsilon^{\mu\nu\rho\sigma}v_\rho S_\sigma+\dfrac{1}{2m_N(1-t/4m_N^2)}\{i(q^\mu v^\nu-q^\nu v^\mu)+2(v^\mu\varepsilon^{\nu\lambda\rho\sigma}-v^\nu\varepsilon^{\mu\lambda\rho\sigma})q_\lambda v_\rho S_\sigma\}$

- Since the Lagrangian (114) was derived from a fully relativistic Lagrangian it defines a Lorentz invariant quantum field theory although it depends explicitly on the arbitrary frame vector v (Ecker and Mojžiš 1996). Reparametrization invariance (Luke and Manohar 1992) is automatically fulfilled.
- The transformation from the original Dirac field Ψ to the velocity–dependent field N_v leads to an unconventional wave function renormalization of N_v that is in general momentum dependent (Ecker and Mojžiš 1997).

Since the theory is Lorentz invariant it must always be possible to express the final amplitudes in a manifestly relativistic form. Of course, this will only be true up to the given order in the chiral expansion one is considering. The general procedure of heavy baryon CHPT for single–nucleon processes can then be summarized as follows.

i. Calculate the heavy baryon amplitudes to a given chiral order with the Lagrangian (114) in a frame defined by the velocity vector v.
ii. Relate those amplitudes to their relativistic counterparts which are independent of v to the order considered. For the special example of the initial nucleon rest frame with $v = p_{\rm in}/m_N$, the translation is given in Table 7 (Ecker and Mojžiš 1997).
iii. Apply wave function renormalization for the external nucleons.

As an application of this procedure, I will now discuss elastic pion–nucleon scattering to $O(p^3)$ in the low–energy expansion. For other applications of CHPT to single–nucleon processes, I refer to the available reviews (Bernard et al. 1995;

Ecker 1995) and conference proceedings (Bernstein and Holstein 1995; Walcher 1998).

3.2 Pion–Nucleon Scattering

Elastic πN scattering is maybe the most intensively studied process of hadron physics, with a long history both in theory and experiment (e.g., Höhler 1983). The systematic CHPT approach is however comparatively new (Gasser et al. 1988). I am going to review here the first complete calculation to $O(p^3)$ by Mojžiš (1998). As for $\pi\pi$ scattering, isospin symmetry is assumed.

A comparison with elastic $\pi\pi$ scattering displays the difficulties of the πN analysis. Although calculations have been performed to next-to-next-to-leading order for both processes, this is only $O(p^3)$ for πN compared to $O(p^6)$ for $\pi\pi$. Of course, this is due to the fact that, unlike for mesons only, every integer order can contribute to the low-energy expansion in the meson-baryon sector. The difference in accuracy also manifests itself in the number of LECs: the numbers are again comparable despite the difference in chiral orders. Finally, while we now know the $\pi\pi$ amplitude to two-loop accuracy the πN amplitude is still not completely known even at the one-loop level as long as the p^4 amplitude has not been calculated.

The amplitude for pion-nucleon scattering

$$\pi^a(q_1) + N(p_1) \to \pi^b(q_2) + N(p_2) \tag{115}$$

can be expressed in terms of four invariant amplitudes $D^\pm$, $B^\pm$:

$$\begin{aligned} T_{ab} &= T^+\delta_{ab} - T^- i\varepsilon_{abc}\tau_c \\ T^\pm &= \overline{u}(p_2)\left[D^\pm(\nu,t) + \frac{i}{2m_N}\sigma^{\mu\nu}q_{2\mu}q_{1\nu}B^\pm(\nu,t)\right]u(p_1) \end{aligned} \tag{116}$$

with

$$\begin{aligned} s &= (p_1+q_1)^2 \quad, t = (q_1-q_2)^2 \quad, \\ u &= (p_1-q_2)^2 \quad, \nu = \frac{s-u}{4m_N} \quad . \end{aligned} \tag{117}$$

With the choice of invariant amplitudes $D^\pm$, $B^\pm$, the low-energy expansion is straightforward: to determine the scattering amplitude to $O(p^n)$, one has to calculate $D^\pm$ to $O(p^n)$ and $B^\pm$ to $O(p^{n-2})$.

In the framework of CHPT, the first systematic calculation of pion-nucleon scattering was performed by Gasser et al. (1988). In heavy baryon CHPT, the pion-nucleon scattering amplitude is not directly obtained in the relativistic form (116) but rather as (Mojžiš 1998)

$$\overline{u}(p_2)P_v^+\left[\alpha^\pm + i\varepsilon^{\mu\nu\rho\sigma}q_{1\mu}q_{2\nu}v_\rho S_\sigma\beta^\pm\right]P_v^+u(p_1) \ . \tag{118}$$

The amplitudes $\alpha^\pm$, $\beta^\pm$ depend on the choice of the velocity v. A natural and convenient choice is the initial nucleon rest frame with $v = p_1/m_N$. In this frame, the relativistic amplitudes can be read off directly from Table 7:

$$\begin{aligned} D^\pm &= \alpha^\pm + \frac{\nu t}{4m_N}\beta^\pm \\ B^\pm &= -m_N\left(1 - \frac{t}{4m_N^2}\right)\beta^\pm \ . \end{aligned} \tag{119}$$

Also the amplitudes $D^\pm$, $B^\pm$ in (119) will depend on the chosen frame. However, as discussed before, they are guaranteed to be Lorentz invariant up to terms of at least $O(p^{n+1})$ if the amplitude (118) has been calculated to $O(p^n)$.

From Eq. (113) one finds that tree–level diagrams with $D = 1, 2, 3$ and one–loop diagrams with $D = 3$ need to be calculated. After proper renormalization, including the nonstandard nucleon wave function renormalization, the final amplitudes depend on the kinematical variables ν, t, m_N, M_π, on the lowest–order LECs F_π, g_A, on four constants of the p^2 Lagrangian and on five combinations of LECs of $O(p^3)$.

The invariant amplitudes $D^\pm$, $B^\pm$ can be projected onto partial–wave amplitudes $f^\pm_{l\pm}(s)$. Threshold parameters are defined as in Eq. (101):

$$\Re e\ f^\pm_{l\pm}(s) = q^{2l}\{a^\pm_{l\pm} + q^2 b^\pm_{l\pm} + O(q^4)\} \ . \tag{120}$$

To confront the chiral amplitude with experiment, Mojžiš (1998) has compared 16 of these threshold parameters with the corresponding values extrapolated from experimental data on the basis of the Karlsruhe–Helsinki phase–shift analysis (Koch and Pietarinen 1980).

Six of the threshold parameters (D and F waves) turn out to be independent of the low–energy constants of $O(p^2)$ and $O(p^3)$. The results are shown in Table 8 and compared with Koch and Pietarinen (1980).

The main conclusion from Table 8 is a definite improvement seen at $O(p^3)$. Since there are no low–energy constants involved (except, of course, M_π, F_π, m_N and g_A), this is clear evidence for the relevance of loop effects. The numbers shown in Table 8 are based on the calculation of Mojžiš (1998), but essentially the same results were obtained by Bernard et al. (1997).

The altogether nine LECs beyond leading order were then fitted by Mojžiš (1998) to the ten remaining threshold parameters, the πN σ–term and the Goldberger–Treiman discrepancy. Referring to Mojžiš (1998) for the details, let me summarize the main results:

- The fit is quite satisfactory although the fitted value of the σ–term tends to be larger than the canonical value (Gasser et al. 1991).
- In many cases, the corrections of $O(p^3)$ are sizable and definitely bigger than what naive chiral order–of–magnitude estimates would suggest.
- The fitted values of the four LECs of $O(p^2)$ agree very well with an independent analysis of Bernard et al. (1997). Moreover, those authors have shown

Table 8. Comparison of two D–wave and four F–wave threshold parameters up to the first, second and third order (the two columns differ by higher–order terms) with (extrapolated) experimental values (Koch and Pietarinen 1980). The theoretical values are based on the calculation of Mojžiš (1998). Units are appropriate powers of GeV^{-1}.

	$O(p)$	$O(p^2)$	$O(p^3)$	HBCHPT $O(p^3)$	exp.
a_{2+}^{+}	0	−48	−35	−36	-36 ± 7
a_{2+}^{-}	0	48	56	56	64 ± 3
a_{3+}^{+}	0	0	226	280	440 ± 140
a_{3-}^{+}	0	14	26	31	160 ± 120
a_{3+}^{-}	0	0	−158	−210	-260 ± 20
a_{3-}^{-}	0	−14	65	57	100 ± 20

that the specific values can be understood on the basis of resonance exchange (baryons and mesons). It seems that the LECs of $O(p^2)$ in the pion–nucleon Lagrangian are under good control, both numerically and conceptually.
– The LECs of $O(p^3)$ are of "natural" magnitude but more work is needed here.

Using the results of Mojžiš (1998), Datta and Pakvasa (1997) have also calculated πN phase shifts near threshold[7]. Again, a clear improvement over tree–level calculations can be seen in most cases. As an example, I reproduce their results for the S_{11} phase shift in Fig. 8.

The main conclusions for the present status of elastic πN scattering are:

1. The results of the first complete analysis (Mojžiš 1998) to $O(p^3)$ in the low–energy expansion are very encouraging.
2. Effects of $O(p^4)$ (still $L \leq 1$) need to be included to check the stability of the expansion.

3.3 Nucleon–Nucleon Interaction

When Weinberg (1990, 1991) investigated the nucleon–nucleon interaction within the chiral framework, he pointed out an obvious clash between the chiral expansion and the existence of nuclear binding. Unlike for the meson–meson interaction

[7] After the School, a new calculation of Fettes et al. (1998) appeared where both threshold parameters and phase shifts are considered.

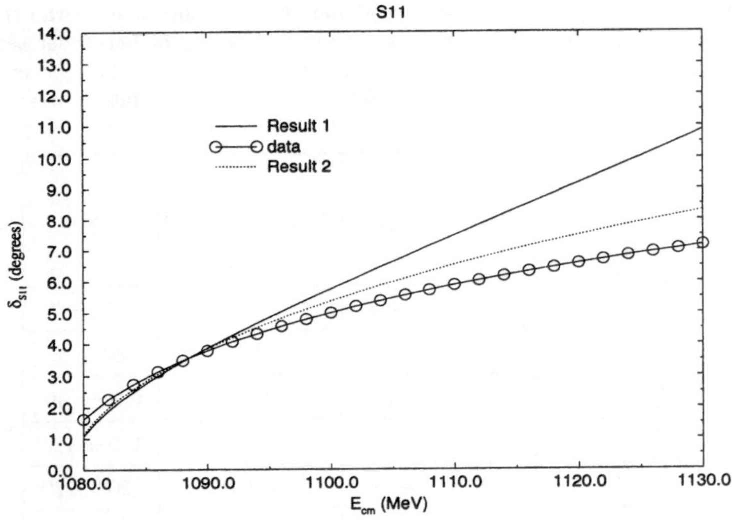

Fig. 8. S_{11} phase shift from Datta and Pakvasa (1997). Solid line: tree–level model with Δ and N^* exchange; dotted line: complete $O(p^3)$ amplitude of Mojžiš (1998). Circles represent the phase shifts extracted from fits to the πN scattering data.

that becomes arbitrarily small for small enough momenta (and meson masses), the perturbative expansion in the NN–system must break down already at low energies. Therefore, the chiral dimension defined in (112) cannot have the same interpretation as for mesonic interactions or for single–nucleon processes.

In heavy baryon CHPT, the problem manifests itself through a seeming infrared divergence associated with the massless propagator of the "light" field N_v. To make the point, we neglect pions for the time being and consider the lowest–order four–nucleon coupling without derivatives ($n = 0$ and $n_B = 4$ in the notation of Eq. (112)). The vertex is characterized by the tree diagram in the first line of Fig. 9. If we now calculate the chiral dimension of the one–loop diagram (second diagram in the first line of the figure) according to (112) we find

$$D = 2L + 2 - \frac{E_B}{2} = 2 \; . \tag{121}$$

However, this result is misleading because the diagram is actually infrared divergent with the propagator (111). Of course, this is an artifact of the approximation made since nucleons are everything else but massless. The way out is to include higher–order corrections in the nucleon propagator. The leading correction is due to $\mathcal{L}_{\pi N}^{(2)}$ in (114). The kinetic terms to this order are

$$\begin{aligned} \mathcal{L}_{\text{kin}} &= \overline{N_v}\left(iv\cdot\nabla + \frac{1}{2m}[(v\cdot\nabla)^2 - \nabla^2]\right)N_v \\ &= \overline{N_v}\left(i\partial_0 + \frac{1}{2m}\partial^2\right)N_v \end{aligned} \tag{122}$$

where the last expression applies for $v = (1,0,0,0)$, which now denotes the center–of–mass system. The corresponding propagator in this frame is

$$\frac{i}{k^0 - \dfrac{\mathbf{k}^2}{2m} + i\varepsilon} . \tag{123}$$

Following Kaplan et al. (1998), we now specialize to NN scattering in the $^1S_0$ channel and denote the incoming momenta as

$$p_{1,2} = \left(\frac{E}{2}, \pm\mathbf{p}\right) , \qquad E = \frac{p^2}{m} + \ldots \tag{124}$$

neglecting higher orders in the expression for the cms–energy E. Including higher orders in derivatives and quark masses, Kaplan et al. (1998) write the general tree amplitude (in d dimensions) for nucleon–nucleon scattering in the $^1S_0$ channel as

$$A_{\text{tree}} = -\left(\frac{\mu}{2}\right)^{4-d} \sum_{n\geq 0} C_{2n}(\mu)p^{2n} =: -\left(\frac{\mu}{2}\right)^{4-d} C(p^2,\mu) . \tag{125}$$

The relevance of the subtraction scale μ will soon become clear. For a general vertex C_{2n} of chiral dimension $2n$, the loop diagram considered before (second diagram in Fig. 9) is easily evaluated (Kaplan et al. 1998) in dimensional regularization:

$$\begin{aligned} I_n &= -i\left(\frac{\mu}{2}\right)^{4-d} \int \frac{d^dk}{(2\pi)^d}\mathbf{k}^{2n} \frac{i}{\frac{E}{2} + k^0 - \frac{\mathbf{k}^2}{2m} + i\varepsilon} \frac{i}{\frac{E}{2} - k^0 - \frac{\mathbf{k}^2}{2m} + i\varepsilon} \\ &= -m(mE)^n(-mE - i\varepsilon)^{\frac{d-3}{2}}\Gamma\left(\frac{3-d}{2}\right)\frac{(\frac{\mu}{2})^{4-d}}{(4\pi)^{\frac{d-1}{2}}} . \end{aligned} \tag{126}$$

The seeming infrared divergence of before now manifests itself as a divergence for $m \to \infty$. The diagram is actually finite for $d = 4$ and clearly of $O(p^{2n+1})$ invalidating the general formula for the chiral dimension that gave $D = 2$ for $n = 0$.

Kaplan et al. (1998) make the point that the diagram would be divergent in $d = 3$ dimensions with

$$I_n \simeq \frac{m(mE)^n\mu}{4\pi(d-3)} \tag{127}$$

near $d = 3$. Although this would not seem to have any great physical significance at first sight, Kaplan et al. (1998) suggest to subtract nevertheless the pole at

$d = 3$ that actually corresponds to a linear ultraviolet divergence in a cutoff regularization. This unconventional subtraction procedure is in line with the observation of other authors (e.g., Lepage 1997; Richardson et al. 1997; Beane et al. 1998) that standard dimensional regularization is not well adapted to the problem at hand.

The one–loop amplitude with the subtraction prescription of Kaplan et al. (1998) is then

$$I_n = -(mE)^n \frac{m}{4\pi}(\mu + ip) \ . \tag{128}$$

Anticipating the following discussion, we now iterate the one–loop diagram and sum the resulting bubble chains to arrive at the final amplitude (Kaplan et al. 1998)

$$A = \frac{-C(p^2,\mu)}{1 + \frac{m}{4\pi}(\mu + ip)C(p^2,\mu)} \ . \tag{129}$$

This amplitude is related to the phase shift as

$$e^{2i\delta} - 1 = \frac{ipm}{2\pi} A \tag{130}$$

or, with the effective range approximation for S–waves in terms of scattering length a and effective range r_0,

$$\begin{aligned} p \cot\delta = ip + \frac{4\pi}{mA} &= -\frac{4\pi}{mC(p^2,\mu)} - \mu \\ &= -\frac{1}{a} + \frac{1}{2} r_0 p^2 + O(p^4) \ . \end{aligned} \tag{131}$$

Note that the (traditional) definition of the scattering length used here has the opposite sign compared to (101) for $\pi\pi$ scattering. With the relations (131), the coefficients C_{2n} can be expressed in terms of $a, r_0, \ldots$:

$$C_0(\mu) = \frac{4\pi}{m}\frac{1}{-\mu + 1/a} \qquad C_2(\mu) = \frac{2\pi r_0}{m}\left(\frac{1}{-\mu + 1/a}\right)^2 \ . \tag{132}$$

It is known from potential scattering (e.g., Goldberger and Watson 1964) that r_0 and the higher–order coefficients in the effective range approximation are bounded by the range of the interaction. This also applies to NN scattering in the ${}^1S_0$ channel: $r_0 \simeq 2.7$ fm $\simeq 2/M_\pi$. On the other hand, the scattering length is sensitive to states near zero binding energy (e.g., Luke and Manohar 1997) and may be much bigger than the interaction range. Therefore, Kaplan et al. (1998) distinguish two scenarios.

- Normal–size scattering length
  In this case, also the scattering length is governed by the range of the interaction. The simplest choice $\mu = 0$ (minimal subtraction) leads to expansion coefficients C_{2n} in (132) in accordance with chiral dimensional analysis. This corresponds to the usual chiral expansion as in the meson or in the single–nucleon sector.

– Large scattering length
In the $^1S_0$ channel of NN scattering, the scattering length is much larger than the interaction range (the situation is similar in the deuteron channel)

$$a = -23.714 \pm 0.013 \text{ fm} \simeq -16/M_\pi \,. \tag{133}$$

With the same choice $\mu = 0$ as before, the coefficients C_{2n} are unnaturally large leading to big cancellations between different orders. Kaplan et al. (1998) therefore suggest to use instead $\mu = O(M_\pi)$ which leads to C_{2n} of natural chiral magnitudes.

The choice $\mu = O(M_\pi)$ immediately explains why we have to sum the iterated loop diagrams that led to amplitude A in (129). Let us consider such a bubble chain graph with coefficients C_{2n} at each four–nucleon vertex. From (132) and the obvious generalization to higher–order coefficients, one obtains $C_{2n} = O(p^{-n-1})$. Altogether, this implies a factor $C_{2n}p^{2n} = O(p^{n-1})$ at each vertex. On the other hand, each loop produces a factor of order $mp/4\pi$ as can be seen from Eq. (128). As a consequence, only the chain graphs with C_0 at each vertex have to be resummed because all such diagrams are of the same order p^{-1}. All other vertices can be treated perturbatively in the usual way.

The chiral expansion of the scattering amplitude (everything still in the $^1S_0$ channel) for $\mu = O(p)$ then takes the form (Kaplan et al. 1998)

$$A = A_{-1} + A_0 + A_1 + \ldots \tag{134}$$

$$A_{-1} = \frac{-C_0}{[1 + \frac{m}{4\pi}(\mu + ip)C_0]} \qquad A_0 = \frac{-C_2 p^2}{[1 + \frac{m}{4\pi}(\mu + ip)C_0]^2} \,. \tag{135}$$

This is also shown pictorially in Fig. 9.

So far, pions have been neglected. Inclusion of pions leaves A_{-1} unchanged but modifies $A_0, A_1, \ldots$. Altogether, to next–to–leading order, $O(p^0)$, the amplitude for NN scattering in the $^1S_0$ channel depends on three parameters: $C_0(M_\pi)$, $C_2(M_\pi)$, $D_2(M_\pi)$. Kaplan et al. (1998) fit these three parameters to the $^1S_0$ phase shift and obtain remarkable agreement with the experimental phase shift all the way up to $p = 300$ MeV. They also apply an analogous procedure to the $^3S_1$ - $^3D_1$ channels (deuteron).

After many attempts during the past years, a systematic low–energy expansion of nucleon–nucleon scattering seems now under control. This is an important step towards unifying the treatment of hadronic interactions at low energies on the basis of chiral symmetry.

Acknowledgements

I want to thank Willi Plessas and the members of his Organizing Committee for all their efforts to continue the successful tradition of the Schladming Winter School. Helpful discussions and email exchange with Jürg Gasser, Harald Grosse, Eduardo de Rafael and Jan Stern are gratefully acknowledged.

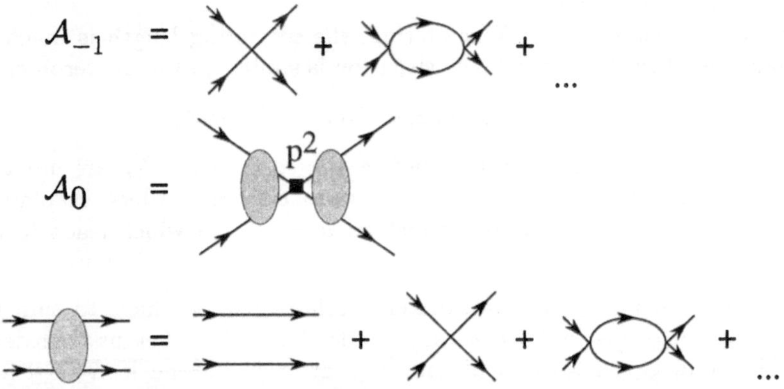

Fig. 9. Feynman graphs contributing to the leading amplitudes for $^1S_0$ nucleon–nucleon scattering (from Kaplan et al. 1998).

References

Adeva, B. et al. (DIRAC) (1994): Proposal to the SPSLC: Lifetime measurement of $\pi^+\pi^-$ atoms to test low energy QCD predictions, CERN/SPSLC/P 284, Dec. 1994

Anisovich, A.V., Leutwyler, H. (1996): Phys. Lett. **B375**, 335

Aoki, S. et al. (CP-PACS) (1998): Nucl. Phys. Proc. Suppl. **60A**, 14

Baillargeon, M., Franzini, P.J. (1995): in Maiani et al. (1995)

Banks, T., Casher, A. (1980): Nucl. Phys. **B169**, 103

Baur, R., Urech, R. (1996): Phys. Rev. **D53**, 6552

Beane, S.R., Cohen, T.D., Phillips, D.R. (1998): Nucl. Phys. **A632**, 445

Bellucci, S., Gasser, J., Sainio, M.E. (1994): Nucl. Phys. **B423**, 80; **B431**, 413 (E)

Bernard, V., Kaiser, N., Kambor, J., Meißner, U.-G. (1992): Nucl. Phys. **B388**, 315

Bernard, V., Kaiser, N., Meißner, U.-G. (1995): Int. J. Mod. Phys. **E4**, 193

Bernard, V., Kaiser, N., Meißner, U.-G. (1997): Nucl. Phys. **A615**, 483

Bernstein, A.M., Holstein, B.R., Eds. (1995): *Chiral Dynamics: Theory and Experiment*, Proc. of the Workshop at MIT, Cambridge, July 1994 (Springer–Verlag, Berlin)

Bijnens, J. (1993): Phys. Lett. **B306**, 343

Bijnens, J., Bruno, C., de Rafael, E. (1993): Nucl. Phys. **B390**, 501

Bijnens, J., Colangelo, G., Gasser, J. (1994): Nucl. Phys. **B427**, 427

Bijnens, J., Ecker, G., Gasser, J. (1995): in Maiani et al. (1995)

Bijnens, J. (1996): Phys. Reports **265**, 369

Bijnens, J., Colangelo, G., Ecker, G., Gasser, J., Sainio, M.E. (1996): Phys. Lett. **B374**, 210

Bijnens, J., Prades, J. (1997): Nucl. Phys. **B490**, 239

Bijnens, J., Talavera, P. (1997): Nucl. Phys. **B489**, 387

Bijnens, J., Colangelo, G., Ecker, G., Gasser, J., Sainio, M.E. (1997): Nucl. Phys. **B508**, 263
Bijnens, J., Colangelo, G., Talavera, P. (1998a): The vector and scalar form factors of the pion to two loops, hep-ph/9805389
Bijnens, J., Colangelo, G., Ecker, G. (1998b): in preparation
Bürgi, U. (1996): Phys. Lett. **B377**, 147; Nucl. Phys. **B479**, 392
Callan, C.G., Coleman, S., Wess, J., Zumino, B. (1969): Phys. Rev. **177**, 2247
Callan, C.G., Dashen, R.F., Gross, D.J. (1976): Phys. Lett. **36B**, 334
Colangelo, G. (1997): Phys. Lett. **B395**, 289
Colangelo, G., Gasser, J., Leutwyler, H., Wanders, G. (1998): in preparation
Coleman, S., Wess, J., Zumino, B. (1969): Phys. Rev. **177**, 2239
Coleman, S., Grossmann, B. (1982): Nucl. Phys. **B203**, 205
Collins, J.C. (1984): *Renormalization* (Cambridge Univ. Press, Cambridge)
Crewther, R.J. (1977): Phys. Lett. **70B**, 349
Dashen, R. (1969): Phys. Rev. **183**, 1245
Datta, A., Pakvasa, S. (1997): Phys. Rev. **D56**, 4322
Donoghue, J.F., Holstein, B.R., Wyler, D. (1993): Phys. Rev. **D47**, 2089
Donoghue, J.F., Pérez, A.F. (1997): Phys. Rev. **D55**, 7075
Dosch, H.G., Narison, S. (1998): Phys. Lett. **B417**, 173
Dumbrajs, O. et al. (1983): Nucl. Phys. **B216**, 277
Ecker, G., Gasser, J., Pich, A., de Rafael, E. (1989): Nucl. Phys. **B321**, 311
Ecker, G. (1994): Phys. Lett. **B336**, 508
Ecker, G. (1995): Prog. Part. Nucl. Phys. **35**, 1
Ecker, G., Mojžiš, M. (1996): Phys. Lett. **B365**, 312
Ecker, G., Mojžiš, M. (1997): Phys. Lett. **B410**, 266
Ecker, G. (1997): Pion–pion and pion–nucleon interactions in chiral perturbation theory, hep-ph/9710560, Contribution to the Workshop on Chiral Dynamics 1997, Mainz, Sept. 1997, to appear in the Proceedings
Fearing, H.W., Scherer, S. (1996): Phys. Rev. **D53**, 315
Fettes, N., Meißner, U.-G., Steininger, S. (1998): Pion–nucleon scattering in chiral perturbation theory I: isospin–symmetric case, hep-ph/9803266
Frishman, Y., Schwimmer, A., Banks, T., Yankielowicz, S. (1981): Nucl. Phys. **B177**, 157
Fuchs, N.H., Sazdjian, H., Stern, J. (1990): Phys. Lett. **B238**, 380
Fuchs, N.H., Sazdjian, H., Stern, J. (1991): Phys. Lett. **B269**, 183
Gasser, J., Leutwyler, H. (1983): Phys. Lett. **B125**, 325
Gasser, J., Leutwyler, H. (1984): Ann. Phys. (N.Y.) **158**, 142
Gasser, J., Leutwyler, H. (1985): Nucl. Phys. **B25**, 465
Gasser, J., Sainio, M.E., Švarc, A. (1988): Nucl. Phys. **B307**, 779
Gasser, J., Leutwyler, H., Sainio, M.E. (1991): Phys. Lett. **B253**, 252, 260
Gell-Mann, M. (1957): Phys. Rev. **106**, 1296
Gell-Mann, M., Oakes, R.J., Renner, B. (1968): Phys. Rev. **175**, 2195
Giménez, V., Giusti, L., Rapuano, F., Talevi, M. (1988): Lattice quark masses: a non-perturbative measurement, hep-lat/9801028
Goldberger, M.L., Watson, K.M. (1964): *Collision Theory* (Wiley, New York)
Goldstone, J. (1961): Nuovo Cimento **19**, 154
Golowich, E., Kambor, J. (1995): Nucl. Phys. **B447**, 373
Golowich, E., Kambor, J. (1997): Two–loop analysis of axialvector current propagators in chiral perturbation theory, hep-ph/9707341

Golterman, M. (1997): Connections between lattice gauge theory and chiral perturbation theory, hep-ph/9710468, Contribution to the Workshop on Chiral Dynamics 1997, Mainz, Sept. 1997, to appear in the Proceedings
Höhler, G. (1983): in Landolt–Börnstein, vol. 9 b2, Ed. H. Schopper (Springer, Berlin)
't Hooft, G. (1976): Phys. Rev. Lett. **37**, 8
't Hooft, G. (1980): in *Recent Developments in Gauge Theories*, G. 't Hooft et al., Eds. (Plenum Press, New York)
Jamin, M. (1998): Nucl. Phys. Proc. Suppl. **64**, 250
Jenkins, E., Manohar, A.V. (1991): Phys. Lett. **B255**, 558
Kambor, J., Wiesendanger, C., Wyler, D. (1996): Nucl. Phys. **B465**, 215
Kaplan, D.B., Manohar, A.V. (1986): Phys. Rev. Lett. **56**, 1994
Kaplan, D.B., Savage, M.J., Wise, M.B. (1998): A new expansion for nucleon–nucleon interactions, nucl-th/9801034; Two–nucleon systems from effective field theory, nucl-th/9802075
Kazakov, D. (1988): Theor. Math. Phys. **75**, 440
Knecht, M., Sazdjian, H., Stern, J., Fuchs, N.H. (1993): Phys. Lett. **B313**, 229
Knecht, M., Moussallam, B., Stern, J., Fuchs, N.H. (1995): Nucl. Phys. **B457**, 513
Knecht, M., Moussallam, B., Stern, J., Fuchs, N.H. (1996): Nucl. Phys. **B471**, 445
Knecht, M., Urech, R. (1997): Virtual photons in low–energy $\pi\pi$ scattering, hep-ph/9709348
Knecht, M., de Rafael, E. (1997): Patterns of spontaneous chiral symmetry breaking in the large–N_c limit of QCD–like theories, hep-ph/9712457
Koch, R., Pietarinen, E. (1980): Nucl. Phys. **A336**, 331
Lee–Franzini, J. (KLOE) (1997): Contribution to the Workshop on Chiral Dynamics 1997, Mainz, Sept. 1997, to appear in the Proceedings
Lellouch, L., de Rafael, E., Taron, J. (1997): Phys. Lett. **B414**, 195
Lepage, G.P. (1997): How to renormalize the Schrödinger equation, Lectures given at the 8th Jorge Andre Swieca Summer School, Sao Paolo, Brazil, Feb. 1997
Leutwyler, H. (1990): Nucl. Phys. **B337**, 108
Leutwyler, H., Smilga, A. (1992): Phys. Rev. **D46**, 5607
Leutwyler, H. (1994): Ann. Phys. (N.Y.) **235**, 165
Leutwyler, H. (1996a): Phys. Lett. **B374**, 181
Leutwyler, H. (1996b): Phys. Lett. **B378**, 313
Leutwyler, H. (1996c): Light quark masses, hep-ph/9609467, Cargèse Lectures 1996
Leutwyler, H. (1997): Probing the quark condensate by means of $\pi\pi$ scattering, hep-ph/9709406, Proceedings of the DAΦCE Workshop, Frascati, Nov. 1996
Lowe, J. (BNL-E865) (1997): Contribution to the Workshop on Chiral Dynamics 1997, Mainz, Sept. 1997, to appear in the Proceedings
Luke, M., Manohar, A.V. (1992): Phys. Lett. **B286**, 348
Luke, M., Manohar, A.V. (1997): Phys. Rev. **D55**, 4129
Lüscher, M. (1997): Theoretical advances in lattice QCD, hep-ph/9711205, Talk given at the 18th Int. Symposium on Lepton–Photon Interactions, Hamburg
Maiani, L., Pancheri, G., Paver, N., Eds. (1995): *The Second DAΦNE Physics Handbook* (INFN, Frascati)
Manohar, A.V., Georgi, H. (1984): Nucl. Phys. **B234**, 189
Meißner, U.-G., Müller, G., Steininger, S. (1997): Phys. Lett. **B406**, 154; B407, 454 (E)
Mojžiš, M. (1998): European Phys. Journal **2**, 181
Moussallam, B. (1997): Nucl. Phys. **B504**, 381 ·

Nambu, Y., Jona-Lasinio, G. (1961): Phys. Rev. **122**, 345
Okubo, S. (1957): Prog. Theor. Phys. **27**, 949
Post, P., Schilcher, K. (1997): Phys. Rev. Lett. **79**, 4088
Prades, J. (1998): Nucl. Phys. Proc. Suppl. **64**, 253
de Rafael, E. (1998): An introduction to sum rules in QCD, hep-ph/9802448, Lectures delivered at Les Houches Summer School 1997
Richardson, K.G., Birse, M.C., McGovern, J.A. (1997): Renormalization and power counting in effective field theories for nucleon–nucleon scattering, hep-ph/9708435
Rosselet, L. et al. (1977): Phys. Rev. **D15**, 574
Schacher, J. (DIRAC) (1997): Contribution to the Workshop on Chiral Dynamics 1997, Mainz, Sept. 1997, to appear in the Proceedings
Sharpe, S.R. (1997): Nucl. Phys. Proc. Suppl. **53**, 181
Shifman, M.A., Vainshtein, A.I., Zakharov, V.I. (1979): Nucl. Phys. **B147**, 385, 447
Stern, J., Sazdjian, H., Fuchs, N.H. (1993): Phys. Rev. **D47**, 3814
Stern, J. (1997): Light quark masses and condensates in QCD, hep-ph/9712438, Contribution to the Workshop on Chiral Dynamics 1997, Mainz, Sept. 1997, to appear in the Proceedings
Stern, J. (1998): Two alternatives of spontaneous chiral symmetry breaking in QCD, hep-ph/9801282, submitted to Phys. Rev. Letters
Urban, P., Ed. (1968): *Particles, Currents, Symmetries*, Proc. of the 7. Int. Universitätswochen für Kernphysik, Schladming, 1968, Acta Phys. Austriaca, Suppl. V (Springer–Verlag, Wien, New York)
Vafa, C., Witten, E. (1984): Nucl. Phys. **B234**, 173; Comm. Math. Phys. **95**, 257
Walcher, T., Ed. (1998): Proceedings of the Workshop on Chiral Dynamics, Mainz, Sept. 1997, in preparation
Weinberg, S. (1966): Phys. Rev. Lett. **17**, 616
Weinberg, S. (1967): Phys. Rev. Lett. **18**, 507
Weinberg, S. (1977): in *A Festschrift for I.I. Rabi*, L. Motz, Ed. (New York Academy of Sciences, N.Y.)
Weinberg, S. (1979): Physica **96A**, 327
Weinberg, S. (1990): Phys. Lett. **B251**, 288
Weinberg, S. (1991): Nucl. Phys. **B363**, 3
Wess, J., Zumino, B. (1971): Phys. Lett. **37B**, 95
Witten, E. (1983): Nucl. Phys. **B223**, 422

Quark Mass Hierarchies, Flavor Mixing and Maximal CP–Violation

Harald Fritzsch

Ludwig–Maximilians–Universität, Sektion Physik,
Theresienstraße 37, D-80333 München, Germany

Abstract. Flavor mixing and the quark mass spectrum are intimately related. In view of the observed strong hierarchy of the quark and lepton masses and of the flavor mixing angles it is argued that the description of flavor mixing must take this into account. One particular interesting way to describe the flavor mixing emerges, which is particularly suited for models of quark mass matrices based on flavor symmetries. We conclude that the unitarity triangle important for B physics should be close to or identical to a rectangular triangle. CP violation is maximal in this sense.

At the magnificent Boston Museum of Fine Arts one can see a big stone brought in from Northern Africa, covered with strange hieroglyphics. More than 2000 years ago it was located in the Great Temple of Amun at the old City of Jebel Barkal in the kingdom of Nubia and is assumed to describe the rulership of king Tanyidamani. The text is written in the Meroitic language, which is still undeciphered. Neither the grammar of that language nor the content of the text on the Stone of Amun is known, only the letters.

In particle physics today one is facing a similar problem, as far as the masses of the leptons and quarks are concerned. After the discovery of the t–quark the spectrum of these masses (apart from the yet unknown neutrino masses) is known. It is a rather wild spectrum, extending over 5 orders of magnitude, from the tiny electron mass to the huge t–mass, but the actual dynamics behind this spectrum remains mysterious. Nature speaks to us in some kind of Meroitic language. The letters of this language, i. e. the masses and flavor mixing parameters, are known, but the grammar and the content of the underlying text are unknown.

Of course, in these lectures I cannot offer a complete solution of the mass problem, but I shall describe what I would like to define as the grammar of patterns and rules, which are not only very simple, but seem to come out very well, if confronted with the experimental results.

The phenomenon of flavor mixing, which is intrinsically linked to CP–violation, is an important ingredient of the Standard Model of Basic Interactions. Yet unlike other features of the Standard Model, e. g. the mixing of the neutral electroweak gauge bosons, it is a phenomenon which can merely be described. A deeper understanding is still lacking, but it is clearly directly linked to the mass spectrum of the quarks – the possible mixing of lepton flavors will not be discussed here. Furthermore there is a general consensus that a deeper dynamical understanding would require to go beyond the physics of the Standard Model.

In my lectures I shall not go thus far. Instead I shall demonstrate that the observed properties of the flavor mixing, combined with our knowledge about the quark mass spectrum, suggest specific symmetry properties which allow to fix the flavor mixing parameters with high precision, thus predicting the outcome of the experiments which will soon be performed at the B–meson factories.

Before we enter the field of fermion mass generation, flavor mixing and CP–violation, let me make some general remarks about the mass issue as it appears today. The gauge interactions of the Standard Model are relevant both for the lefthanded (L) and righthanded (R) fermion fields. Chirality is conserved by the gauge interaction – a lefthanded quark, after interacting with a gauge boson, e. g. a W–boson or a gluon, stays lefthanded. A CP–transformation turns a left-handed quark into a righthanded antiquark, but the interaction with the gauge bosons is unaffected. Thus the gauge sector of the Standard Model can be divided into two disjoint worlds, the world of L–fermions and of R–fermions. Formally the gauge interactions do not provide a bridge between those two sectors.

In reality the situation is more complex, which can be observed in particular by looking at the strong interactions. In the limit in which the quarks are taken to be massless (limit of chiral $SU(n)_L \times SU(n)_R$) the world of QCD can also be divided up into the world of L–quarks and of R–quarks. However nonperturbative effects generate a non–zero value for the v. e. v. of $\bar{q}_R q_L$:

$$< 0 \mid \bar{q}_R q_L + h.c. \mid 0 > \neq 0, \qquad (1)$$

which is of order Λ_c (Λ_c: QCD scale).

Thus there exists a strong correlation between the lefthanded and righthanded fields, which is responsible for the mass generation of the bound states like the proton or the ρ–meson. These masses are due to the dynamical breaking of the chiral symmetry.

A consequence of this symmetry breaking is that the matrix elements of the axial vector currents acquire a pole at $q^2 = 0$ (q: momentum transfer), due to the massless pseudoscalar mesons which serve as the corresponding Goldstone particles.

In the Standard Model of the electroweak interaction the masses are introduced by the coupling of the gauge fields and fermions to the scalar field φ whose neutral component φ° acquires a non–zero v.e.v.:

$$< 0 \mid \varphi^\circ \mid 0 > = \frac{1}{\sqrt{2}} v. \qquad (2)$$

In order to reproduce the observed gauge boson masses, one needs to have $v \cong$ 246 GeV.

The quark and lepton masses are introduced by the coupling of the fermions to φ, which is described by a coupling constant which is a free parameter and varies for the different fermions in proportion to the masses. These couplings of the type

$$\lambda \cdot \bar{\psi}_R \psi_L \varphi + \text{h.c.} \qquad (3)$$

provide a correlation between the L–world and the R–world. The v.e.v. of φ, multiplied with λ_Z, describes the corresponding fermion mass. Since the coupling constants λ can be complex, the CP–symmetry will be violated, if there are more than two families of fermions, and if flavor mixing is present.

In the Standard Model the fermion masses are introduced via the spontaneous symmetry breaking in order to ensure the renormalizability of the underlying gauge theory. However, it can be seen from a more general point of view that the introduction of the fermion masses in the electroweak gauge theory is a dynamical issue, unlike the introduction of the quark masses in QCD. Let us consider a "Gedankenexperiment", the process $\bar{t}t \rightarrow W^+W^-$, in which both incoming quarks are polarized. In the center of mass frame we prepare the outgoing W–bosons in a $J = 1$ wave by colliding both a t_L–quark and a $\bar{t}_R$–quark. The tree-diagrams describing the process are either the formation of a virtual γ or Z, decaying into the W–pair, or the exchange of a b–quark in the t–channel, leading to the production of the W–pair. Both diagrams, if considered in isolation, lead to a cross section which violates the unitarity bound for $J = 1$ at high energy, but the coherent sum does not. This is the famous gauge theory cancellation.

The dynamical aspect of the t–mass enters, if we study the W^+W^-–production in the $J = 0$ wave by considering the process $t_L\bar{t}_L \rightarrow W^+W^-$. Since $\bar{t}_L$–quarks do not interact with the W–bosons, the cross section in the $J = 0$ wave would vanish for massless t–quarks. However, due to the non–zero t–mass a t–quark prepared in the center–of–mass system with its spin opposite to its momentum has a righthanded component, and the scattering amplitude in the s–wave is proportional to $m_t \cdot \sqrt{s}$. Thus unitarity is violated at high energy.

In the Standard Model this problem is avoided, since there is a cancellation in the $J = 0$ channel provided by the scalar "Higgs"–particle. The coupling of the latter to the t–quark is proportional to m_t. Hence the cancellation is present, no matter how large m_t is.

This simple "Gedankenexperiment" shows the general condition: The cross section for the reaction $\bar{t}t \rightarrow W^+W$ in the s–wave must be finite at high energies. This requires a new dynamics besides the one provided by the quarks and electroweak gauge bosons. It could be either the addition of a new scalar particle, as in the Standard Model, or a string of resonances in the $J = 0$ channel, generated by new types of interactions or, perhaps, a new substructure of the leptons and quarks. At present we do not know, which possibility is realized, but in general it is implied that the lepton and quark masses are more than just kinematical quantities. They must play an essential rôle in the dynamics. For this reason one should expect that the fermion masses, especially the t–mass, are linked in a specific way to the masses of the W–bosons.

After these introductory remarks about the rôle of the lepton and quark masses in the electroweak gauge theory, let me turn to the main topic of these lectures, the connection between quark masses and the mixing of the quark flavors. According to the standard electroweak theory one is dealing with three

$SU(2)_w$–doublets:

$$\begin{pmatrix} u' \\ d' \end{pmatrix}_L \begin{pmatrix} c' \\ s' \end{pmatrix}_L \begin{pmatrix} t' \\ b' \end{pmatrix}_L \tag{4}$$

where $u', d' \ldots$ stand for certain superpositions of the corresponding mass eigenstates. In terms of mass eigenstates the charged weak currents are given by:

$$\overline{(u,c,t)_L} \begin{pmatrix} V_{ud} & V_{us} & V_{ub} \\ V_{cd} & V_{cs} & V_{cb} \\ V_{td} & V_{ts} & V_{tb} \end{pmatrix} \begin{pmatrix} d \\ s \\ s \end{pmatrix}_L . \tag{5}$$

This generalizes the standard Cabibbo–type rotation between the first and second family [1]. The matrix elements V_{ij} are the elements of the CKM matrix [2]. In general they are complex numbers. Their absolute values are measurable quantities. For example, $|V_{cb}|$ primarily determines the lifetime of B mesons. The phases of V_{ij}, however, are not physical, like the phases of quark fields. A phase transformation of the u quark ($u \to u\, e^{i\alpha}$), for example, leaves the quark mass term invariant but changes the elements in the first row of V (i.e., $V_{uj} \to V_{uj}\, e^{-i\alpha}$). Only a common phase transformation of all quark fields leaves all elements of V invariant, thus there is a five-fold freedom to adjust the phases of V_{ij}.

In general the unitary matrix V depends on nine parameters. Note that in the absence of complex phases V would consist of only three independent parameters, corresponding to three (Euler) rotation angles. Hence one can describe the complex matrix V by three angles and six phases. Due to the freedom in redefining the quark field phases, five of the six phases in V can be absorbed and we arrive at the well-known result that the CKM matrix V can be parametrized in terms of three rotation angles and one CP-violating phase [2].

The standard parametrization of the CKM matrix is given as follows:

$$V_{ij} = \begin{pmatrix} c_{12}\, c_{13} & s_{12}\, c_{13} & s_{13}\, e^{-i\delta_{13}} \\ -s_{12}\, c_{23} - c_{12}\, s_{23}\, s_{13} e^{i\delta_{13}} & c_{12}\, c_{23} - s_{12}\, s_{23}\, s_{13}\, e^{i\delta_{13}} & s_{23}\, c_{13} \\ s_{12}\, s_{23} - c_{12}\, c_{23}\, s_{13}\, e^{i\delta_{13}} & -c_{12}\, s_{23} - s_{12} c_{23} s_{13}\, e^{i\delta_{13}} & c_{23}\, c_{13} \end{pmatrix} \tag{6}$$

Here s_{12} stands for $\sin\Theta_{12}, c_{12}$ for $\cos\Theta_{12}$ etc. Since the observed mixing angles are small the three angles Θ_{12}, Θ_{23} and Θ_{13} are related in a good approximation to the moduli of specific V–elements as follows:

$$|\, V_{us} \,| \cong s_{12}\,, \quad |\, V_{ub} \,| \cong s_{13}\,, \quad |\, V_{cb} \,| \cong s_{23}\,. \tag{7}$$

The experiments give [3]:

$$\Theta_{12} \cong 12.7^\circ\,, \quad \Theta_{13} \cong 0.18^\circ\,, \quad \Theta_{23} \cong 2.25^\circ\,. \tag{8}$$

(Here we have given the central values of these angles for illustration, without indicating the errors. The phase δ_{13} will be discussed later).

Another way to describe the flavor mixing matrix is to follow Wolfenstein [4] and to use the modulus of V_{us} as an expansion parameter:

$$V = \begin{pmatrix} 1-\frac{1}{2}\lambda^2 & \lambda & A\lambda^3(\rho - i\eta) \\ -\lambda & 1-\frac{1}{2}\lambda^2 & A\lambda^2 \\ A\lambda^3(1-\rho-i\eta) & -A\lambda^2 & 1 \end{pmatrix} + O\left(\lambda^4\right) \tag{9}$$

The central values of the parameters are:

$$\lambda = 0.2205\,, \quad A = 0.806\,, \quad |\, \rho - i\eta \,| = 0.36\,. \tag{10}$$

When the standard parametrization of the CKM–matrix in terms of the angles Θ_{ij} was introduced years ago by a number of authors including this one [5], the large value of the t–mass was not known. Thus the striking mass hierarchy exhibited in the quark mass spectrum was not explicitly taken into account. But the flavor mixing and the mass spectrum are intimately related to each other, and the question arises whether the standard way of describing the flavor mixing is the best way in doing so. We shall discuss this issue below. The same question can be asked for the other description proposed in the literature, e. g. the original one given by Kobayashi and Maskawa [2] or the one given recently [6].

Adopting a particular parametrization of flavor mixing is arbitrary and not directly a physical issue. Nevertheless it is quite likely that the actual values of flavor mixing parameters (including the strength of CP violation), once they are known with high precision, will give interesting information about the physics beyond the standard model. Probably at this point it will turn out that a particular description of the CKM matrix is more useful and transparent than the others. For this reason, let me first analyze all possible parametrizations and point out their respective advantages and disadvantages.

The question about how many different ways to describe V may exist was raised some time ago [7]. Below we shall reconsider this problem and give a complete analysis.

If the flavor mixing matrix V is first assumed to be a real orthogonal matrix, it can in general be written as a product of three matrices R_{12}, R_{23} and R_{31}, which describe simple rotations in the (1,2), (2,3) and (3,1) planes:

$$R_{12}(\theta) = \begin{pmatrix} c_\theta & s_\theta & 0 \\ -s_\theta & c_\theta & 0 \\ 0 & 0 & 1 \end{pmatrix} ,$$

$$R_{23}(\sigma) = \begin{pmatrix} 1 & 0 & 0 \\ 0 & c_\sigma & s_\sigma \\ 0 & -s_\sigma & c_\sigma \end{pmatrix} ,$$

$$R_{31}(\tau) = \begin{pmatrix} c_\tau & 0 & s_\tau \\ 0 & 1 & 0 \\ -s_\tau & 0 & c_\tau \end{pmatrix} , \tag{11}$$

where $s_\theta \equiv \sin\theta$, $c_\theta \equiv \cos\theta$, etc. Clearly these rotation matrices do not commute with each other. There exist twelve different ways to arrange products of these matrices such that the most general orthogonal matrix R can be obtained. Note that the matrix $R_{ij}^{-1}(\omega)$ plays an equivalent role as $R_{ij}(\omega)$ in constructing R, because of $R_{ij}^{-1}(\omega) = R_{ij}(-\omega)$. Note also that $R_{ij}(\omega)R_{ij}(\omega') = R_{ij}(\omega+\omega')$ holds, thus the product $R_{ij}(\omega)R_{ij}(\omega')R_{kl}(\omega'')$ or $R_{kl}(\omega'')R_{ij}(\omega)R_{ij}(\omega')$ cannot cover the whole space of a 3×3 orthogonal matrix and should be excluded. Explicitly the twelve different forms of R read as

$$
\begin{aligned}
(1)\ R &= R_{12}(\theta)\ R_{23}(\sigma)\ R_{12}(\theta')\ ,\\
(2)\ R &= R_{12}(\theta)\ R_{31}(\tau)\ R_{12}(\theta')\ ,\\
(3)\ R &= R_{23}(\sigma)\ R_{12}(\theta)\ R_{23}(\sigma')\ ,\\
(4)\ R &= R_{23}(\sigma)\ R_{31}(\tau)\ R_{23}(\sigma')\ ,\\
(5)\ R &= R_{31}(\tau)\ R_{12}(\theta)\ R_{31}(\tau')\ ,\\
(6)\ R &= R_{31}(\tau)\ R_{23}(\sigma)\ R_{31}(\tau')\ ,
\end{aligned}
$$

in which a rotation in the (i,j) plane occurs twice; and

$$
\begin{aligned}
(7)\ R &= R_{12}(\theta)\ R_{23}(\sigma)\ R_{31}(\tau)\ ,\\
(8)\ R &= R_{12}(\theta)\ R_{31}(\tau)\ R_{23}(\sigma)\ ,\\
(9)\ R &= R_{23}(\sigma)\ R_{12}(\theta)\ R_{31}(\tau)\ ,\\
(10)\ R &= R_{23}(\sigma)\ R_{31}(\tau)\ R_{12}(\theta)\ ,\\
(11)\ R &= R_{31}(\tau)\ R_{12}(\theta)\ R_{23}(\sigma)\ ,\\
(12)\ R &= R_{31}(\tau)\ R_{23}(\sigma)\ R_{12}(\theta)\ ,
\end{aligned}
$$

where all three R_{ij} are present.

Although all the above twelve combinations represent the most general orthogonal matrices, only nine of them are structurally different. The reason is that the products $R_{ij}R_{kl}R_{ij}$ and $R_{ij}R_{mn}R_{ij}$ (with $ij \neq kl \neq mn$) are correlated with each other, leading essentially to the same form for R. Indeed it is straightforward to see the correlation between patterns (1), (3), (5) and (2), (4), (6), respectively, as follows:

$$
\begin{aligned}
R_{12}(\theta)\ R_{31}(\tau)\ R_{12}(\theta') &= R_{12}(\theta+\pi/2)\ R_{23}(\sigma=\tau)\ R_{12}(\theta'-\pi/2)\ ,\\
R_{23}(\sigma)\ R_{31}(\tau)\ R_{23}(\sigma') &= R_{23}(\sigma-\pi/2)\ R_{12}(\theta=\tau)\ R_{23}(\sigma'+\pi/2)\ ,\\
R_{31}(\tau)\ R_{23}(\sigma)\ R_{31}(\tau') &= R_{31}(\tau+\pi/2)\ R_{12}(\theta=\sigma)\ R_{31}(\tau'-\pi/2)\ . \qquad (12)
\end{aligned}
$$

Thus the orthogonal matrices (2), (4) and (6) need not be treated as independent choices. We then draw the conclusion that there exist *nine* different forms for the orthogonal matrix R, i.e., patterns (1), (3) and (5) as well as (7) - (12).

We proceed to include the CP-violating phase, denoted by φ, in the above rotation matrices. The resultant matrices should be unitary such that a unitary flavor mixing matrix can be finally produced. There are several different ways for φ to enter R_{12}, e.g.,

$$R_{12}(\theta,\varphi) = \begin{pmatrix} c_\theta & s_\theta\, e^{+i\varphi} & 0 \\ -s_\theta\, e^{-i\varphi} & c_\theta & 0 \\ 0 & 0 & 1 \end{pmatrix} ,$$

or

$$R_{12}(\theta,\varphi) = \begin{pmatrix} c_\theta & s_\theta & 0 \\ -s_\theta & c_\theta & 0 \\ 0 & 0 & e^{-i\varphi} \end{pmatrix} ,$$

or

$$R_{12}(\theta,\varphi) = \begin{pmatrix} c_\theta\, e^{+i\varphi} & s_\theta & 0 \\ -s_\theta & c_\theta\, e^{-i\varphi} & 0 \\ 0 & 0 & 1 \end{pmatrix} . \tag{13}$$

Similarly one may introduce a phase parameter into R_{23} or R_{31}. Then the CKM matrix V can be constructed, as a product of three rotation matrices, by use of one complex R_{ij} and two real ones. Note that the location of the CP-violating phase in V can be arranged by redefining the quark field phases, thus it does not play an essential role in classifying different parametrizations. We find that it is always possible to locate the phase parameter φ in a 2×2 submatrix of V, in which each element is a sum of two terms with the relative phase φ. The remaining five elements of V are real in such a phase assignment. Accordingly we arrive at the nine distinctive parametrizations of the CKM matrix V listed in Table 1, where the complex rotation matrices $R_{12}(\theta,\varphi)$, $R_{23}(\sigma,\varphi)$ and $R_{31}(\tau,\varphi)$ are obtained directly from the real ones in Eq. (11) with the replacement $1 \rightarrow e^{-i\varphi}$. These nine possibilities have been discussed recently in ref. [8] (see also ref. [9]).

One can see that *P2* and *P3* correspond to the cases given in [2] and [5], although different notations for the CP-violating phase and three mixing angles are adopted here. The latter is indeed equivalent to the "standard" parametrization advocated by the Particle Data Group (see also ref. [3]). This can be seen clearly if one makes three transformations of quark field phases: $c \rightarrow c\, e^{-i\varphi}$, $t \rightarrow t\, e^{-i\varphi}$, and $b \rightarrow b\, e^{-i\varphi}$. In addition, *P1* is just the one discussed by Xing and the author in ref. [6].

From a mathematical point of view, all nine different parametrizations are equivalent. However this is not the case if we apply our considerations to the quarks and their mass spectrum. It is well–known that both the observed quark mass spectrum and the observed values of the flavor mixing parameters exhibit a striking hierarchical structure. The latter can be understood in a natural way as the consequence of a specific pattern of chiral symmetries whose breaking causes the towers of different masses to appear step by step [10], [11], [12]. Such a chiral evolution of the mass matrices leads, as argued in ref. [11], to a specific way to introduce and describe the flavor mixing.

In the limit $m_u = m_d = 0$, which is close to the real world, since $m_u/m_t \ll 1$ and $m_d/m_b \ll 1$, the flavor mixing is merely a rotation between the t–c and b–s

Table 1. Classification of different parametrizations for the flavor mixing matrix.

Parametrization	Useful relations
P1: $V = R_{12}(\theta)\ R_{23}(\sigma,\varphi)\ R_{12}^{-1}(\theta')$ $\begin{pmatrix} s_\theta s_{\theta'} c_\sigma + c_\theta c_{\theta'} e^{-i\varphi} & s_\theta c_{\theta'} c_\sigma - c_\theta s_{\theta'} e^{-i\varphi} & s_\theta s_\sigma \\ c_\theta s_{\theta'} c_\sigma - s_\theta c_{\theta'} e^{-i\varphi} & c_\theta c_{\theta'} c_\sigma + s_\theta s_{\theta'} e^{-i\varphi} & c_\theta s_\sigma \\ -s_{\theta'} s_\sigma & -c_{\theta'} s_\sigma & c_\sigma \end{pmatrix}$	$\mathcal{J} = s_\theta c_\theta s_{\theta'} c_{\theta'} s_\sigma^2 c_\sigma \sin\varphi$ $\tan\theta = \|V_{ub}/V_{cb}\|$ $\tan\theta' = \|V_{td}/V_{ts}\|$ $\cos\sigma = \|V_{tb}\|$
P2: $V = R_{23}(\sigma)\ R_{12}(\theta,\varphi)\ R_{23}^{-1}(\sigma')$ $\begin{pmatrix} c_\theta & s_\theta c_{\sigma'} & -s_\theta s_{\sigma'} \\ -s_\theta c_\sigma & c_\theta c_\sigma c_{\sigma'} + s_\sigma s_{\sigma'} e^{-i\varphi} & -c_\theta c_\sigma s_{\sigma'} + s_\sigma c_{\sigma'} e^{-i\varphi} \\ s_\theta s_\sigma & -c_\theta s_\sigma c_{\sigma'} + c_\sigma s_{\sigma'} e^{-i\varphi} & c_\theta s_\sigma s_{\sigma'} + c_\sigma c_{\sigma'} e^{-i\varphi} \end{pmatrix}$	$\mathcal{J} = s_\theta^2 c_\theta s_\sigma c_\sigma s_{\sigma'} c_{\sigma'} \sin\varphi$ $\cos\theta = \|V_{ud}\|$ $\tan\sigma = \|V_{td}/V_{cd}\|$ $\tan\sigma' = \|V_{ub}/V_{us}\|$
P3: $V = R_{23}(\sigma)\ R_{31}(\tau,\varphi)\ R_{12}(\theta)$ $\begin{pmatrix} c_\theta c_\tau & s_\theta c_\tau & s_\tau \\ -c_\theta s_\sigma s_\tau - s_\theta c_\sigma e^{-i\varphi} & -s_\theta s_\sigma s_\tau + c_\theta c_\sigma e^{-i\varphi} & s_\sigma c_\tau \\ -c_\theta c_\sigma s_\tau + s_\theta s_\sigma e^{-i\varphi} & -s_\theta c_\sigma s_\tau - c_\theta s_\sigma e^{-i\varphi} & c_\sigma c_\tau \end{pmatrix}$	$\mathcal{J} = s_\theta c_\theta s_\sigma c_\sigma s_\tau c_\tau^2 \sin\varphi$ $\tan\theta = \|V_{us}/V_{ud}\|$ $\tan\sigma = \|V_{cb}/V_{tb}\|$ $\sin\tau = \|V_{ub}\|$
P4: $V = R_{12}(\theta)\ R_{31}(\tau,\varphi)\ R_{23}^{-1}(\sigma)$ $\begin{pmatrix} c_\theta c_\tau & c_\theta s_\sigma s_\tau + s_\theta c_\sigma e^{-i\varphi} & c_\theta c_\sigma s_\tau - s_\theta s_\sigma e^{-i\varphi} \\ -s_\theta c_\tau & -s_\theta s_\sigma s_\tau + c_\theta c_\sigma e^{-i\varphi} & -s_\theta c_\sigma s_\tau - c_\theta s_\sigma e^{-i\varphi} \\ -s_\tau & s_\sigma c_\tau & c_\sigma c_\tau \end{pmatrix}$	$\mathcal{J} = s_\theta c_\theta s_\sigma c_\sigma s_\tau c_\tau^2 \sin\varphi$ $\tan\theta = \|V_{cd}/V_{ud}\|$ $\tan\sigma = \|V_{ts}/V_{tb}\|$ $\sin\tau = \|V_{td}\|$
P5: $V = R_{31}(\tau)\ R_{12}(\theta,\varphi)\ R_{31}^{-1}(\tau')$ $\begin{pmatrix} c_\theta c_\tau c_{\tau'} + s_\tau s_{\tau'} e^{-i\varphi} & s_\theta c_\tau & -c_\theta c_\tau s_{\tau'} + s_\tau c_{\tau'} e^{-i\varphi} \\ -s_\theta c_{\tau'} & c_\theta & s_\theta s_{\tau'} \\ -c_\theta s_\tau c_{\tau'} + c_\tau s_{\tau'} e^{-i\varphi} & -s_\theta s_\tau & c_\theta s_\tau s_{\tau'} + c_\tau c_{\tau'} e^{-i\varphi} \end{pmatrix}$	$\mathcal{J} = s_\theta^2 c_\theta s_\tau c_\tau s_{\tau'} c_{\tau'} \sin\varphi$ $\cos\theta = \|V_{cs}\|$ $\tan\tau = \|V_{ts}/V_{us}\|$ $\tan\tau' = \|V_{cb}/V_{cd}\|$
P6: $V = R_{12}(\theta)\ R_{23}(\sigma,\varphi)\ R_{31}(\tau)$ $\begin{pmatrix} -s_\theta s_\sigma s_\tau + c_\theta c_\tau e^{-i\varphi} & s_\theta c_\sigma & s_\theta s_\sigma c_\tau + c_\theta s_\tau e^{-i\varphi} \\ -c_\theta s_\sigma s_\tau - s_\theta c_\tau e^{-i\varphi} & c_\theta c_\sigma & c_\theta s_\sigma c_\tau - s_\theta s_\tau e^{-i\varphi} \\ -c_\sigma s_\tau & -s_\sigma & c_\sigma c_\tau \end{pmatrix}$	$\mathcal{J} = s_\theta c_\theta s_\sigma c_\sigma^2 s_\tau c_\tau \sin\varphi$ $\tan\theta = \|V_{us}/V_{cs}\|$ $\sin\sigma = \|V_{ts}\|$ $\tan\tau = \|V_{td}/V_{tb}\|$
P7: $V = R_{23}(\sigma)\ R_{12}(\theta,\varphi)\ R_{31}^{-1}(\tau)$ $\begin{pmatrix} c_\theta c_\tau & s_\theta & -c_\theta s_\tau \\ -s_\theta c_\sigma c_\tau + s_\sigma s_\tau e^{-i\varphi} & c_\theta c_\sigma & s_\theta c_\sigma s_\tau + s_\sigma c_\tau e^{-i\varphi} \\ s_\theta s_\sigma c_\tau + c_\sigma s_\tau e^{-i\varphi} & -c_\theta s_\sigma & -s_\theta s_\sigma s_\tau + c_\sigma c_\tau e^{-i\varphi} \end{pmatrix}$	$\mathcal{J} = s_\theta c_\theta^2 s_\sigma c_\sigma s_\tau c_\tau \sin\varphi$ $\sin\theta = \|V_{us}\|$ $\tan\sigma = \|V_{ts}/V_{cs}\|$ $\tan\tau = \|V_{ub}/V_{ud}\|$
P8: $V = R_{31}(\tau)\ R_{12}(\theta,\varphi)\ R_{23}(\sigma)$ $\begin{pmatrix} c_\theta c_\tau & s_\theta c_\sigma c_\tau - s_\sigma s_\tau e^{-i\varphi} & s_\theta s_\sigma c_\tau + c_\sigma s_\tau e^{-i\varphi} \\ -s_\theta & c_\theta c_\sigma & c_\theta s_\sigma \\ -c_\theta s_\tau & -s_\theta c_\sigma s_\tau - s_\sigma c_\tau e^{-i\varphi} & -s_\theta s_\sigma s_\tau + c_\sigma c_\tau e^{-i\varphi} \end{pmatrix}$	$\mathcal{J} = s_\theta c_\theta^2 s_\sigma c_\sigma s_\tau c_\tau \sin\varphi$ $\sin\theta = \|V_{cd}\|$ $\tan\sigma = \|V_{cb}/V_{cs}\|$ $\tan\tau = \|V_{td}/V_{ud}\|$
P9: $V = R_{31}(\tau)\ R_{23}(\sigma,\varphi)\ R_{12}^{-1}(\theta)$ $\begin{pmatrix} -s_\theta s_\sigma s_\tau + c_\theta c_\tau e^{-i\varphi} & -c_\theta s_\sigma s_\tau - s_\theta c_\tau e^{-i\varphi} & c_\sigma s_\tau \\ s_\theta c_\sigma & c_\theta c_\sigma & s_\sigma \\ -s_\theta s_\sigma c_\tau - c_\theta s_\tau e^{-i\varphi} & -c_\theta s_\sigma c_\tau + s_\theta s_\tau e^{-i\varphi} & c_\sigma c_\tau \end{pmatrix}$	$\mathcal{J} = s_\theta c_\theta s_\sigma c_\sigma^2 s_\tau c_\tau \sin\varphi$ $\tan\theta = \|V_{cd}/V_{cs}\|$ $\sin\sigma = \|V_{cb}\|$ $\tan\tau = \|V_{ub}/V_{tb}\|$

systems, described by one rotation angle. No complex phase is present; i.e., CP violation is absent. This rotation angle is expected to change very little, once m_u and m_d are introduced as tiny perturbations. A sensible parametrization should make use of this feature. This implies that the rotation matrix R_{23} appears exactly once in the description of the CKM matrix V, eliminating *P2* (in which R_{23} appears twice) and *P5* (where R_{23} is absent). This leaves us with seven parametrizations of the flavor mixing matrix.

The list can be reduced further by considering the location of the phase φ. In the limit $m_u = m_d = 0$, the phase must disappear in the weak transition elements V_{tb}, V_{ts}, V_{cb} and V_{cs}. In *P7* and *P8*, however, φ appears particularly in V_{tb}. Thus these two parametrizations should be eliminated, leaving us with five parametrizations (i.e., *P1*, *P3*, *P4*, *P6* and *P9*). In the same limit, the phase φ appears in the V_{ts} element of *P3* and the V_{cb} element of *P4*. Hence these two parametrizations should also be eliminated. Then we are left with three parametrizations, *P1*, *P6* and *P9*. As expected, these are the parametrizations containing the complex rotation matrix $R_{23}(\sigma, \varphi)$. We stress that the "standard" parametrization [3] (equivalent to *P3*) does not obey the above constraints and should be dismissed.

Among the remaining three parametrizations, *P1* is singled out by the fact that the CP-violating phase φ appears only in the 2×2 submatrix of V describing the weak transitions among the light quarks. This is precisely the phase where the phase φ should appear, not in any of the weak transition elements involving the heavy quarks t and b.

In the parametrization *P6* or *P9*, the complex phase φ appears in V_{cb} or V_{ts}, but this phase factor is multiplied by a product of $\sin\theta$ and $\sin\tau$, i.e., it is of second order of the weak mixing angles. Hence the imaginary parts of these elements are not exactly vanishing, but very small in magnitude.

In our view the best possibility to describe the flavor mixing in the standard model is to adopt the parametrization *P1*. As discussed in ref. [6], this parametrization has a number of significant advantages in addition to that mentioned above. Especially it is well suited for specific models of quark mass matrices.

In the following part I shall show that the parametrization *P1* follows automatically, if we impose the constraints from the chiral symmetries and the hierarchical structure of the mass eigenvalues. We take the point of view that the quark mass eigenvalues are dynamical entities, and one could change their values in order to study certain symmetry limits, as it is done in QCD. In the standard electroweak model, in which the quark mass matrices are given by the coupling of a scalar field to various quark fields, this can certainly be done by adjusting the related coupling constants. Whether it is possible in reality is an open question. It is well–known that the quark mass matrices can always be made hermitian by a suitable transformation of the right–handed fields. Without loss of generality, we shall suppose in this paper that the quark mass matrices are hermitian. In the limit where the masses of the u and d quarks are set to zero, the quark mass matrix $\tilde{M}$ (for both charge $+2/3$ and charge $-1/3$ sectors) can be arranged such that its elements $\tilde{M}_{i1}$ and $\tilde{M}_{1i}$ $(i = 1, 2, 3)$ are all zero [10], [11]. Thus the quark mass matrices have the form

$$\tilde{M} = \begin{pmatrix} 0 & 0 & 0 \\ 0 & \tilde{C} & \tilde{B} \\ 0 & \tilde{B}^* & \tilde{A} \end{pmatrix} . \tag{14}$$

The observed mass hierarchy is incorporated into this structure by denoting the entry which is of the order of the t-quark or b-quark mass by $\tilde{A}$, with $\tilde{A} \gg \tilde{C}, |\tilde{B}|$. It can easily be seen (see, e.g., ref. [13]) that the complex phases in the mass matrices (14) can be rotated away by subjecting both $\tilde{M}_\mathrm{u}$ and $\tilde{M}_\mathrm{d}$ to the same unitary transformation. Thus we shall take $\tilde{B}$ to be real for both up- and down-quark sectors. As expected, CP violation cannot arise at this stage. The diagonalization of the mass matrices leads to a mixing between the second and third families, described by an angle $\tilde{\theta}$. The flavor mixing matrix is then given by

$$\tilde{V} = \begin{pmatrix} 1 & 0 & 0 \\ 0 & \tilde{c} & \tilde{s} \\ 0 & -\tilde{s} & \tilde{c} \end{pmatrix} , \tag{15}$$

where $\tilde{s} \equiv \sin\tilde{\theta}$ and $\tilde{c} \equiv \cos\tilde{\theta}$. In view of the fact that the limit $m_u = m_d = 0$ is not far from reality, the angle $\tilde{\theta}$ is essentially given by the observed value of $|V_{cb}|$ $(= 0.039 \pm 0.002$ [14]); i.e., $\tilde{\theta} = 2.24° \pm 0.12°$.

At the next and final stage of the chiral evolution of the mass matrices, the masses of the u and d quarks are introduced. The Hermitian mass matrices have in general the form:

$$M = \begin{pmatrix} E & D & F \\ D^* & C & B \\ F^* & B^* & A \end{pmatrix} \tag{16}$$

with $A \gg C, |B| \gg E, |D|, |F|$. By a common unitary transformation of the up- and down-type quark fields, one can always arrange the mass matrices M_u and M_d in such a way that $F_\mathrm{u} = F_\mathrm{d} = 0$; i.e.,

$$M = \begin{pmatrix} E & D & 0 \\ D^* & C & B \\ 0 & B^* & A \end{pmatrix} . \tag{17}$$

This can easily be seen as follows. If phases are neglected, the two symmetric mass matrices M_u and M_d can be transformed by an orthogonal transformation matrix O, which can be described by three angles such that they assume the form (17). The condition $F_\mathrm{u} = F_\mathrm{d} = 0$ gives two constraints for the three angles of O. If complex phases are allowed in M_u and M_d, the condition $F_\mathrm{u} = F_\mathrm{u}^* = F_\mathrm{d} = F_\mathrm{d}^* = 0$ imposes four constraints, which can also be fulfilled, if M_u and M_d are subjected to a common unitary transformation matrix U. The latter depends on nine parameters. Three of them are not suitable for our purpose, since they are just diagonal phases; but the remaining six can be chosen such that the vanishing of F_u and F_d results.

The basis in which the mass matrices take the form (17) is a basis in the space of quark flavors, which in our view is of special interest. It is a basis in which the mass matrices exhibit two texture zeros, for both up- and down-type quark sectors. These, however, do not imply special relations among mass eigenvalues and flavor mixing parameters (as pointed out above). In this basis the mixing is of the "nearest neighbour" form, since the (1,3) and (3,1) elements of $M_{\rm u}$ and $M_{\rm d}$ vanish; no direct mixing between the heavy t (or b) quark and the light u (or d) quark is present (see also ref. [15]). In certain models (see, e.g. ref. ([15], [16]), this basis is indeed of particular interest, but we shall proceed without relying on a special texture models for the mass matrices.

A mass matrix of the type (17) can in the absence of complex phases be diagonalized by a rotation matrix, described only by two angles in the hierarchy limit of quark masses [15]. At first the off-diagonal element B is rotated away by a rotation between the second and third families (angle θ_{23}); at the second step the element D is rotated away by a transformation of the first and second families (angle θ_{12}). No rotation between the first and third families is required to an excellent degree of accuracy [15], [16]. The rotation matrix for this sequence takes the form

$$R = R_{12}R_{23} = \begin{pmatrix} c_{12} & s_{12} & 0 \\ -s_{12} & c_{12} & 0 \\ 0 & 0 & 1 \end{pmatrix} \begin{pmatrix} 1 & 0 & 0 \\ 0 & c_{23} & s_{23} \\ 0 & -s_{23} & c_{23} \end{pmatrix}$$

$$= \begin{pmatrix} c_{12} & s_{12}c_{23} & s_{12}s_{23} \\ -s_{12} & c_{12}c_{23} & c_{12}s_{23} \\ 0 & -s_{23} & c_{23} \end{pmatrix} , \tag{18}$$

where $c_{12} \equiv \cos\theta_{12}$, $s_{12} \equiv \sin\theta_{12}$, etc. The flavor mixing matrix V is the product of two such matrices, one describing the rotation among the up-type quarks, and the other describing the rotation among the down-type quarks:

$$V = R^{\rm u}_{12}R^{\rm u}_{23}(R^{\rm d}_{23})^{-1}(R^{\rm d}_{12})^{-1} . \tag{19}$$

Note that V itself is exact, since a rotation between the first and third families can always be incorporated and absorbed through redefining the relevant rotation matrices. The product $R^{\rm u}_{23}(R^{\rm d}_{23})^{-1}$ can be written as a rotation matrix described by a single angle θ. In the limit $m_u = m_d = 0$, this is just the angle $\tilde{\theta}$ encountered in Eq. (15). The angle which describes the $R^{\rm u}_{12}$ rotation shall be denoted by $\theta_{\rm u}$; the corresponding angle for the $R^{\rm d}_{12}$ rotation by $\theta_{\rm d}$. Thus in the absence of CP-violating phases the flavor mixing matrix takes the following specific form:

$$V = \begin{pmatrix} c_{\rm u} & s_{\rm u} & 0 \\ -s_{\rm u} & c_{\rm u} & 0 \\ 0 & 0 & 1 \end{pmatrix} \begin{pmatrix} 1 & 0 & 0 \\ 0 & c & s \\ 0 & -s & c \end{pmatrix} \begin{pmatrix} c_{\rm d} & -s_{\rm d} & 0 \\ s_{\rm d} & c_{\rm d} & 0 \\ 0 & 0 & 1 \end{pmatrix}$$

$$= \begin{pmatrix} s_{\rm u}s_{\rm d}c + c_{\rm u}c_{\rm d} & s_{\rm u}c_{\rm d}c - c_{\rm u}s_{\rm d} & s_{\rm u}s \\ c_{\rm u}s_{\rm d}c - s_{\rm u}c_{\rm d} & c_{\rm u}c_{\rm d}c + s_{\rm u}s_{\rm d} & c_{\rm u}s \\ -s_{\rm d}s & -c_{\rm d}s & c \end{pmatrix} , \tag{20}$$

where $c_{\rm u} \equiv \cos\theta_{\rm u}$, $s_{\rm u} \equiv \sin\theta_{\rm u}$, etc.

We proceed by including the phase parameters of the quark mass matrices in Eq. (17). Each mass matrix has in general two complex phases. But it can easily be seen that, by suitable rephasing of the quark fields, the flavor mixing matrix can finally be written in terms of only a single phase φ as follows [6]:

$$V = \begin{pmatrix} c_{\rm u} & s_{\rm u} & 0 \\ -s_{\rm u} & c_{\rm u} & 0 \\ 0 & 0 & 1 \end{pmatrix} \begin{pmatrix} e^{-i\varphi} & 0 & 0 \\ 0 & c & s \\ 0 & -s & c \end{pmatrix} \begin{pmatrix} c_{\rm d} & -s_{\rm d} & 0 \\ s_{\rm d} & c_{\rm d} & 0 \\ 0 & 0 & 1 \end{pmatrix}$$
$$= \begin{pmatrix} s_{\rm u}s_{\rm d}c + c_{\rm u}c_{\rm d}e^{-i\varphi} & s_{\rm u}c_{\rm d}c - c_{\rm u}s_{\rm d}e^{-i\varphi} & s_{\rm u}s \\ c_{\rm u}s_{\rm d}c - s_{\rm u}c_{\rm d}e^{-i\varphi} & c_{\rm u}c_{\rm d}c + s_{\rm u}s_{\rm d}e^{-i\varphi} & c_{\rm u}s \\ -s_{\rm d}s & -c_{\rm d}s & c \end{pmatrix} . \qquad (21)$$

Note that the three angles $\theta_{\rm u}$, $\theta_{\rm d}$ and θ in Eq. (21) can all be arranged to lie in the first quadrant through a suitable redefinition of quark field phases. Consequently all $s_{\rm u}$, $s_{\rm d}$, s and $c_{\rm u}$, $c_{\rm d}$, c are positive. The phase φ can in general take values from 0 to 2π; and CP violation is present in weak interactions if $\varphi \neq 0, \pi$ and 2π.

This representation of the flavor mixing matrix, in comparison with all other parametrizations discussed previously, has a number of interesting features which in our view make it very attractive and provide strong arguments for its use in future discussions of flavor mixing phenomena, in particular, those in B-meson physics. We shall discuss them below.

a) The flavor mixing matrix V in Eq. (21) follows directly from the chiral expansion of the mass matrices. Thus it naturally takes into account the hierarchical structure of the quark mass spectrum.

b) The complex phase describing CP violation (φ) appears only in the (1,1), (1,2), (2,1) and (2,2) elements of V, i.e., in the elements involving only the quarks of the first and second families. This is a natural description of CP violation since in our hierarchical approach CP violation is not directly linked to the third family, but rather to the first and second ones, and in particular to the mass terms of the u and d quarks.

It is instructive to consider the special case $s_{\rm u} = s_{\rm d} = s = 0$. Then the flavor mixing matrix V takes the form

$$V = \begin{pmatrix} e^{-i\varphi} & 0 & 0 \\ 0 & 1 & 0 \\ 0 & 0 & 1 \end{pmatrix} . \qquad (22)$$

This matrix describes a phase change in the weak transition between u and d, while no phase change is present in the transitions between c and s as well as t and b. Of course, this effect can be absorbed in a phase change of the u- and d-quark fields, and no CP violation is present. Once the angles $\theta_{\rm u}$, $\theta_{\rm d}$ and θ are introduced, however, CP violation arises. It is due to a phase change in the weak transition between u' and d', where u' and d' are the rotated quark fields, obtained by applying the corresponding rotation matrices given in Eq. (21) to

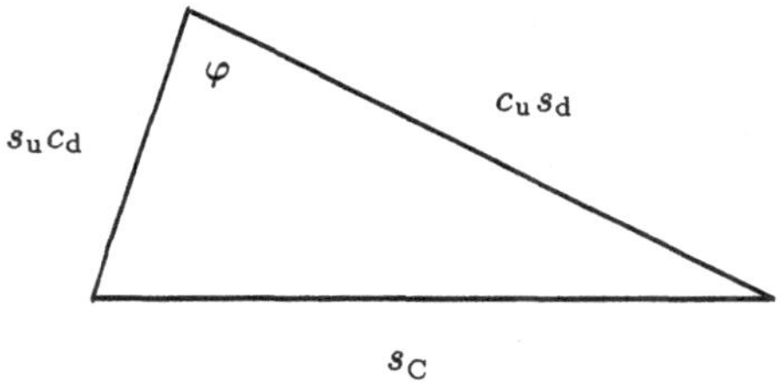

Fig. 1. The LQ–triangle in the complex plane.

the quark mass eigenstates (u': mainly u, small admixture of c; d': mainly d, small admixture of s).

Since the mixing matrix elements involving t or b quark are real in the representation (21), one can find that the phase parameter of B^0_q-$\bar{B}^0_q$ mixing ($q = d$ or s), dominated by the box-diagram contributions in the standard model [3], is essentially unity:

$$\left(\frac{q}{p}\right)_{B_q} = \frac{V^*_{tb} V_{tq}}{V_{tb} V^*_{tq}} = 1 . \qquad (23)$$

In most of other parametrizations of the flavor mixing matrix, however, the two rephasing-variant quantities $(q/p)_{B_d}$ and $(q/p)_{B_s}$ take different (maybe complex) values.

c) The dynamics of flavor mixing can easily be interpreted by considering certain limiting cases in Eq. (21). In the limit $\theta \rightarrow 0$ (i.e., $s \rightarrow 0$ and $c \rightarrow 1$), the flavor mixing is, of course, just a mixing between the first and second families, described by only one mixing angle (the Cabibbo angle $\theta_{\rm C}$). It is a special and essential feature of the representation (21) that the Cabibbo angle is *not* a basic angle, used in the parametrization. The matrix element V_{us} (or V_{cd}) is indeed a superposition of two terms including a phase. This feature arises naturally in our hierarchical approach, but it is not new. In many models of specific textures of mass matrices, it is indeed the case that the Cabibbo-type transition V_{us} (or V_{cd}) is a superposition of several terms. At first, it was obtained in the discussion of the two-family mixing [17].

In the limit $\theta = 0$ considered here, one has $|V_{us}| = |V_{cd}| = \sin\theta_{\rm C} \equiv s_{\rm C}$ and

$$s_{\rm C} = \left| s_{\rm u} c_{\rm d} - c_{\rm u} s_{\rm d} e^{-{\rm i}\varphi} \right| . \qquad (24)$$

This relation describes a triangle in the complex plane, as illustrated in Fig. 1, which we shall denote as the "LQ– triangle" ("light quark triangle"). This triangle is a feature of the mixing of the first two families. Explicitly one has (for $s = 0$):

$$\tan\theta_{\rm C} = \sqrt{\frac{\tan^2\theta_{\rm u} + \tan^2\theta_{\rm d} - 2\tan\theta_{\rm u}\tan\theta_{\rm d}\cos\varphi}{1 + \tan^2\theta_{\rm u}\tan^2\theta_{\rm d} + 2\tan\theta_{\rm u}\tan\theta_{\rm d}\cos\varphi}} . \qquad (25)$$

Certainly the flavor mixing matrix V cannot accommodate CP violation in this limit. However, the existence of φ seems necessary in order to make Eq. (25) compatible with current data, as one can see below.

d) The three mixing angles θ, $\theta_{\rm u}$ and $\theta_{\rm d}$ have a precise physical meaning. The angle θ describes the mixing between the second and third families, which is generated by the off-diagonal terms $B_{\rm u}$ and $B_{\rm d}$ in the up and down mass matrices of Eq. (17). We shall refer to this mixing involving t and b as the "heavy quark mixing". The angle $\theta_{\rm u}$, however, solely describes the u-c mixing, corresponding to the $D_{\rm u}$ term in $M_{\rm u}$. We shall denote this as the "u-channel mixing". The angle $\theta_{\rm d}$ solely describes the d-s mixing, corresponding to the $D_{\rm d}$ term in $M_{\rm d}$; this will be denoted as the "d-channel mixing". Thus there exists an asymmetry between the mixing of the first and second families and that of the second and third families, which in our view reflects interesting details of the underlying dynamics of flavor mixing. The heavy quark mixing is a combined effect, involving both charge $+2/3$ and charge $-1/3$ quarks, while the u- or d-channel mixing (described by the angle $\theta_{\rm u}$ or $\theta_{\rm d}$) proceeds solely in the charge $+2/3$ or charge $-1/3$ sector. Therefore an experimental determination of these two angles would allow to draw interesting conclusions about the amount and perhaps the underlying pattern of the u- or d-channel mixing.

e) The three angles θ, $\theta_{\rm u}$ and $\theta_{\rm d}$ are related in a very simple way to observable quantities of B-meson physics. For example, θ is related to the rate of the semileptonic decay $B \rightarrow D^* l \nu_l$; $\theta_{\rm u}$ is associated with the ratio of the decay rate of $B \rightarrow (\pi, \rho) l \nu_l$ to that of $B \rightarrow D^* l \nu_l$; and $\theta_{\rm d}$ can be determined from the ratio of the mass difference between two B_d mass eigenstates to that between two B_s mass eigenstates. We find the following exact relations:

$$\sin\theta \;=\; |V_{cb}| \sqrt{1 + \left|\frac{V_{ub}}{V_{cb}}\right|^2} \;, \tag{26}$$

and

$$\begin{aligned} \tan\theta_{\rm u} &= \left|\frac{V_{ub}}{V_{cb}}\right| \;, \\ \tan\theta_{\rm d} &= \left|\frac{V_{td}}{V_{ts}}\right| \;. \end{aligned} \tag{27}$$

These simple results make the parametrization (21) uniquely favorable for the study of B-meson physics.

By use of current data on $|V_{ub}|$ and $|V_{cb}|$, i.e., $|V_{cb}| = 0.039 \pm 0.002$ [14] and $|V_{ub}/V_{cb}| = 0.08 \pm 0.02$ [3], we obtain $\theta_{\rm u} = 4.57° \pm 1.14°$ and $\theta = 2.25° \pm 0.12°$. Taking $|V_{td}| = (8.6 \pm 2.1) \times 10^{-3}$, which was obtained from the analysis of current data on B^0_d-$\bar{B}^0_d$ mixing, we get $|V_{td}/V_{ts}| = 0.22 \pm 0.07$, i.e., $\theta_{\rm d} = 12.7° \pm 3.8°$. Both the heavy quark mixing angle θ and the u-channel mixing angle $\theta_{\rm u}$ are relatively small. The smallness of θ implies that Eqs. (24) and (25) are valid to a high degree of precision (of order $1 - c \approx 0.001$).

Recently a global fit of these angles was made [18], with rather small uncertainties for the angles and the phase φ. One finds:

$$\begin{aligned} \theta &= (2.30 \pm 0.09)^\circ \; , \quad \theta_{\rm u} = (4.87 \pm 0.86)^\circ \; , \\ \theta_{\rm d} &= (11.71 \pm 1.09)^\circ \; , \quad \varphi = (91.1 \pm 11.8)^\circ \; , \end{aligned} \tag{28}$$

These values are consistent with the ones given above, however, the errors are smaller.

f) According to Eq. (22), as well as Eq. (21), the phase φ is a phase difference between the contributions to V_{us} (or V_{cd}) from the u-channel mixing and the d-channel mixing. Therefore φ is given by the relative phase of $D_{\rm d}$ and $D_{\rm u}$ in the quark mass matrices (17), if the phases of $B_{\rm u}$ and $B_{\rm d}$ are absent or negligible.

The phase φ is not likely to be 0° or 180°, according to the experimental values given above, even though the measurement of CP violation in K^0-$\bar{K}^0$ mixing is not taken into account. For $\varphi = 0^\circ$, one finds $\tan\theta_{\rm C} = 0.14 \pm 0.08$; and for $\varphi = 180^\circ$, one gets $\tan\theta_{\rm C} = 0.30 \pm 0.08$. Both cases are barely consistent with the value of $\tan\theta_{\rm C}$ obtained from experiments ($\tan\theta_{\rm C} \approx |V_{us}/V_{ud}| \approx 0.226$).

g) The CP-violating phase φ in the flavor mixing matrix V can be determined from $|V_{us}|$ ($= 0.2205 \pm 0.0018$) through the following formula, obtained easily from Eq. (21):

$$\varphi = \arccos\left(\frac{s_{\rm u}^2 c_{\rm d}^2 c^2 + c_{\rm u}^2 s_{\rm d}^2 - |V_{us}|^2}{2 s_{\rm u} c_{\rm u} s_{\rm d} c_{\rm d} c}\right) . \tag{29}$$

The two-fold ambiguity associated with the value of φ, coming from $\cos\varphi = \cos(2\pi - \varphi)$, is removed if one takes $\sin\varphi > 0$ into account (this is required by current data on CP violation in K^0-$\bar{K}^0$ mixing (i.e., ϵ_K). More precise measurements of the angles $\theta_{\rm u}$ and $\theta_{\rm d}$ in the forthcoming experiments of B physics will remarkably reduce the uncertainty of φ to be determined from Eq. (29). This approach is of course complementary to the direct determination of φ from CP asymmetries in some weak B-meson decays into hadronic CP eigenstates. As mentioned above, the phase φ appears to be very close to 90°.

h) It is well-known that CP violation in the flavor mixing matrix V can be described in a way which is invariant with respect to phase changes by a universal quantity $\mathcal{J}$ [19]:

$$\text{Im}\left(V_{il} V_{jm} V_{im}^* V_{jl}^*\right) = \mathcal{J} \sum_{k,n=1}^{3} \left(\epsilon_{ijk} \epsilon_{lmn}\right) . \tag{30}$$

In the parametrization (21), $\mathcal{J}$ reads

$$\mathcal{J} = s_{\rm u} c_{\rm u} s_{\rm d} c_{\rm d} s^2 c \sin\varphi . \tag{31}$$

Obviously $\varphi = 90^\circ$ leads to the maximal value of $\mathcal{J}$. Indeed $\varphi = 90^\circ$, a particularly interesting case for CP violation, is quite consistent with current data. This possibility exists if $0.202 \le \tan\theta_{\rm d} \le 0.222$, or $11.4^\circ \le \theta_{\rm d} \le 12.5^\circ$. In this case the mixing term $D_{\rm d}$ in Eq. (17) can be taken to be real, and the term $D_{\rm u}$

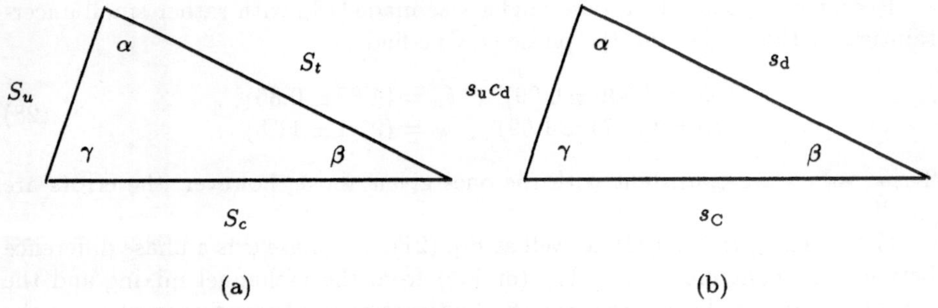

Fig. 2. The unitarity triangle (a) and its rescaled counterpart (b) in the complex plane.

to be imaginary, if $\mathrm{Im}(B_u) = \mathrm{Im}(B_d) = 0$ is assumed. Since in our description of the flavor mixing the complex phase φ is related in a simple way to the phases of the quark mass terms, the case $\varphi = 90°$ is especially interesting. It can hardly be an accident, and this case should be studied further. The possibility that the phase φ describing CP violation in the standard model is given by the algebraic number $\pi/2$ should be taken seriously. It may provide a useful clue towards a deeper understanding of the origin of CP violation and of the dynamical origin of the fermion masses.

In reference [20] the case $\varphi = 90°$ has been denoted as "maximal" CP violation. It implies in our framework that in the complex plane the u-channel and d-channel mixings are perpendicular to each other. In this special case (as well as $\theta \to 0$), we have

$$\tan^2\theta_C = \frac{\tan^2\theta_u + \tan^2\theta_d}{1 + \tan^2\theta_u \tan^2\theta_d} . \tag{32}$$

To a good approximation (with the relative error $\sim 2\%$), one finds $s_C^2 \approx s_u^2 + s_d^2$.

i) At future B-meson factories, the study of CP violation will concentrate on measurements of the unitarity triangle

$$S_u + S_c + S_t = 0 , \tag{33}$$

where $S_i \equiv V_{id}V_{ib}^*$ in the complex plane (see Fig. 2(a)). The inner angles of this triangle are denoted as usual:

$$\begin{aligned} \alpha &\equiv \arg(-S_t S_u^*) , \\ \beta &\equiv \arg(-S_c S_t^*) , \\ \gamma &\equiv \arg(-S_u S_c^*) . \end{aligned} \tag{34}$$

In terms of the parameters θ, θ_u, θ_d and φ, we obtain

$$\sin(2\alpha) = \frac{2c_u c_d \sin\varphi\,(s_u s_d c + c_u c_d \cos\varphi)}{s_u^2 s_d^2 c^2 + c_u^2 c_d^2 + 2 s_u c_u s_d c_d c \cos\varphi} ,$$

$$\sin(2\beta) = \frac{2s_{\rm u}c_{\rm d}\sin\varphi\,(c_{\rm u}s_{\rm d}c - s_{\rm u}c_{\rm d}\cos\varphi)}{c_{\rm u}^2 s_{\rm d}^2 c^2 + s_{\rm u}^2 c_{\rm d}^2 - 2s_{\rm u}c_{\rm u}s_{\rm d}c_{\rm d}c\cos\varphi} \,. \tag{35}$$

To an excellent degree of accuracy, one finds $\alpha \approx \varphi$. In order to illustrate how accurate this relation is, let us input the central values of θ, $\theta_{\rm u}$ and $\theta_{\rm d}$ (i.e., $\theta = 2.25°$, $\theta_{\rm u} = 4.57°$ and $\theta_{\rm d} = 12.7°$) to Eq. (35). Then one arrives at $\varphi - \alpha \approx 1°$ as well as $\sin(2\alpha) \approx 0.34$ and $\sin(2\beta) \approx 0.65$. It is expected that $\sin(2\alpha)$ and $\sin(2\beta)$ will be directly measured from the CP asymmetries in $B_d \rightarrow \pi^+\pi^-$ and $B_d \rightarrow J/\psi K_S$ modes at a B-meson factory.

Note that the three sides of the unitarity triangle can be rescaled by $|V_{cb}|$. In a very good approximation (with the relative error $\sim 2\%$), one arrives at

$$|S_u| \;:\; |S_c| \;:\; |S_t| \;\approx\; s_{\rm u}c_{\rm d} \;:\; s_{\rm C} \;:\; s_{\rm d} \,. \tag{36}$$

Equivalently, one can obtain

$$s_\alpha \;:\; s_\beta \;:\; s_\gamma \;\approx\; s_{\rm C} \;:\; s_{\rm u}c_{\rm d} \;:\; s_{\rm d} \,, \tag{37}$$

where $s_\alpha \equiv \sin\alpha$, etc. The rescaled unitarity triangle is shown in Fig. 2(b). Comparing this triangle with the LQ–triangle in Fig. 1, we find that they are indeed congruent with each other to a high degree of accuracy. The congruent relation between these two triangles is particularly interesting, since the LQ–triangle is essentially a feature of the physics of the first two quark families, while the unitarity triangle is linked to all three families. In this connection it is of special interest to note that in models which specify the textures of the mass matrices the Cabibbo triangle and hence three inner angles of the unitarity triangle can be fixed by the spectrum of the light quark masses and the CP-violating phase φ (see, e.g. ref. [20]).

j) It is worth pointing out that the u-channel and d-channel mixing angles are related to the Wolfenstein parameters [4] in a simple way:

$$\tan\theta_{\rm u} = \left|\frac{V_{ub}}{V_{cb}}\right| \;\approx\; \lambda\sqrt{\rho^2+\eta^2} \,,$$
$$\tan\theta_{\rm d} = \left|\frac{V_{td}}{V_{ts}}\right| \;\approx\; \lambda\sqrt{(1-\rho)^2+\eta^2} \,, \tag{38}$$

where $\lambda \approx s_{\rm C}$ measures the magnitude of V_{us}. Note that the CP-violating parameter η is linked to φ through

$$\sin\varphi \;\approx\; \frac{\eta}{\sqrt{\rho^2+\eta^2}\sqrt{(1-\rho)^2+\eta^2}} \tag{39}$$

in the lowest-order approximation. Then $\varphi = 90°$ implies $\eta^2 \approx \rho(1-\rho)$, on the condition $0 < \rho < 1$. In this interesting case, of course, the flavor mixing matrix can fully be described in terms of only three independent parameters.

k) Compared with the standard parametrization of the flavor mixing matrix V our parametrization has an additional advantage: the renormalization-group evolution of V, from the weak scale to an arbitrary high energy scale, is to a very good approximation associated only with the angle θ. This can easily be seen if one keeps the t and b Yukawa couplings only and neglects possible threshold effect in the one-loop renormalization-group equations of the Yukawa matrices [21]. Thus the parameters $\theta_{\rm u}$, $\theta_{\rm d}$ and φ are essentially independent of the energy scale, while θ does depend on it and will change if the underlying scale is shifted, say from the weak scale ($\sim 10^2$ GeV) to the grand unified theory scale (of order 10^{16} GeV). In short, the heavy quark mixing is subject to renormalization-group effects; but the u- and d-channel mixings are not, likewise the phase φ describing CP violation and the LQ–triangle as a whole.

We have presented a new description of the flavor mixing phenomenon, which is based on the phenomenological fact that the quark mass spectrum exhibits a clear hierarchy pattern. This leads uniquely to the interpretation of the flavor mixing in terms of a heavy quark mixing, followed by the u-channel and d-channel mixings. The complex phase φ, describing the relative orientation of the u-channel mixing and the d-channel mixing in the complex plane, signifies CP violation, which is a phenomenon primarily linked to the physics of the first two families. The Cabibbo angle is not a basic mixing parameter, but given by a superposition of two terms involving the complex phase φ. The experimental data suggest that the phase φ, which is directly linked to the phases of the quark mass terms, is close to 90°. This opens the possibility to interpret CP violation as a maximal effect, in a similar way as parity violation.

Our description of flavor mixing has many clear advantages compared with other descriptions. We propose that it should be used in the future description of flavor mixing and CP violation, in particular, for the studies of quark mass matrices and B-meson physics.

The description of the flavor mixing phenomenon given above is of special interest if for the u- and d-channel mixings specific quark mass textures are used. In that case one often finds (see, e. g., ref. [22]) apart from small corrections

$$\tan\theta_{\rm d} = \sqrt{\frac{m_d}{m_s}}\,,$$
$$\tan\theta_{\rm u} = \sqrt{\frac{m_u}{m_c}}\,. \tag{40}$$

The experimental value for $\tan\theta_u$ given by the ratio $|V_{ub}/V_{cb}|$ is in agreement with the observed value for $(m_u/m_c)^{1/2} \approx 0.07$, but the errors for both $(m_u/m_c)^{1/2}$ and $|V_{ub}/V_{cb}|$ are the same (about 25%). Thus from the underlying texture no new information is obtained.

This is not true for the angle $\theta_{\rm d}$, whose experimental value is due to a large uncertainty.: $\theta_{\rm d} = 12.7° \pm 3.8°$. (The analysis given in [18] indicates, however, that the uncertainty for $\theta_{\rm d}$ may be less). If $\theta_{\rm d}$ is given indeed by $(m_d/m_s)^{1/2}$, which is known to a high accuracy, we would know $\theta_{\rm d}$ and therefore all four parameters of the CKM matrix with high precision.

As emphasized in ref. [20], the phase angle φ is very close to $90°$, implying that the LQ–triangle and the unitarity triangle are essentially rectangular triangles. In particular the angle β which is likely to be measured soon in the study of the reaction $B^0 \rightarrow J/\psi K_S^0$ is expected to be close to $20°$.

It will be very interesting to see whether the angles $\theta_{\rm d}$ and $\theta_{\rm u}$ are indeed given by the square roots of the light quark mass ration m_d/m_s and m_u/m_c, which imply that the phase φ is close to or exactly $90°$. This would mean that the light quarks play the most important rôle in the dynamics of flavor mixing and CP violation and that a small window has been opened allowing the first view across the physics landscape beyond the mountain chain of the Standard Model.

References

[1] N. Cabibbo, Phys. Rev. Lett. **10** (1963) 531.
[2] M. Kobayashi and T. Maskawa, Prog. Theor. Phys. **49** (1973) 652.
[3] Particle Data Group, R.M. Barnett et al., Phys. Rev. **D54** (1996) 94.
[4] L. Wolfenstein, Phys. Rev. Lett **51** (1983) 1945.
[5] L. Maiani, in *Proc. 1977 Int. Symp. on Lepton and Photon Interactions at High Energies* (DESY, Hamburg), (1997) 867; L.L. Chau and W.Y. Keung, Phys. Rev. Lett. **53** (1984) 1802; H. Fritzsch, Phys. Rev. **D32** (1985) 3058; H. Harari and M. Leurer, Phys. Lett. **B181** (1986) 123; H. Fritzsch, J. Plankl Phys. Rev. **D35** (1987) 1732.
[6] H. Fritzsch and Z.Z. Xing, Phys. Lett. **B413** (1997) 396.
[7] C. Jarlskog, in *CP Violation*, edited by C. Jarlskog (World Scientific), (1989) 3.
[8] H. Fritzsch and Z.Z. Xing, Phys. Rev.**D57** (1989) 594.
[9] A. Rasin, Report No. hep-ph/9708216 (unpublished).
[10] H. Fritzsch, Phys. Lett. **B184** (1987) 391.
[11] H. Fritzsch, Phys. Lett. **B189** (1987) 191.
[12] L.J. Hall, and S. Weinberg, Phys. Rev.**D48** (1993) 979.
[13] H. Lehmann, C. Newton and T.T. Wu, Phys. Lett. **B384** (1996) 249.
[14] M. Neubert, Int. J. Mod. Phys. **A11** (1996) 4173.
[15] H. Fritzsch, Nucl. Phys.**B155** (1979) 189.
[16] S. Dimopoulos, L.J. Hall and S. Raby, Phys. Rev. Lett. **68** (1992) 1984; R. Barbieri, L.J. Hall and A. Romanino, Phys. Lett. **B401** (1997) 47.
[17] H. Fritzsch, Phys. Lett **B70** (1977) 436; ibid. **B73** (1978) 317.
[18] F. Parodi, R. Roudeau and A. Stocchi, Report No. hep-ph/9802289.
[19] C. Jarlskog, Phys. Rev. Lett. **55** (1985) 1039.
[20] H. Fritzsch and Z.Z. Xing, Phys. Lett. **B353** (1995) 114.
[21] K.S. Babu and Q. Shafi, Phys. Rev. **D47** (1993) 5004; and references therein.
[22] H. Fritzsch and Z.Z. Xing, in preparation.

Duality in Quantum Field Theory (and String Theory)

Luis Álvarez-Gaumé

Theory Division, CERN, CH-1211 Geneva 23, Switzerland.

Abstract. These lectures give an introduction to duality in Quantum Field Theory. We discuss the phases of gauge theories and the implications of the electric-magnetic duality transformation to describe the mechanism of confinement. We review the exact results of $N = 1$ supersymmetric QCD and the Seiberg-Witten solution of $N = 2$ super Yang-Mills. Some of its extensions to String Theory are also briefly discussed.

1 The Duality Symmetry

From a historical point of view we can say that many of the fundamental concepts of twentieth century physics have Maxwell's equations at its origin. In particular some of the symmetries that have led to our understanding of the fundamental interactions in terms of relativistic quantum field theories have their roots in the equations describing electromagnetism. As we will now describe, the most basic form of the duality symmetry also appears in the source-free Maxwell equations:

$$\nabla \cdot (\mathbf{E} + i\,\mathbf{B}) = 0\,,$$
$$\frac{\partial}{\partial t}(\mathbf{E} + i\,\mathbf{B}) + i\nabla \times (\mathbf{E} + i\,\mathbf{B}) = 0. \tag{1}$$

These equations are invariant under Lorentz transformations, and making all of physics compatible with these symmetries led Einstein to formulate the Theory of Relativity. Other important symmetries of (1) are conformal and gauge invariance, which have later played important roles in our understanding of phase transitions and critical phenomena, and in the formulation of the fundamental interactions in terms of gauge theories. In these lectures however we will study the implications of yet another symmetry hidden in (1): duality. The simplest form of duality is the invariance of (1) under the interchange of electric and magnetic fields:

$$\mathbf{B} \to \mathbf{E}$$
$$\mathbf{E} \to -\mathbf{B}. \tag{2}$$

In fact, the vacuum Maxwell equations (1) admit a continuous $SO(2)$ transformation symmetry

$$(\mathbf{E} + i\,\mathbf{B}) \to e^{i\phi}(\mathbf{E} + i\,\mathbf{B})\,. \tag{3}$$

If we include ordinary electric sources the equations (1.1) become:

$$\begin{aligned} \nabla \cdot (\mathbf{E} + i\,\mathbf{B}) &= q \\ \frac{\partial}{\partial t}(\mathbf{E} + i\,\mathbf{B}) + i\nabla \times (\mathbf{E} + i\,\mathbf{B}) &= \mathbf{j}_e\,. \end{aligned} \tag{4}$$

In presence of matter, the duality symmetry is not valid. To keep it, magnetic sources have to be introduced:

$$\begin{aligned} \nabla \cdot (\mathbf{E} + i\,\mathbf{B}) &= (q + ig) \\ \frac{\partial}{\partial t}(\mathbf{E} + i\,\mathbf{B}) + i\nabla \times (\mathbf{E} + i\,\mathbf{B}) &= (\mathbf{j}_e + i\,\mathbf{j}_m)\,. \end{aligned} \tag{5}$$

Now the duality symmetry is restored if at the same time we also rotate the electric and magnetic charges

$$(q + ig) \to e^{i\phi}(q + ig)\,. \tag{6}$$

The complete physical meaning of the duality symmetry is still not clear, but a lot of work has been dedicated in recent years to understand the implications of this type of symmetry. We will focus mainly on the applications to Quantum Field Theory. In the final sections, we will briefly review some of the applications to String Theory, where duality makes striking and profound predictions.

2 Dirac's charge quantization

From the classical point of view the inclusion of magnetic charges is not particularly problematic. Since the Maxwell equations, and the Lorentz equations of motion for electric and magnetic charges only involve the electric and magnetic field, the classical theory can accommodate any values for the electric and magnetic charges.

However, when we try to make a consistent quantum theory including monopoles, deep consequences are obtained. Dirac obtained his celebrated quantization condition precisely by studying the consistency conditions for a quantum theory in the presence of electric and magnetic charges (Dirac 1931). We derive it here by the quantization of the angular momentum, since it allows to extend it to the case of dyons, *i.e.*, particles that carry both electric and magnetic charges.

Consider a non-relativistic charge q in the vicinity of a magnetic monopole of strength g, situated at the origin. The charge q experiences a force $m\ddot{\mathbf{r}} = q\dot{\mathbf{r}} \times \mathbf{B}$, where $\mathbf{B}$ is the monopole field given by $\mathbf{B} = g\mathbf{r}/4\pi r^3$. The change in the orbital angular momentum of the electric charge under the effect of this force is given by

$$\begin{aligned} &\frac{d}{dt}\,(\,m\mathbf{r} \times \dot{\mathbf{r}}) = m\mathbf{r} \times \ddot{\mathbf{r}} \\ &= \frac{qg}{4\pi r^3}\,\mathbf{r} \times (\dot{\mathbf{r}} \times \mathbf{r}) = \frac{d}{dt}\left(\frac{qg}{4\pi}\,\frac{\mathbf{r}}{r}\right)\,. \end{aligned} \tag{7}$$

Hence, the total conserved angular momentum of the system is

$$\mathbf{J} = \mathbf{r} \times m\dot{\mathbf{r}} - \frac{qg}{4\pi}\frac{\mathbf{r}}{r}. \tag{8}$$

The second term on the right hand side (henceforth denoted by $\mathbf{J}_{em}$) is the contribution coming from the electromagnetic field. This term can be directly computed by using the fact that the momentum density of an electromagnetic field is given by its Poynting vector, $\mathbf{E} \times \mathbf{B}$, and hence its contribution to the angular momentum is given by

$$\mathbf{J}_{em} = \int d^3x\, \mathbf{r} \times (\mathbf{E} \times \mathbf{B}) = \frac{g}{4\pi}\int d^3x\, \mathbf{r} \times \left(\mathbf{E} \times \frac{\mathbf{r}}{r^3}\right).$$

In components,

$$\begin{aligned} J^i_{em} &= \frac{g}{4\pi}\int d^3x E^j \partial_j(\hat{x}^i) \\ &= \frac{g}{4\pi}\int_{S^2} \hat{x}^i \mathbf{E}\cdot d\mathbf{s} - \frac{g}{4\pi}\int d^3x \nabla\cdot\mathbf{E}\hat{x}^i. \end{aligned} \tag{9}$$

When the separation between the electric and magnetic charges is negligible compared to their distance from the boundary S^2, the contribution of the first integral to $\mathbf{J}_{em}$ vanishes by spherical symmetry. We are therefore left with

$$\mathbf{J}_{em} = -\frac{gq}{4\pi}\hat{r}. \tag{10}$$

Returning to equation (8), if we assume that orbital angular momentum is quantized. Then it follows that

$$\frac{qg}{4\pi} = \frac{1}{2}n, \tag{11}$$

where n is an integer. Equation (11) is the Dirac's charge quantization condition. It implies that if there exists a magnetic monopole of charge g somewhere in the universe, then all electric charges are quantized in units of $2\pi/g$. If we have a number of purely electric charges q_i and purely magnetic charges g_j, then any pair of them will satisfy a quantization condition:

$$q_i g_j = 2\pi n_{ij} \tag{12}$$

Thus, any electric charge is an integral multiple of $2\pi/g_j$. For a given g_j, let these charges have n_{0j} as the highest common factor. Then, all the electric charges are multiples of $q_0 = n_{0j}2\pi/g_j$. Similar considerations apply to the quantization of the magnetic charge.

Till now, we have only dealt with particles that carry either an electric or a magnetic charge. Consider now two dyons of charges (q_1, g_1) and (q_2, g_2). For this system, we can repeat the calculation of $\mathbf{J}_{em}$ by following the steps in (9), where now the electromagnetic fields are split as $\mathbf{E} = \mathbf{E}_1 + \mathbf{E}_2$ and $\mathbf{B} = \mathbf{B}_1 + \mathbf{B}_2$. The answer is easily found to be

$$\mathbf{J}_{em} = -\frac{1}{4\pi}\left(q_1 g_2 - q_2 g_1\right)\hat{r} \tag{13}$$

The charge quantization condition is thus generalized to

$$\frac{q_1 g_2 - q_2 g_1}{4\pi} = \frac{1}{2} n_{12} \tag{14}$$

This is referred to as the Dirac-Schwinger-Zwanziger condition (Schwinger 1966).

3 A charge lattice and the $SL(2, \mathbf{Z})$ group

In the previous section we derived the quantization of the electric charge of particles without magnetic charge, in terms of some smallest electric charge q_0. For a dyon (q_n, g_n), this gives $q_0 g_n = 2\pi n$. Thus, the smallest magnetic charge the dyon can have is $g_0 = 2\pi m_0/q_0$, with m_0 a positive integer dependent on the detailed theory considered. For two dyons of the same magnetic charge g_0 and electric charges q_1 and q_2, the quantization condition implies $q_1 - q_2 = nq_0$, with n a multiple of m_0. Therefore, although the difference of electric charges is quantized, the individual charges are still arbitrary. It introduces a new parameter θ that contributes to the electric charge of any dyon with magnetic charge g_0 by

$$q = q_0\left(n_e + \frac{\theta}{2\pi}\right). \tag{15}$$

Observe that the parameter $\theta + 2\pi$ gives the same electric charges as the parameter θ by shifting $n_e \to n_e + 1$. Thus, we look at the parameter θ as an angular variable.

This arbitrariness in the electric charge of dyons through the θ parameter can be fixed if the theory is CP invariant. Under a CP transformation $(q, g) \to (-q, g)$. If the theory is CP invariant, the existence of a state (q, g_0) necessarily leads to the existence of $(-q, g_0)$. Applying the quantization condition to this pair, we get $2q = q_0 \times \mathit{integer}$. This implies that $q = nq_0$ or $q = (n + \frac{1}{2})q_0$. If $\theta \neq 0, \pi$, the theory is not CP invariant. It indicates that the θ parameter is a source of CP violation. Later on we will identify θ with the instanton angle.

One can see that the general solution of the Dirac-Schwinger-Zwanziger condition (14) is

$$q = q_0\left(n_e + \frac{\theta}{2\pi} n_m\right) \tag{16}$$

$$g = n_m g_0 \tag{17}$$

with n_e and n_m integer numbers. These equations can be expressed in terms of the complex number

$$q + ig = q_0(n_e + n_m \tau), \tag{18}$$

where

$$\tau \equiv \frac{\theta}{2\pi} + \frac{2\pi i m_0}{q_0^2}. \tag{19}$$

Observe that this definition only includes intrinsic parameters of the theory, and that the imaginary part of τ is positive definite. This complex parameter will play an important role in supersymmetric gauge theories. Thus, physical states with electric and magnetic charges (q, g) are located on a discrete two dimensional lattice with periods q_0 and $q_0\tau$, and are represented by the corresponding vector (n_m, n_e) (see the figure below).

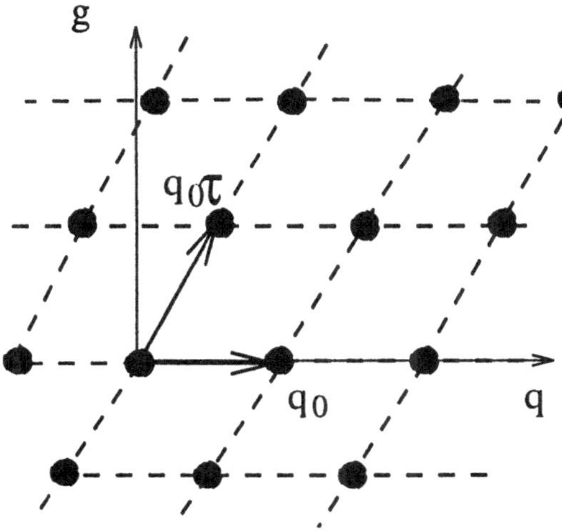

Notice that the lattice of charges obtained from the quantization condition breaks the classical duality symmetry group $SO(2)$ that rotated the electric and magnetic charges (6). But another symmetry group arises at quantum level. Given a lattice as in the above figure we can describe it in terms of different fundamental cells. Different choices correspond to transforming the electric and magnetic numbers (n_m, n_e) by a two-by-two matrix:

$$(n_m, n_e) \to (n_m, n_e) \begin{pmatrix} \alpha & \beta \\ \gamma & \delta \end{pmatrix}^{-1}, \tag{20}$$

with $\alpha, \beta, \gamma, \delta \in \mathbf{Z}$ satisfying $\alpha\delta - \beta\gamma = 1$. This transformation leaves invariant the Dirac-Schwinger-Zwanziger quantization condition (14). Hence the duality transformations are elements of the discrete group $SL(2, \mathbf{Z})$. Its action on the charge lattice can be implemented by modular transformations of the parameter τ:

$$\tau \to \frac{\alpha\tau + \beta}{\gamma\tau + \delta} \tag{21}$$

This transformations preserve the sign of the imaginary part of τ, and are generated just by the action of two elements:

$$T: \quad \tau \rightarrow \tau + 1\,, \tag{22}$$

$$S: \quad \tau \rightarrow \frac{-1}{\tau}\,. \tag{23}$$

The effect of T is to shift $\theta \rightarrow \theta + 2\pi$. Its action is well understood: it just maps the charge lattice (n_m, n_e) to $(n_m, n_e - n_m)$. As physics is 2π-periodic in θ, it is a symmetry of the theory. Then, if the state $(1,0)$ is in the physical spectrum, the state $(1, n_e)$, with any integer n_e, is also a physical state.

The effect of S is less trivial. If we take $\theta = 0$ just for simplicity, the S action is $q_0 \rightarrow g_0$ and sends the lattice vector (n_m, n_e) to the lattice vector $(-n_e, n_m)$. So it interchanges the electric and magnetic roles. In terms of coupling constants, it represents the transformation $\tau \rightarrow -1/\tau$, implying the exchange between the weak and strong coupling regimes. In this respect the duality symmetry could provide a new source of information on non-perturbative physics.

If we claim that the S transformation is also a symmetry of the theory we have full $SL(2, \mathbf{Z})$ symmetry. It implies the existence of any state (n_m, n_e) in the physical spectrum, with n_m and n_e relatively co-prime, just from the knowledge that there are the physical states $\pm(0,1)$ and $\pm(1,0)$. There are some examples of theories 'duality invariant', for instance the $SU(2)$ gauge theory with $N = 4$ supersymmetry and the $SU(2)$ gauge theory with $N = 2$ supersymmetry and four flavours (Vafa and Witten 1994).

A priori however there is no physical reason to impose S-invariance in contrast with T-invariance. The stable physical spectrum may not be $SL(2, \mathbf{Z})$ invariant. But if the theory still admits somehow magnetic monopoles, we could apply the S-transformation as a change of variables of the theory, where a magnetic state is mapped to an electric state in terms of the dual variables. It could be convenient for several reasons: Maybe there are some physical phenomena where the magnetic monopoles become relevant degrees of freedom; this is the case for the mechanism of confinement, as we will see below. The other reason could be the difficulty in the computation of some dynamical effects in terms of the original electric variables because of the large value of the electric coupling q_0. The S-transformation sends q_0 to $1/q_0$. In terms of the dual magnetic variables, the physics is weakly coupled.

Just by general arguments we have learned a good deal of information about the duality transformations. Next we have to see where such concepts appear in quantum field theory.

4 The Higgs Phase

4.1 The Higgs mechanism and mass gap

We start considering that the relevant degrees of freedom at large distances of some theory in 3+1 dimensions are reduced to an abelian Higgs model:

$$\mathcal{L}(\phi^*, \phi, A_\mu) = -\frac{1}{4}F_{\mu\nu}F^{\mu\nu} + (D_\mu\phi)^*(D^\mu\phi) - \frac{\lambda}{2}(\phi^*\phi - M^2)^2 , \tag{24}$$

where

$$\begin{aligned} F_{\mu\nu} &= \partial_\mu A_\nu - \partial_\nu A_\mu , \\ D_\mu\phi &= (\partial_\mu + iqA_\mu)\phi , \end{aligned} \tag{25}$$

and q is the electric charge of the particle ϕ.

An important physical example of a theory described at large distances by the effective Lagrangian (24) (in its nonrelativistic approximation) is a superconductor. Sound waves of a solid material causes complicated deviations from the ideal lattice of the material. Conducting electrons interact with the quantums of those sound waves, called phonons. For electrons near the Fermi surface, their interactions with the phonons create an attractive force. This force can be strong enough to cause bound states of two electrons with opposite spin, called Cooper pairs. The lowest state is a scalar particle with charge $q = -2e$, which is represented by ϕ in (24). To understand the basic features of a superconductor we only need to consider its relevant self-interactions and the interaction with the electromagnetic field resulting from its electric charge q. This is the dynamics which is encoded in the effective Lagrangian (24). The values of the parameters λ and M^2 are dependent of the temperature T, and in general contribute to increase the energy of the system. To have an stable ground state, we require $\lambda(T) > 0$ for any value of the temperature. But the function $M^2(T)$ does not need to be negative for all T. In fact, when the temperature T drops below a critical value T_c, the function $M^2(T)$ becomes positive. In such situation, the ground state reaches its minimal energy when the Higgs particle condenses,

$$|\langle\phi\rangle| = M . \tag{26}$$

If we make perturbation theory around this minimum,

$$\phi(x) = M + \varphi(x), \tag{27}$$

with vanishing external electromagnetic fields, we find that there is a mass gap between the ground state and the first excited levels. There are particles of spin one with mass square

$$\mathcal{M}_V^2 = 2qM^2, \tag{28}$$

which corresponds to the inverse of the penetration depth of static electromagnetic fields in the superconductor. There are also spin zero particles with mass square

$$\mathcal{M}^{\epsilon}{}_{H} = 2\lambda M^2. \tag{29}$$

So perturbation theory already shows a quite different behaviour of the Higgs theory from the Coulomb theory. There is only one real massive scalar field and the electromagnetic interaction becomes short-ranged, with the photon correlator being exponentially suppressed. This is a distinction that must survive nonperturbatively. But up to now, the above does not yet distinguish a Higgs theory from just any non-gauge theory with massive vector particles. There is yet another new phenomena in the Higgs mode which shows the spontaneous symmetry breaking of the $U(1)$ gauge theory.

4.2 Vortex tubes and flux quantization

We have seen that the Higgs condensation produces the electromagnetic interactions to be short-range. Ignoring boundary effects in the material, the electric and magnetic fields are zero inside the superconductor. This phenomena is called the Meissner effect.

If we turn on an external magnetic field $\mathbf{H}_0$ beyond some critical value, one finds that small regions in the superconductor make a transition to a 'non-superconducting' state. Stable magnetic flux tubes are allowed along the material, with a transverse size of the order of the inverse of the mass gap. Their magnetic flux satisfies a quantization rule that can be understood only by a combination of the spontaneous symmetry breaking of the $U(1)$ gauge symmetry and some topological arguments.

Parametrize the complex Higgs field by

$$\phi(x) = \rho(x)e^{i\chi(x)}, \tag{30}$$

and perform fluctuations around the configuration which minimizes the energy. *I.e.*, we consider that $\rho(x) \simeq M$ nearly everywhere, but at some points ρ may be zero. At such points χ needs not be well defined and therefore in all the rest of space χ could be multivalued. For instance, if we take a closed contour C around a zero of $\rho(x)$, then following χ around C could give values that run from 0 to $2\pi n$, with n an integer number, instead of coming back to zero. These are exactly the field configurations that produce the quantized magnetic flux tubes (Nielsen and Olesen 1973).

Consider a two-dimensional plane, cut somewhere through a superconducting piece of material, with polar coordinates (r, θ) and work in the time-like $A_0 = 0$. To have a finite energy per unit length static configuration we should demand that

$$\begin{aligned} \phi(x) &\to M e^{i\chi(\theta)} \\ A_i(x) &\to \frac{\text{const}}{r} \end{aligned} \tag{31}$$

for $r \to \infty$. Obviously, to keep the fields single valued, we must have

$$\chi(2\pi) = \chi(0) + 2\pi n. \tag{32}$$

If $n \neq 0$, it is clear that at some point of the two-dimensional plane we should have that the continuous field ϕ vanishes. Such field configurations do not correspond to the ground state.

We solve the field equations with the boundary conditions (31) and (32) fixed, and minimize the energy. We find stable vortex tubes with non-trivial magnetic flux through the two-dimensional plane. To see this, perform a singular gauge transformation [1]

$$\begin{aligned} \phi(x) &\to e^{iq\Lambda(x)}\phi(x) \\ A_\mu(x) &\to A_\mu(x) - \partial_\mu \Lambda(x)\,, \end{aligned} \tag{33}$$

with $\Lambda = 2\pi n\theta/q$. We compute the magnetic flux in such a gauge and we find

$$\Phi = \oint A_\mu dx^\mu = \Lambda(2\pi) - \Lambda(0) = \frac{2\pi n}{q}. \tag{34}$$

It is important to realize that such field configurations, called Abrikosov vortices, are stable. The vortex tube cannot break since it cannot have an end point: as the magnetic flux is quantized, we would have to be able to deform continuously the singular gauge transformation Λ to zero, something obviously not possible for $n \neq 0$. Physically this is the statement that the magnetic flux is conserved, a consequence of the Maxwell equations. Mathematically it means that for $n \neq 0$ the function $\chi(\theta)$ belongs to a nontrivial homotopy class of the fundamental group $\Pi_1(U(1)) = \mathbf{Z}$.

The existence of these macroscopic stable objects can be used as another characterization of the Higgs phase. They should survive beyond perturbation theory.

4.3 Magnetic monopoles and permanent magnetic confinement

The magnetic flux conservation in the abelian Higgs model tells us that the theory does not include magnetic monopoles. But it is significant that the magnetic flux is precisely a multiple of the quantum of magnetic charge $2\pi/q$ found by Dirac. If we imagine the effective gauge theory (24) enriched somehow by magnetic monopoles, they would form end points of the vortex tubes. The energy per unit length, *i.e.* the string tension σ, of these flux tubes is of the order of the scale of the Higgs condensation,

$$\sigma \sim M^2. \tag{35}$$

It implies that the total energy of a system composed of a monopole and an antimonopole, with a convenient magnetic flux tube attached between them, would be at least proportional to the separation length of the monopoles. In other words: magnetic monopoles in the Higgs phase are permanently confined.

[1] Singular in the sense of being not well defined in all space.

5 The Georgi-Glashow model and the Coulomb phase

The Georgi-Glashow model is a Yang-Mills-Higgs system which contains a Higgs multiplet ϕ^a ($a = 1,2,3$) transforming as a vector in the adjoint representation of the gauge group $SO(3)$, and the gauge fields $W_\mu = W_\mu^a T^a$. Here, T^a are the hermitian generators of $SO(3)$ satisfying $[T^a, T^b] = if^{abc}T^c$. In the adjoint representation, we have $(T^a)_{bc} = -if^a_{bc}$ and, for $SO(3)$, $f^{abc} = \epsilon^{abc}$. The field strength of W_μ and the covariant derivative on ϕ^a are defined by

$$\begin{aligned} G_{\mu\nu} &= \partial_\mu W_\nu - \partial_\nu W_\mu + ie[W_\mu, W_\nu]\,, \\ D_\mu \phi^a &= \partial_\mu \phi^a - e\epsilon^{abc} W_\mu^b \phi^c\,. \end{aligned} \tag{36}$$

The minimal Lagrangian is then given by

$$\begin{aligned} \mathcal{L} = &-\frac{1}{4} G^a_{\mu\nu} G^{a\mu\nu} \\ &+ \frac{1}{2} D^\mu \phi^a D_\mu \phi^a - V(\phi)\,, \end{aligned} \tag{37}$$

where

$$V(\phi) = \frac{\lambda}{4}\left(\phi^a\phi^a - a^2\right)^2. \tag{38}$$

The equations of motion following from this Lagrangian are

$$\begin{aligned} (D_\nu G^{\mu\nu})^a &= -e\,\epsilon^{abc}\,\phi^b\,(D^\mu\phi)^c\,, \\ D^\mu D_\mu \phi^a &= -\lambda\phi^a(\phi^2 - a^2)\,. \end{aligned} \tag{39}$$

The gauge field strength also satisfies the Bianchi identity

$$D_\nu \widetilde{G}^{\mu\nu a} = 0\,. \tag{40}$$

Let us find the vacuum configurations in this theory. Introducing non-abelian electric and magnetic fields, $G^{0i}_a = -\mathcal{E}^i_a$ and $G^{ij}_a = -\epsilon^{ij}{}_k \mathcal{B}^k_a$, the energy density is written as

$$\begin{aligned} \theta_{00} = &\frac{1}{2}\left((\mathcal{E}^i_a)^2 + (\mathcal{B}^i_a)^2 \right. \\ &+ \left.(D^0\phi_a)^2 + (D^i\phi_a)^2\right) + V(\phi)\,. \end{aligned} \tag{41}$$

Note that $\theta_{00} \geq 0$, and it vanishes only if

$$G^{\mu\nu}_a = 0, \quad D_\mu\phi = 0, \quad V(\phi) = 0\,. \tag{42}$$

The first equation implies that in the vacuum, W^a_μ is pure gauge and the last two equations define the Higgs vacuum. The structure of the space of vacua is determined by $V(\phi) = 0$ which solves to $\phi^a = \phi^a_{vac}$ such that $|\phi_{vac}| = a$. The space of Higgs vacua is therefore a two-sphere (S^2) of radius a in field space. To formulate a perturbation theory, we have to choose one of these vacua and hence, break the gauge symmetry spontaneously The part of the symmetry which keeps this vacuum invariant, still survives and the corresponding unbroken generator is $\phi^c_{vac}T^c/a$. The gauge boson associated with this generator is $A_\mu = \phi^c_{vac}W^c_\mu/a$ and the electric charge operator for this surviving $U(1)$ is given by

$$Q = e\frac{\phi^c_{vac}T^c}{a} . \tag{43}$$

If the group is compact, this charge is quantized. The perturbative spectrum of the theory can be found by expanding ϕ^a around the chosen vacuum as

$$\phi^a = \phi^a_{vac} + \phi'^a .$$

A convenient choice is $\phi^c_{vac} = \delta^{c3}a$. The perturbative spectrum (which becomes manifest after choosing an appropriate unitary gauge) consists of a massive Higgs of spin zero with a square mass

$$\mathcal{M}^2_H = 2\lambda a^2, \tag{44}$$

a massless photon, corresponding to the $U(1)$ gauge boson A^3_μ, and two charged massive W-bosons, A^1_μ and A^2_μ, with square mass

$$\mathcal{M}^2_W = e^2a^2. \tag{45}$$

This mass spectrum is realistic as long as we are at weak coupling, $e^2 \sim \lambda \ll 1$. At strong coupling, nonperturbative effects could change significantly eqs. (44) and (45). But the fact that there is an unbroken subgroup of the gauge symmetry ensures that there is some massless gauge boson, with a long range interaction. This is the characteristic of the Coulomb phase.

6 The 't Hooft-Polyakov monopoles

Let us look for time-independent, finite energy solutions in the Georgi-Glashow model. Finiteness of energy requires that as $r \to \infty$, the energy density θ_{00} given by (41) must approach zero faster than $1/r^3$. This means that as $r \to \infty$, our solution must go over to a Higgs vacuum defined by (42). In the following, we will first assume that such a finite energy solution exists and show that it can have a monopole charge related to its soliton number which is, in turn, determined by the associated Higgs vacuum. This result is proven without having to deal with any particular solution explicitly. Next, we will describe the 't Hooft-Polyakov ansatz for explicitly constructing one such monopole solution, where we will also comment on the existence of Dyonic solutions. In the last two subsections we will derive the Bogomol'nyi bound and the Witten effect.

6.1 The Topological nature of the magnetic charge

For convenience, in this subsection we will use the vector notation for the $SO(3)$ gauge group indices and not for the spatial indices.

Let $\boldsymbol{\phi}_{vac}$ denote the field $\boldsymbol{\phi}$ in a Higgs vacuum. It then satisfies the equations

$$\begin{aligned} \boldsymbol{\phi}_{vac} \cdot \boldsymbol{\phi}_{vac} &= a^2 \,, \\ \partial_\mu \boldsymbol{\phi}_{vac} - e\, \mathbf{W}_\mu \times \boldsymbol{\phi}_{vac} &= 0 \,, \end{aligned} \tag{46}$$

which can be solved for $\mathbf{W}_\mu$. The most general solution is given by

$$\mathbf{W}_\mu = \frac{1}{ea^2}\, \boldsymbol{\phi}_{vac} \times \partial_\mu \boldsymbol{\phi}_{vac} + \frac{1}{a} \boldsymbol{\phi}_{vac} A_\mu \,. \tag{47}$$

To see that this actually solves (46), note that $\partial_\mu \boldsymbol{\phi}_{vac} \cdot \boldsymbol{\phi}_{vac} = 0$, so that

$$\begin{aligned} &\frac{1}{ea^2} (\boldsymbol{\phi}_{vac} \times \partial_\mu \boldsymbol{\phi}_{vac}) \times \boldsymbol{\phi}_{vac} = \\ &\frac{1}{ea^2} \left(\partial_\mu \boldsymbol{\phi}_{vac} a^2 - \boldsymbol{\phi}_{vac} (\boldsymbol{\phi}_{vac} \cdot \partial_\mu \boldsymbol{\phi}_{vac}) \right) = \frac{1}{e} \partial_\mu \boldsymbol{\phi}_{vac} \,. \end{aligned} \tag{48}$$

The first term on the right-hand side of Eq. (47) is the particular solution, and $\boldsymbol{\phi}_{vac} A_\mu$ is the general solution to the homogeneous equation. Using this solution, we can now compute the field strength tensor $\mathbf{G}_{\mu\nu}$. The field strength $F_{\mu\nu}$ corresponding to the unbroken part of the gauge group can be identified as

$$\begin{aligned} F_{\mu\nu} = \frac{1}{a} \boldsymbol{\phi}_{vac} \cdot \mathbf{G}_{\mu\nu} &= \partial_\mu A_\nu - \partial_\nu A_\mu \\ &+ \frac{1}{a^3 e} \boldsymbol{\phi}_{vac} \cdot (\partial_\mu \boldsymbol{\phi}_{vac} \times \partial_\nu \boldsymbol{\phi}_{vac}) \,. \end{aligned} \tag{49}$$

Using the equations of motion in the Higgs vacuum it follows that

$$\partial_\mu F^{\mu\nu} = 0 \,, \qquad \partial_\mu \widetilde{F}^{\mu\nu} = 0 \,.$$

This confirms that $F_{\mu\nu}$ is a valid $U(1)$ field strength tensor. The magnetic field is given by $B^i = -\frac{1}{2}\epsilon^{ijk} F_{jk}$. Let us now consider a static, finite energy solution and a surface Σ enclosing the core of the solution. We take Σ to be far enough so that, on it, the solution is already in the Higgs vacuum. We can now use the magnetic field in the Higgs vacuum to calculate the magnetic charge g_Σ associated with our solution:

$$\begin{aligned} g_\Sigma &= \int_\Sigma B^i ds^i \\ &= -\frac{1}{2ea^3} \int_\Sigma \epsilon_{ijk}\, \boldsymbol{\phi}_{vac} \cdot \left(\partial^j \boldsymbol{\phi}_{vac} \times \partial^k \boldsymbol{\phi}_{vac} \right) ds^i \,. \end{aligned} \tag{50}$$

It turns out that the expression on the right hand side is a topological quantity as we explain below: Since $\phi^2 = a$; the manifold of Higgs vacua ($\mathcal{M}_0$) has the topology of S^2. The field $\boldsymbol{\phi}_{vac}$ defines a map from Σ into $\mathcal{M}_0$. Since Σ is also an

S^2, the map $\phi_{vac} : \Sigma \to \mathcal{M}_0$ is characterized by its homotopy group $\pi_2(S^2)$. In other words, ϕ_{vac} is characterized by an integer ν (the winding number) which counts the number of times it wraps Σ around $\mathcal{M}_0$. In terms of the map ϕ_{vac}, this integer is given by

$$\nu = \frac{1}{4\pi a^3} \int_\Sigma \frac{1}{2} \epsilon_{ijk} \phi_{vac} \cdot \left(\partial^j \phi_{vac} \times \partial^k \phi_{vac} \right) ds^i \,. \tag{51}$$

Comparing this with the expression for magnetic charge, we get the important result

$$g_\Sigma = \frac{-4\pi\nu}{e} \,. \tag{52}$$

Hence, the winding number of the soliton determines its monopole charge. Note that the above equation differs from the Dirac quantization condition by a factor of 2. This is because the smallest electric charge which could exist in our model is $e/2$ for an spinorial representation of $SU(2)$, the universal covering group of $SO(3)$. Then, in this model $m_0 = 2$.

6.2 The 't Hooft-Polyakov ansatz

Now we describe an ansatz proposed by 't Hooft ('t Hooft 1974) and Polyakov (Polyakov 1974) for constructing a monopole solution in the Georgi-Glashow model. For a spherically symmetric, parity-invariant, static solution of finite energy, they proposed:

$$\begin{aligned} \phi^a &= \frac{x^a}{er^2} H(aer) \,, \\ W_i^a &= -\epsilon_{ij}^a \frac{x^j}{er^2} \left(1 - K(aer)\right) \,, \\ W_0^a &= 0 \,. \end{aligned} \tag{53}$$

For the non-trivial Higgs vacuum at $r \to \infty$, they chose $\phi_{vac}^c = ax^c/r = a\hat{x}^c$. Note that this maps an S^2 at spatial infinity onto the vacuum manifold with a unit winding number. The asymptotic behaviour of the functions $H(aer)$ and $K(aer)$ are determined by the Higgs vacuum as $r \to \infty$ and regularity at $r = 0$. Explicitly, defining $\xi = aer$, we have: as $\xi \to \infty$, $H \sim \xi$, $K \to 0$ and as $\xi \to 0$, $H \sim \xi$, $(K - 1) \sim \xi$. The mass of this solution can be parametrized as

$$\mathcal{M} = \frac{4\pi a}{e} f\left(\lambda/e^2\right).$$

For this ansatz, the equations of motion reduce to two coupled equations for K and H which have been solved exactly only in certain limits. For $r \to 0$, one gets $H \to ec_1 r^2$ and $K = 1 + ec_2 r^2$ which shows that the fields are non-singular at $r = 0$. For $r \to \infty$, we get $H \to \xi + c_3 exp(-a\sqrt{2\lambda} r)$ and $K \to c_4 \xi exp(-\xi)$ which leads to $W_i^a \approx -\epsilon_{ij}^a x^j / er^2$. Once again, defining $F_{ij} = \phi^c G_{ij}^c / a$, the magnetic field turns out to be $B^i = -x^i / er^3$. The associated monopole charge is

$g = -4\pi/e$, as expected from the unit winding number of the solution. It should be mentioned that 't Hooft's definition of the Abelian field strength tensor is slightly different but, at large distances, it reduces to the form given above.

Note that in the above monopole solution, the presence of the Dirac string is not obvious. To extract the Dirac string, we have to perform a singular gauge transformation on this solution which rotates the non-trivial Higgs vacuum $\phi^c_{vac} = a\hat{x}^c$ into the trivial vacuum $\phi^c_{vac} = a\delta^{c3}$. In the process, the gauge field develops a Dirac string singularity which now serves as the source of the magnetic charge ('t Hooft 1974).

The 't Hooft-Polyakov monopole carries one unit of magnetic charge and no electric charge. The Georgi-Glashow model also admits solutions which carry both magnetic as well as electric charges. An ansatz for constructing such a solution was proposed by Julia and Zee (Julia and Zee 1975). In this ansatz, ϕ^a and W^a_i have exactly the same form as in the 't Hooft-Polyakov ansatz, but W^a_0 is no longer zero: $W^a_0 = x^a J(aer)/er^2$. This serves as the source for the electric charge of the dyon. It turns out that the dyon electric charge depends on a continuous parameter and, at the classical level, does not satisfy the quantization condition. However, semiclassical arguments show that, in CP invariant theories, and at the quantum level, the dyon electric charge is quantized as $q = ne$. This can be easily understood if we recognize that a monopole is not invariant under a gauge transformation which is, of course, a symmetry of the equations of motion. To treat the associated zero-mode properly, the gauge degree of freedom should be regarded as a collective coordinate. Upon quantization, this collective coordinate leads to the existence of electrically charged states for the monopole with discrete charges. In the presence of a CP violating term in the Lagrangian, the situation is more subtle as we will discuss later. In the next subsection, we describe a limit in which the equations of motion can be solved exactly for the 'tHooft-Polyakov and the Julia-Zee ansatz. This is the limit in which the soliton mass saturates the Bogomol'nyi bound.

6.3 The Bogomol'nyi bound and the BPS states

In this subsection, we derive the Bogomol'nyi bound (Bogomol'nyi 1976) on the mass of a dyon in term of its electric and magnetic charges which are the sources for $F^{\mu\nu} = \boldsymbol{\phi} \cdot \mathbf{G}^{\mu\nu}/a$. Using the Bianchi identity (40) and the first equation in (39), we can write the charges as

$$
\begin{aligned}
g &\equiv \int_{S^2_\infty} B_i dS^i = \frac{1}{a}\int \mathcal{B}^a_i (D^i\phi)^a d^3x\,, \\
q &\equiv \int_{S^2_\infty} E_i dS^i = \frac{1}{a}\int \mathcal{E}^a_i (D^i\phi)^a d^3x\,.
\end{aligned}
\tag{54}
$$

Now, in the center of mass frame, the dyon mass is given by

$$
\mathcal{M} \equiv \int d^3x\theta_{00} = \int d^3x \left(\frac{1}{2}\left[(\mathcal{E}^a_k)^2 + (\mathcal{B}^a_k)^2\right.\right.
$$

$$+ \ (D_k\phi^a)^2 + (D_0\phi^a)^2] + V(\phi)) \ , \tag{55}$$

where $\theta_{\mu\nu}$ is the energy momentum tensor. Using (54) and some algebra we obtain

$$\begin{aligned}\mathcal{M} = \int d^3x \ \Big(& \frac{1}{2} [(\mathcal{E}^a_k - D_k\phi^a \sin\theta)^2 \\ + \ & (\mathcal{B}^a_k - D_k\phi^a \cos\theta)^2 + (D_0\phi^a)^2] \\ + \ & V(\phi)) + a(q\sin\theta + g\cos\theta) \, ,\end{aligned} \tag{56}$$

where θ is an arbitrary angle. Since the terms in the first line are positive, we can write $\mathcal{M} \geq (q\sin\theta + g\cos\theta)$. This bound is maximized for $\tan\theta = q/g$. Thus we get the Bogomol'nyi bound on the dyon mass as

$$\mathcal{M} \geq a\sqrt{g^2 + q^2} \, . \tag{57}$$

For the 't Hooft-Polyakov solution, we have $q = 0$, and thus, $\mathcal{M} \geq a|g|$. But $|g| = 4\pi/e$ and $\mathcal{M}_W = ae = aq$, so that

$$\mathcal{M} \geq a\frac{4\pi}{e} = \frac{4\pi}{e^2}\mathcal{M}_W = \frac{4\pi}{q^2}\mathcal{M}_W = \frac{\nu}{\alpha}\mathcal{M}_W \, .$$

Here, α is the fine structure constant and $\nu = 1$ or $1/4$, depending on whether the electron charge is q or $q/2$. Since α is small ($\sim 1/137$ for electromagnetism), the above relation implies that the monopole is much heavier than the W-bosons associated with the symmetry breaking.

From (56) it is clear that the bound is not saturated unless $\lambda \to 0$, so that $V(\phi) = 0$. This is the Bogomol'nyi-Prasad-Sommerfield (BPS) limit of the theory (Bogomol'nyi 1976), (Prasad and Sommerfield 1975). Note that, in this limit, $\phi^2_{vac} = a^2$ is no longer determined by the theory and, therefore, has to be imposed as a boundary condition on the Higgs field. Moreover, in this limit, the Higgs scalar becomes massless. Now, to saturate the bound we set

$$\begin{aligned} D_0\phi^a &= 0 \, , \\ \mathcal{E}^a_k &= (D_k\phi)^a \sin\theta \, , \\ \mathcal{B}^a_k &= (D_k\phi)^a \cos\theta \, , \end{aligned} \tag{58}$$

where $\tan\theta = q/g$. In the BPS limit, one can use the 't Hooft-Polyakov (or the Julia-Zee) ansatz either in (39), or in (58), to obtain the exact monopole (or dyon) solutions (Bogomol'nyi 1976), (Prasad and Sommerfield 1975). These solutions automatically saturate the Bogomol'nyi bound and are referred to as the BPS states. Also, note that in the BPS limit, all the perturbative excitations of the theory saturate this bound and, therefore, belong to the BPS spectrum. As we will see later, BPS states appear in a very natural way in theories with $N = 2$ supersymmetry.

6.4 The θ parameter and the Witten effect

In this section we will show that in the presence of a θ-term in the Lagrangian, the magnetic charge of a particle always contributes to its electric charge in the way given by formula (16) (Witten 1979).

To study the effect of CP violation, we consider the Georgi-Glashow model with an additional θ-term as the only source of CP violation:

$$\mathcal{L} = -\frac{1}{4} F^a_{\mu\nu} F^{a\mu\nu} + \frac{1}{2}(D_\mu \phi^a)^2 - \lambda(\phi^2 - a^2)^2 \\ + \frac{\theta e^2}{32\pi^2} F^a_{\mu\nu} \tilde{F}^{a\mu\nu} . \tag{59}$$

Here, $\tilde{F}^{a\mu\nu} = \frac{1}{2}\epsilon^{\mu\nu\rho\sigma} F^a_{\rho\sigma}$. The presence of the θ-term does not affect the equations of motion but changes the physics since the theory is no longer CP invariant. We want to construct the electric charge operator in this theory. The theory has an $SO(3)$ gauge symmetry but the electric charge is associated with an unbroken $U(1)$ which keeps the Higgs vacuum invariant. Hence, we define an operator N which implements a gauge rotation around the $\hat{\phi}$ direction with gauge parameter $\Lambda^a = \phi^a/a$. These transformations correspond to the electric charge. Under N, a vector v^a and the gauge fields A^a_μ transform as

$$\delta v^a = \frac{1}{a}\,\epsilon^{abc}\phi^b v^c\,, \quad \delta A^a_\mu = \frac{1}{ea}\, D_\mu \phi^a .$$

Clearly, ϕ^a is kept invariant. At large distances, where $|\phi| = a$, the operator $e^{2\pi i N}$ is a 2π-rotation about $\hat{\phi}$ and therefore $\exp(2\pi i N) = 1$. Elsewhere, the rotation angle is $2\pi|\phi|/a$. However, by Gauss' law, if the gauge transformation is 1 at ∞, it leaves the physical states invariant. Thus, it is only the large distance behaviour of the transformation which matters and the eigenvalues of N are quantized in integer units. Now, we use Noether's formula to compute N:

$$N = \int d^3x \left(\frac{\delta \mathcal{L}}{\delta \partial_0 A^a_i}\, \delta A^a_i + \frac{\delta \mathcal{L}}{\delta \partial_0 \phi^a}\, \delta \phi^a \right) .$$

Since $\delta \boldsymbol{\phi} = 0$, only the gauge part (which also includes the θ-term) contributes:

$$\frac{\delta}{\delta \partial_0 A^a_i}\, \left(F^a_{\mu\nu} F^{a\mu\nu}\right) = 4F^{aoi} = -4\mathcal{E}^{ai}\,,$$

$$\frac{\delta}{\delta \partial_0 A^a_i}\, \left(\tilde{F}^a_{\mu\nu} F^{a\mu\nu}\right) = 2\epsilon^{ijk} F^a_{jk} = -4\mathcal{B}^{ai} .$$

Thus,

$$N = \frac{1}{ae} \int d^3x D_i \boldsymbol{\phi} \cdot \boldsymbol{\mathcal{E}}^i - \frac{\theta e}{8\pi^2 a} \int d^3x\, D_i \boldsymbol{\phi} \cdot \boldsymbol{\mathcal{B}}^i \\ = \frac{1}{e} Q_e - \frac{\theta e}{8\pi^2} Q_m\,,$$

where we have used (54). Here, Q_e and Q_m are the electric and magnetic charge operators with eigenvalues q and g, respectively, and N is quantized in integer units. This leads to the following formula for the electric charge:

$$q = ne + \frac{\theta e^2}{8\pi^2} g \, .$$

For the 't Hooft-Polyakov monopole, $n = 1$, $g = -4\pi/e$, and therefore, $q = e(1 - \theta/2\pi)$. For a general dyonic solution we get

$$g = \frac{4\pi}{e} n_m , \qquad q = n_e e + \frac{\theta e}{2\pi} n_m \, , \tag{60}$$

and we recover, (16) and (17) for $q_0 = e$. In the presence of a θ-term, a magnetic monopole always carries an electric charge which is not an integral multiple of some basic unit. In section III we introduced the charge lattice of periods e and $e\tau$. In this parametrization, the Bogomol'nyi bound (57) takes the form

$$\mathcal{M} \geq \sqrt{2} |ae(n_e + n_m \tau)| \, . \tag{61}$$

Notice that for a BPS state, equation (61) implies that its mass is proportional to the distance of its lattice point from the origin.

7 The Confining phase

7.1 The abelian projection

In non-abelian gauge theories, gauge fixing is a subject full of interesting surprises (ghosts, phantom solitons, ...) which often obscure the physical content of the theory ('t Hooft 1994).

't Hooft gave a qualitative program to overcome these difficulties and provided a scenario that explains confinement in a gauge theory. The idea is to perform the gauge fixing procedure in two steps. In the first one a unitary gauge is chosen for the non-abelian degrees of freedom. It reduces the non-abelian gauge symmetry to the maximal abelian subgroup of the gauge group. Here one gets particle gauge singularities. [2] This procedure is called the abelian projection ('t Hooft 1994). In this way, the dynamics of the Yang-Mills theory will be reduced to an abelian gauge theory with certain additional degrees of freedom.

We need a field that transforms without derivatives under gauge transformations. An example is a real field, X, in the adjoint representation of $SU(N)$,

$$X \to \Omega X \Omega^{-1} . \tag{62}$$

Such a field can always be found; take for instance $X^a = G^a_{12}$. We will use the field X to implement the unitary gauge condition which will carry us to the

[2] We will discuss the physical meaning of them later on.

abelian projection of the $SU(N)$ gauge group. The gauge is fixed by requiring that X be diagonal:

$$X = \begin{pmatrix} \lambda_1 & & 0 \\ & \ddots & \\ 0 & & \lambda_N \end{pmatrix} . \tag{63}$$

The eigenvalues of the matrix X are gauge invariant. Generically they are all different, and the gauge condition (63) leaves an abelian $U(1)^{N-1}$ gauge symmetry. It corresponds to the subgroup generated by the gauge transformations

$$\Omega = \begin{pmatrix} e^{i\omega_1} & & 0 \\ & \ddots & \\ 0 & & e^{i\omega_1} \end{pmatrix}, \quad \sum_{i=1}^{N} \omega_i = 0 . \tag{64}$$

There is also a discrete subgroup of transformations which still leave X in diagonal form. It is the Weyl group of $SU(N)$, which corresponds to permutations of the eigenvalues λ_i. We also fix the Weyl group with the convention $\lambda_1 > \lambda_2 \cdots \lambda_N$.

At this stage, we have an abelian $U(1)$ gauge theory with $N-1$ photons, $N(N-1)$ charged vector particles and some additional degrees of freedom that will appear presently.

7.2 The nature of the gauge singularities

So far we assumed that the eigenvalues λ_i coincide nowhere. But there are some gauge field configurations that produce two consecutive eigenvalues to coincide at some spacetime points

$$\lambda_i = \lambda_{i+1} = \lambda, \quad \text{for certain } i. \tag{65}$$

These spacetime points are 'singular' points of the abelian projection. The $SU(2)$ gauge subgroup corresponding to the 2×2 block matrix with coinciding eigenvalues leaves invariant the gauge-fixing condition (63).

Let us consider the vicinity of such a point. Prior to the complete gauge-fixing we may take X to be

$$X = \begin{pmatrix} D_1 & 0 & 0 & 0 \\ 0 & \lambda + \epsilon_3 & \epsilon_1 - i\epsilon_2 & 0 \\ 0 & \epsilon_1 + i\epsilon_2 & \lambda - \epsilon_3 & 0 \\ 0 & 0 & 0 & D_2 \end{pmatrix}, \tag{66}$$

where D_1 and D_2 may safely be considered to be diagonalized because the other eigenvalues do not coincide. With respect to that $SU(2)$ subgroup of $SU(N)$ that corresponds to rotations among the ith and $i+1$st components, the three fields $\epsilon_a(x)$ form an isovector. We may write the central block as

$$\lambda I_2 + \epsilon_a \sigma^a , \tag{67}$$

where σ^a are the Pauli matrices.

Consider static fields configurations. The points of space where the two eigenvalues coincide correspond to the points $\mathbf{x}_0$ that satisfy

$$\epsilon^a(\mathbf{x}_0) = 0\,. \tag{68}$$

These three equations define a single space point, and then the singularity is particle-like. Which is its physical interpretation?.

By analyticity we have $\epsilon^a \sim (x - x_0)^a$, and our gauge condition corresponds to rotating the isovector ϵ^a such that

$$\epsilon = \begin{pmatrix} 0 \\ 0 \\ |\epsilon_3| \end{pmatrix}. \tag{69}$$

From the previous sections, we know that the zero-point of ϵ^a at $\mathbf{x}_0$ behaves as a magnetic charge with respect to the remaining $U(1) \subset SU(2)$ rotations. We realize that those gauge field configurations that produce such gauge 'singularities' correspond to magnetic monopoles.

The non-abelian $SU(N)$ gauge theory is topologically such that it can be cast into a $U(1)^{N-1}$ abelian gauge theory, which will feature not only electrically charged particles but also magnetic monopoles.

7.3 The phases of the Yang-Mills vacuum

We can now give a qualitative description of the possible phases of the Yang-Mills vacuum. It is only the dynamics which, as a function of the microscopic bare parameters, determines in which phase the Yang-Mills vacuum is actually realized.

Classically, the Yang-Mills Lagrangian is scale invariant. One can write down field configurations with magnetic charge and arbitrarily low energy. But quantum corrections are likely to violate their masslessness. If dynamics simply chooses to give a positive mass to the monopoles, we are in a Higgs or Coulomb phase. We must look for the magnetic vortex tubes to figure out if we are in a Higgs phase. It will be a signal that the ordinary Higgs mechanism has taken place in the abelian gauge formulation of the Yang-Mills theory. The role of the dynamically generated Higgs field could be taken by some scalar composite charge operator respecting the $U(1)^{N-1}$ gauge symmetries. There is also the possibility that no Higgs phenomenon occurs at all in the abelian sector, or that some $U(1)$ gauge symmetries are not spontaneously broken. In this case we are in the Coulomb phase, with some massless photons, or in a mixed Coulomb-Higgs phase.

There is yet a third possibility, however. Maybe the quantum corrections give a formally negative mass squared for the monopole: a magnetically charged object condenses. We apply an 'electric-magnetic dual transformation' to write an effective Lagrangian which encodes the relevant magnetic degrees of freedom in the infrared limit. In such effective Lagrangian, the Higgs mechanism takes place in term of dual variables. We are in a dual Higgs phase. We have electric flux tubes with finite energy per unit of length. There is a confining potential between electrically charged objects, like quarks.

In 1994, Seiberg and Witten gave a quantitative proof that such dynamical mechanism of color confinement takes place in $N = 2$ super-QCD (SQCD) broken to $N = 1$ (Seiberg and Witten 1994a), giving a non-trivial realization of 't Hooft scenario. When $N = 2$ SQCD is softly broken to $N = 0$ the same mechanism of confinement persists (Álvarez-Gaumé et al. 1996), (Álvarez-Gaume et al. 1997).

7.4 Oblique confinement

For simplicity let us consider an $SU(2)$ gauge group. We have seen that for a non-zero CP violating parameter θ, the physical electric charge of a particle with electric (resp. magnetic) number n_e (resp. n_m) is:

$$q = (n_e + \frac{\theta}{2\pi} n_m)e. \tag{70}$$

Dyons with large electric charges may have larger self-energies contributing positively to their mass squared. If the state (n_e, n_m) condenses at $\theta \simeq 0$, it is likely that the state $(n_e - 1, n_m)$ condenses at $\theta \simeq 2\pi$. It suggests that there is a phase transition around $\theta \simeq \pi$. Such first order phase transitions have been observed in softly broken $N = 2$ SQCD to $N = 0$ (Evans et al. 1997).

't Hooft propossed a new condensation mode at $\theta \simeq \pi$ ('t Hooft 1994). He imagined the possibility that a bound state of the dyons (n_e, n_m) and $(n_e - 1, n_m)$, with zero electric charge at $\theta = \pi$, could be formed. Its smaller electric charge could favor its condensation, leading to what he called an oblique confinement mode. These oblique modes have also been observed in softly broken $N = 2$ SQCD with matter (Álvarez-Gaumé et al. 1996), (Álvarez-Gaume et al. 1997).

8 The Higgs/confining phase

In the previous section we have characterized the confining phase as the dual of the Higgs phase, *i.e.*, the physical states are gauge singlets made by the electric degrees of freedom bound by stable electric flux tubes. A good gauge invariant order parameter measuring such behaviour is the Wilson loop (Wilson 1974):

$$W(C) = \mathrm{Tr}\,\exp\left(ig \oint_C dx^\mu A_\mu\right). \tag{71}$$

For $SU(N)$ Yang-Mills in the confining phase, for contours C, the Wilson loop obeys the area law,

$$\langle W(C)\rangle \sim \exp(-\sigma \cdot (\text{area})), \tag{72}$$

with σ the string tension of the electric flux tube.

But dynamical matter fields in the fundamental representation immediately create a problem in identifying the confining phase of the theory through the Wilson loop. The criterion used for confinement in the pure gauge theory, the energy between static sources, no longer works. Even if the energy starts increasing as the sources separate, it eventually becomes favorable to produce a particle-antiparticle pair out of the vacuum. This pair shields the gauge charge of the sources, and the energy stops growing. So, even in a theory that 'looks' very confining, our signal fails, and the perimeter law replaces (72),

$$\langle W(C) \rangle =\sim \exp(-\Lambda \cdot (\text{perimeter})). \tag{73}$$

If some scalar field is in the fundamental representation of the gauge group, there is no distinction at all between the confinement phases and the Higgs phase. Using the scalar field in the fundamental representation one can built gauge invariant interpolating operators for all possible physical states. As the vacuum expectation value of the Higgs field in the fundamental representation continuously changes from large values to smaller ones, the spectrum of all physical states, and all other measurable quantities, change smoothly (Fradkin and Shenker 1979). There is no gauge invariant operator which can distinguish between the Higgs or confining phases. We are in a Higgs/confining phase.

In supersymmetric gauge theories, it is common to have scalar fields in the fundamental representation of the gauge group, the scalar quarks. In such situation, when the theory is not in the Coulomb phase, we will see that the theory is presented in a Higgs/confining phase. We could take the phase description which is more appropriate for the theory. For instance, if the theory is in the weak coupling region, it is better to realize it in the Higgs phase; if the theory is in the strong coupling region, it is better to think in a confining phase.

9 Supersymmetry

9.1 The supersymmetry algebra and its massless representations

The $N = 1$ supersymmetry algebra is written as (Wess and Bagger 1992)

$$\begin{aligned} \{Q_\alpha, \bar{Q}_{\dot\alpha}\} &= 2\sigma^\mu_{\alpha\dot\alpha} P_\mu \\ \{Q_\alpha, Q_\beta\} &= 0\,, \; \{\bar{Q}_{\dot\alpha}\,, \bar{Q}_{\dot\beta}\} = 0\,. \end{aligned} \tag{74}$$

Here, Q and $\bar{Q}$ are the supersymmetry generators and transform as spin $1/2$ operators, $\alpha, \dot\alpha = 1, 2$. Moreover, the supersymmetry generators commute with the momentum operator P_μ and hence, with P^2. Therefore, all states in a given representation of the algebra have the same mass. For a theory to be supersymmetric, it is necessary that its particle content forms a representation of the above algebra. The irreducible representations of (74) can be obtained using Wigner's method.

For massless states, we can always go to a frame where $P^\mu = E(1,0,0,1)$. Then the supersymmetry algebra becomes

$$\{Q_\alpha, \bar{Q}_{\dot\alpha}\} = \begin{pmatrix} 0 & 0 \\ 0 & 4E \end{pmatrix}.$$

In a unitary theory the norm of a state is always positive. Since Q_α and $\bar{Q}_{\dot\alpha}$ are conjugate to each other, and $\{Q_1, \bar{Q}_{\dot 1}\} = 0$, it follows that $Q_1|phys> = \bar{Q}_{\dot 1}|phys> = 0$. As for the other generators, it is convenient to rescale them as

$$a = \frac{1}{2\sqrt{E}} Q_2\,, \qquad a^\dagger = \frac{1}{2\sqrt{E}} \bar{Q}_{\dot 2}\,.$$

Then, the supersymmetry algebra takes the form

$$\{a, a^\dagger\} = 1\,, \quad \{a, a\} = 0\,, \quad \{a^\dagger, a^\dagger\} = 0\,.$$

This is a Clifford algebra with 2 fermionic generators and has a 2-dimensional representation. From the point of view of the angular momentum algebra, a is a rising operator and $a^\dagger$ is a lowering operator for the helicity of massless states. We choose the vacuum such that $J_3|\Omega_\lambda> = \lambda|\Omega_\lambda>$ and $a|\Omega_\lambda> = 0$. Then

$$J_3(a^\dagger|\Omega_\lambda>) = (\lambda - \frac{1}{2})(a^\dagger|\Omega_\lambda>). \tag{75}$$

The irreducible representations are not necessarily CPT invariant. Therefore, if we want to assign physical states to these representations, we have to supplement them with their CPT conjugates $|-\lambda>_{CPT}$. If a representation is CPT self-conjugate, it is left unchanged. Thus, from a Clifford vacuum with helicity $\lambda = 1/2$ we obtain the $N = 1$ supermultiplet

$$\begin{pmatrix} \{\ |1/2>, & |-1/2>_{CPT}\ \} \\ \{\ |0>, & |0>_{CPT}\ \} \end{pmatrix} \tag{76}$$

which contains a Weyl spinor ψ and a complex scalar ϕ. It is called the scalar multiplet.

The other relevant representation of a renormalizable quantum field theory is the vector multiplet. It is constructed from a Clifford vacuum with helicity $\lambda = 1$:

$$\begin{pmatrix} \{\ |1>, & |-1>_{CPT}\ \} \\ \{\ |1/2>, & |-1/2>_{CPT}\ \} \end{pmatrix}. \tag{77}$$

It contains a vector A_μ and a Weyl spinor λ.

9.2 Superspace and superfields

To make supersymmetry linearly realized it is convenient to use the superspace formalism and superfields (Salam and Strathdee 1974). Superspace is obtained by adding four spinor degrees of freedom $\theta^\alpha, \bar{\theta}_{\dot{\alpha}}$ to the spacetime coordinates x^μ. Under the supersymmetry transformations implemented by the operator $\xi^\alpha Q_\alpha + \bar{\xi}_{\dot{\alpha}} \bar{Q}^{\dot{\alpha}}$ with transformation parameters ξ and $\bar{\xi}$, the superspace coordinates transform as

$$\begin{aligned} x^\mu \to x'^\mu &= x^\mu + i\theta\sigma^\mu\bar{\xi} - i\xi\sigma^\mu\bar{\theta}\,, \\ \theta \to \theta' &= \theta + \xi\,, \\ \bar{\theta} \to \bar{\theta}' &= \bar{\theta} + \bar{\xi}\,. \end{aligned} \tag{78}$$

These transformations can easily be obtained by the following representation of the supercharges acting on (x, θ):

$$\begin{aligned} Q_\alpha &= \frac{\partial}{\partial\theta^\alpha} - i\sigma^\mu_{\alpha\dot{\alpha}}\bar{\theta}^{\dot{\alpha}}\,\partial_\mu\,, \\ \bar{Q}_{\dot{\alpha}} &= -\frac{\partial}{\partial\bar{\theta}^{\dot{\alpha}}} + i\theta^\alpha\sigma^\mu_{\alpha\dot{\alpha}}\,\partial_\mu\,. \end{aligned} \tag{79}$$

These satisfy $\{Q_\alpha, \bar{Q}_{\dot{\alpha}}\} = 2i\sigma^\mu_{\alpha\dot{\alpha}}\,\partial_\mu$. Moreover, using the chain rule, it is easy to see that $\partial/\partial x^\mu$ is invariant under (78) but not $\partial/\partial\theta$ and $\partial/\partial\bar{\theta}$. Therefore, we introduce the super-covariant derivatives

$$\begin{aligned} D_\alpha &= \frac{\partial}{\partial\theta^\alpha} + i\sigma^\mu_{\alpha\dot{\alpha}}\,\partial_\mu\,, \\ \bar{D}_{\dot{\alpha}} &= -\frac{\partial}{\partial\bar{\theta}^{\dot{\alpha}}} - i\sigma^\mu_{\alpha\dot{\alpha}}\theta^\alpha\,\partial_\mu\,. \end{aligned} \tag{80}$$

They satisfy $\{D_\alpha, \bar{D}_{\dot{\alpha}}\} = -2i\sigma^\mu_{\alpha\dot{\alpha}}\,\partial_\mu$ and anti-commute with Q and $\bar{Q}$.

The quantum fields transform as components of a superfield defined on superspace, $F(x, \theta, \bar{\theta})$. Since the θ-variables are anti-commuting, the Taylor expansion of $F(x, \theta, \bar{\theta})$ in $(\theta, \bar{\theta})$ is finite, indicating that the supersymmetry representations are finite dimensional. The coefficients of the expansion are the component fields.

To have irreducible representations we must impose supersymmetric invariant constraints on the superfields. The scalar multiplet (76) is represented by a chiral scalar superfield, Φ, satisfying the chiral constraint

$$\bar{D}_{\dot{\alpha}}\Phi = 0\,. \tag{81}$$

Note that for $y^\mu = x^\mu + i\theta\sigma^\mu\bar{\theta}$, we have $\bar{D}_{\dot{\alpha}} y^\mu = 0$, $\bar{D}_{\dot{\alpha}}\theta^\beta = 0$. Therefore, any function of (y, θ) is a chiral superfield. It can be shown that this also is a necessary condition. Hence, any chiral superfield can be expanded as

$$\Phi(y, \theta) = \phi(y) + \sqrt{2}\theta\psi(y) + \theta\theta F(y)\,. \tag{82}$$

Here, ψ and ϕ are the fermionic and scalar components, respectively, and F is an auxiliary field, linear and homogeneous. Similarly, an anti-chiral superfield is defined by $D_\alpha \Phi^\dagger = 0$ and can be expanded as

$$\Phi^\dagger(y^\dagger, \bar\theta) = \phi^\dagger(y^\dagger) + \sqrt{2}\bar\theta\bar\psi(y^\dagger) + \bar\theta\bar\theta F^\dagger(y^\dagger) , \tag{83}$$

where $y^{\mu\dagger} = x^\mu - i\theta\sigma^\mu\bar\theta$.

The vector multiplet (77) is represented off-shell by a real scalar superfield

$$V = V^\dagger . \tag{84}$$

In local quantum field theories, spin one massless particles carry gauge symmetries (Weinberg 1996). These symmetries commute with the supersymmetry transformations. For a vector superfield, many of its component fields can be gauged away using the abelian gauge transformation $V \to V + \Lambda + \Lambda^\dagger$, where Λ ($\Lambda^\dagger$) are chiral (antichiral) superfields. In the Wess-Zumino gauge (Wess and Bagger 1992), it becomes

$$V = -\theta\sigma^\mu\bar\theta A_\mu + i\theta^2\bar\theta\bar\lambda - i\bar\theta^2\theta\lambda + \frac{1}{2}\theta^2\bar\theta^2 D .$$

In this gauge, $V^2 = \frac{1}{2}A_\mu A^\mu \theta^2\bar\theta^2$ and $V^3 = 0$. The Wess-Zumino gauge breaks supersymmetry, but not the gauge symmetry of the abelian gauge field A_μ. The Abelian superfield gauge field strength is defined by

$$W_\alpha = -\frac{1}{4}\bar D^2 D_\alpha V , \qquad \bar W_{\dot\alpha} = -\frac{1}{4} D^2 \bar D_{\dot\alpha} V .$$

It can be verified that W_α is a chiral superfield. Since it is gauge invariant, it can be computed in the Wess-Zumino gauge,

$$\begin{aligned} W_\alpha = & -i\lambda_\alpha(y) + \theta_\alpha D - \frac{i}{2}(\sigma^\mu\bar\sigma^\nu\theta)_\alpha F_{\mu\nu} \\ & + \theta^2(\sigma^\mu\partial_\mu\bar\lambda)_\alpha , \end{aligned} \tag{85}$$

where $F_{\mu\nu} = \partial_\mu A_\nu - \partial_\nu A_\mu$.

In the non-Abelian case, V belongs to the adjoint representation of the gauge group: $V = V_A T^A$, where, $T^{A\dagger} = T^A$. The gauge transformations are now implemented by

$$e^{-2V} \to e^{-i\Lambda^\dagger} e^{-2V} e^{i\Lambda} ,$$

where $\Lambda = \Lambda_A T^A$ is a chiral superfield. The non-Abelian gauge field strength is defined by

$$W_\alpha = \frac{1}{8}\bar D^2 e^{2V} D_\alpha e^{-2V}$$

and transforms as

$$W_\alpha \to W'_\alpha = e^{-i\Lambda} W_\alpha e^{i\Lambda} .$$

In components, in the WZ gauge it takes the form

$$W^a_\alpha = -i\lambda^a_\alpha + \theta_\alpha D^a - \frac{i}{2}(\sigma^\mu \bar{\sigma}^\nu \theta)_\alpha F^a_{\mu\nu} \\ + \theta^2 \sigma^\mu D_\mu \bar{\lambda}^a \,, \tag{86}$$

where

$$F^a_{\mu\nu} = \partial_\mu A^a_\nu - \partial_\nu A^a_\mu + f^{abc} A^b_\mu A^c_\nu \,, \\ D_\mu \bar{\lambda}^a = \partial_\mu \bar{\lambda}^a + f^{abc} A^b_\mu \bar{\lambda}^c \,.$$

Now we are ready to construct supersymmetric Lagrangians in terms of superfields.

9.3 Supersymmetric Lagrangians

Clearly, any function of superfields is, by itself, a superfield. Under supersymmetry, the superfield transforms as $\delta F = (\xi Q + \bar{\xi}\bar{Q})F$, from which the transformation of the component fields can be obtained. Note that the coefficient of the $\theta^2\overline{\theta}^2$ component is the field component of highest dimension in the multiplet. Then, its variation under supersymmetry is always a total derivative of other components. Thus, ignoring surface terms, the spacetime integral of this component is invariant under supersymmetry. This tells us that a supersymmetric Lagrangian density may be constructed as the highest dimension component of an appropriate superfield.

Let us first consider the product of a chiral and an anti-chiral superfield $\Phi^\dagger\Phi$. This is a general superfield and its highest component can be computed using (82) as

$$\Phi^\dagger\Phi\,|_{\theta^2\bar{\theta}^2} = -\frac{1}{4}\phi^\dagger\Box\phi - \frac{1}{4}\Box\phi^\dagger\phi + \frac{1}{2}\partial_\mu\phi^\dagger\partial^\mu\phi \\ -\frac{i}{2}\psi\sigma^\mu\partial_\mu\bar{\psi} + \frac{i}{2}\partial_\mu\psi\sigma^\mu\bar{\psi} + F^\dagger F \,. \tag{87}$$

Dropping some total derivatives we get the free field Lagrangian for a massless scalar and a massless fermion with an auxiliary field.

The product of chiral superfields is a chiral superfield. In general, any arbitrary function of chiral superfields is a chiral superfield:

$$\mathcal{W}(\Phi_i) = \mathcal{W}(\phi_i + \sqrt{2}\theta\psi_i + \theta\theta F_i) \\ = \mathcal{W}(\phi_i) + \frac{\partial\mathcal{W}}{\partial\phi_i}\sqrt{2}\theta\psi_i \\ + \theta\theta\left(\frac{\partial\mathcal{W}}{\partial\phi_i}F_i - \frac{1}{2}\frac{\partial^2\mathcal{W}}{\partial\phi_i\phi_j}\psi_i\psi_j\right) . \tag{88}$$

$\mathcal{W}$ is referred to as the superpotential. Moreover, the space of the chiral fields Φ may have a non-trivial metric g^{ij} in which case the scalar kinetic term, for example, takes the form $g^{ij}\partial_\mu\phi_i^\dagger\partial^\mu\phi_j$, with appropriate modifications for other terms. In such cases, the free field Lagrangian above has to be replaced by a non-linear σ-model (Zumino 1979). Thus, the most general $N = 1$ supersymmetric Lagrangian for the scalar multiplet is given by

$$\mathcal{L} = \int d^4\theta\, K(\Phi, \Phi^\dagger) + \int d^2\theta \mathcal{W}(\Phi) + \int d^2\bar{\theta}\bar{\mathcal{W}}(\Phi^\dagger)\,.$$

Note that the θ-integrals pick up the highest component of the superfield and in our conventions, $\int d^2\theta\theta^2 = 1$ and $\int d^2\bar{\theta}\bar{\theta}^2 = 1$. In terms of the non-holomorphic function $K(\phi, \phi^\dagger)$, the metric in field space is given by $g^{ij} = \partial^2 K/\partial\phi_i\partial\phi_j^\dagger$, *i.e.*, the target space for chiral superfields is always a Kähler space. For this reason, the function $K(\Phi, \Phi^\dagger)$ is referred to as the Kähler potential.

Remember that the super-field strength W_α is a chiral superfield spinor. Using the normalization $\mathrm{Tr}(T^aT^b) = \frac{1}{2}\delta^{ab}$, we have that

$$\begin{aligned}\mathrm{Tr}(W^\alpha W_\alpha)\,|_{\theta\theta} &= -i\lambda^a\sigma^\mu D_\mu\bar{\lambda}^a + \frac{1}{2}D^aD^a \\ &\quad -\frac{1}{4}F^{a\mu\nu}F^a_{\mu\nu} + \frac{i}{8}\epsilon^{\mu\nu\rho\sigma}F^a_{\mu\nu}F^a_{\rho\sigma}\,.\end{aligned} \tag{89}$$

The first three terms are real and the last one is pure imaginary. It means that we can include the gauge coupling constant and the θ parameter in the Lagrangian in a compact form

$$\begin{aligned}\mathcal{L} &= \frac{1}{4\pi}\mathrm{Im}\left(\tau\,\mathrm{Tr}\int d^2\theta\, W^\alpha W_\alpha\right) \\ &= -\frac{1}{4g^2}F^a_{\mu\nu}F^{a\mu\nu} + \frac{\theta}{32\pi^2}F^a_{\mu\nu}\tilde{F}^{a\mu\nu} \\ &\quad + \frac{1}{g^2}(\frac{1}{2}D^aD^a - i\lambda^a\sigma^\mu D_\mu\bar{\lambda}^a)\,,\end{aligned} \tag{90}$$

where $\tau = \theta/2\pi + 4\pi i/g^2$.

We now include matter fields by the introduction of the chiral superfield Φ in a given representation of the gauge group in which the generators are the matrices T^a_{ij}. The kinetic energy term $\Phi^\dagger\Phi$ is invariant under global gauge transformations $\Phi' = e^{-i\Lambda}\Phi$. In the local case, to insure that Φ' remains a chiral superfield, Λ has to be a chiral superfield. The supersymmetric gauge invariant kinetic energy term is then given by $\Phi^\dagger e^{-2V}\Phi$. We are now in a position to write down the full N=1 supersymmetric gauge invariant Lagrangian as

$$\begin{aligned}\mathcal{L} &= \frac{1}{8\pi}\,\mathrm{Im}\left(\tau\mathrm{Tr}\int d^2\theta\, W^\alpha W_\alpha\right) \\ &\quad + \int d^2\theta d^2\bar{\theta}\,(\Phi^\dagger e^{-2V}\Phi) + \int d^2\theta\,\mathcal{W} + \int d^2\bar{\theta}\,\bar{\mathcal{W}}\,.\end{aligned} \tag{91}$$

Note that since each term is separately invariant, the relative normalization between the scalar part and the Yang-Mills part is not fixed by $N = 1$ supersymmetry. In fact, under loop effects, by virtue of the perturbative non-renormalization theorem (Grisaru et al. 1979), only the term with the complete superspace integral $\int d^2\theta d^2\bar{\theta}$ gets an overall renormalization factor $Z(\mu, g(\mu))$, with μ the renormalization scale and $g(\mu)$ the renormalized gauge coupling constant. Observe the unique dependence on $Re(\tau)$ in Z, breaking the holomorphic τ-dependence of the Lagrangian $\mathcal{L}$. But quantities as the superpotential $\mathcal{W}$ are renormalization group invariant under perturbation theory (Grisaru et al. 1979) (we will see dynamically generated superpotentials by nonperturbative effects).

In terms of component fields, the Lagrangian (91) becomes

$$\begin{aligned}\mathcal{L} &= -\frac{1}{4g^2}F^a_{\mu\nu}F^{a\mu\nu} + \frac{\theta}{32\pi^2}F^a_{\mu\nu}\widetilde{F}^{a\mu\nu} \\ &- \frac{i}{g^2}\lambda^a\sigma^\mu D_\mu\bar{\lambda}^a + \frac{1}{2g^2}D^aD^a \\ &+ (\partial_\mu\phi - iA^a_\mu T^a\phi)^\dagger(\partial^\mu\phi - iA^{a\mu}T^a\phi) - D^a\phi^\dagger T^a\phi \\ &- i\bar{\psi}\bar{\sigma}^\mu(\partial_\mu\psi - iA^a_\mu T^a\psi) + F^\dagger F \\ &+ \left(-i\sqrt{2}\,\phi^\dagger T^a\lambda^a\psi + \frac{\partial\mathcal{W}}{\partial\phi}F - \frac{1}{2}\frac{\partial^2\mathcal{W}}{\partial\phi\partial\phi}\psi\psi + h.c.\right).\end{aligned} \tag{92}$$

Here, $\mathcal{W}$ denotes the scalar component of the superpotential. The auxiliary fields F and D^a can be eliminated by using their equations of motion:

$$F = \frac{\partial\mathcal{W}}{\partial\phi} \tag{93}$$

$$D^a = g^2(\phi^\dagger T^a\phi). \tag{94}$$

The terms involving these fields, thus, give rise to the scalar potential

$$V = |F|^2 + \frac{1}{2g^2}D^aD^a. \tag{95}$$

Using the supersymmetry algebra (74) it is not difficult to see that the hamiltonian $P^0 = H$ is a positive semi-definite operator, $\langle H\rangle \geq 0$, and that the ground state has zero energy if and only if it is supersymmetric invariant. At the level of local fields, the equation (95) means that the supersymmetric ground state configuration is such that

$$F = D^a = 0. \tag{96}$$

9.4 *R*-symmetry

The supercharges Q_α and $\overline{Q}_{\dot\alpha}$ are complex spinors. In the supersymmetry algebra (74) there is a $U(1)$ symmetry associated to the phase of the supercharges:

$$\begin{aligned} Q &\to Q' = e^{i\beta} Q \\ \overline{Q} &\to \overline{Q}' = e^{-i\beta}\overline{Q}. \end{aligned} \tag{97}$$

This symmetry is called the *R*-symmetry. It plays an important role in the study of supersymmetric gauge theories.

In terms of superspace, the *R*-symmetry is introduced through the superfield generator $(\theta Q + \bar\theta \overline{Q})$. Then, it rotates the phase of the superspace components θ and $\overline{\theta}$, in the opposite way as Q and $\overline{Q}$. It gives different *R*-charges for the component fields of a superfield. Consider that the chiral superfield Φ has *R*-charge n,

$$\Phi(x,\theta) \to \Phi'(x,\theta) = e^{in\beta}\Phi(x, e^{-i\beta}\theta)\,. \tag{98}$$

In terms of its component fields we have that:

$$\begin{aligned} \phi &\to \phi' = e^{in\beta}\phi\,, \\ \psi &\to \psi' = e^{i(n-1)\beta}\psi\,, \\ F &\to F' = e^{i(n-2)\beta}F\,. \end{aligned}$$

Since $d^2(e^{-i\beta}\theta) = e^{2i\beta}d^2\theta$, we derive that the superpotential has *R*-charge two,

$$\mathcal{W}(\Phi) \to \mathcal{W}(\Phi',\theta) = e^{2i\beta}\mathcal{W}(\Phi, e^{-i\beta}\theta)\,, \tag{99}$$

and that the Kähler potential is *R*-neutral.

10 The uses of supersymmetry

10.1 Flat directions and super-Higgs mechanism

We have seen that the field configuration of the supersymmetric ground state are those corresponding to zero energy. To find them we solve (96). Consider a supersymmetric gauge theory with gauge group G, and matter superfields Φ_i in the representation $R(f)$ of G. The classical equations of motion of the D^a $(a = 1, \ldots, \dim G)$ auxiliary fields give

$$D^a = \sum_i \phi_i^\dagger T_i^a \phi_i. \tag{100}$$

The solutions of $D^a = 0$ usually lead to the concept of flat directions. They play an important role in the analysis of SUSY theories. These flat directions may be lifted by F-terms in the Lagrangian, as for instance mass terms.

As an illustrative example of flat directions and some of its consequences, consider the $SU(2)$ gauge group, one chiral superfield Q in the fundamental representation of $SU(2)$ and another chiral superfield $\tilde{Q}$ in the anti-fundamental representation of $SU(2)$. This is supersymmetric QCD (SQCD) with one massless flavor. In this particular case, the equation (100) becomes

$$D^a = q^\dagger \sigma^a q - \tilde{q}\sigma^a \tilde{q}^\dagger . \tag{101}$$

The equations $D^a = 0$ have the general solution (up to gauge and global symmetry transformations)

$$q = \tilde{q}^\dagger = \begin{pmatrix} a \\ 0 \end{pmatrix}, \quad a \text{ arbitrary}. \tag{102}$$

The scalar superpartners of the fermionic quarks, called squarks, play the role of Higgs fields. As these are in the fundamental representation of the gauge group, $SU(2)$ is completely broken by the super-Higgs mechanism (for $a \neq 0$). It is just the supersymmetric generalization of the familiar Higgs mechanism: three real scalars are eaten by the gluon, in the adjoint representation, and three Weyl spinor combinations of the quark spinors are eaten by the gluino to form a massive Dirac spinor in the adjoint of $SU(2)$. Gluons and gluinos acquire the classical square mass

$$\mathcal{M}_g^2 = 2g_0^2|a|^2, \tag{103}$$

where g_0 is the bare gauge coupling. We see that the theory is in the Higgs/confining phase. But there is no mass gap; it remains a massless superfield. Its corresponding massless scalar must move along some flat direction of the classical potential. This flat direction is given by the arbitrary value of the real number $|a|$. This degeneracy is not unphysical, as in the spontaneous breaking of a symmetry. When we move along the supersymmetric flat direction the physical observables change, as for instance the gluon mass (103). Different values of $|a|$ correspond to physically inequivalent vacua. The space they expand is called the moduli space. It would be nice to have a gauge invariant parametrization of such an additional parameter of the gauge invariant vacuum. It can only come from the vacuum expectation value of some gauge invariant operator, since it is an independent new classical parameter which does not appear in the bare Lagrangian. The simplest choice is to take the following gauge invariant chiral superfield:

$$M = Q\tilde{Q}. \tag{104}$$

Classically, its vacuum expectation value is

$$\langle M \rangle = |a|^2, \tag{105}$$

a gauge invariant statement and a good parametrization of the flat direction.

There is one consequence of the flat directions in supersymmetric gauge theories that, when combined with the property of holomorphy, will be important to obtain exact results in supersymmetric theories. SQCD depends of the complex coupling $\tau(\mu) = \theta(\mu)/2\pi + 4\pi i/g^2(\mu)$ at scale μ. The angle $\theta(\mu)$ measures the strength of CP violation at scale μ. By asymptotic freedom, the theory is weakly coupled at scales higher than the dynamically generated scale $|\Lambda|$, which is defined by

$$\Lambda \equiv \mu_0 e^{\frac{2\pi i \tau(\mu_0)}{b_0}}, \tag{106}$$

where μ_0 is the ultraviolet cut-off where the bare parameter $\tau_0 = \tau(\mu_0)$ is defined, and b_0 is the one-loop coefficient of the beta function,

$$\mu \frac{\partial g}{\partial \mu}(\mu) = g\left(-b_0(g^2/16\pi^2) + \mathcal{O}(g^4)\right) . \tag{107}$$

The complex parameter Λ is renormalization group invariant in the scheme of the Wilsonian effective actions, where holomorphy is not lost (see below). Observe also that the bare instanton angle θ_0 plays the role of the complex phase of Λ^{b_0}.

At scales $\mu \leq \mathcal{M}_g$, all the gluons decouple and the relevant degrees of freedom are those of the 'meson' M. Its self-interactions are completely determined by the 'microscopic' degrees of freedom of the super-gluons and super-quarks. We must perform a matching condition for the physics at some scale of order $\mathcal{M}_g$; the renormalization group will secure the physical equivalence at the other energies. If $\mathcal{M}_g \gg \Lambda$, this matching takes place at weak coupling, where perturbation theory in the gauge coupling g is reliable, and we can trust the semiclassical arguments, like those leading to formulae (103) and (105).

So far we have shown the existence of a flat direction at the classical level. When quantum corrections are included, the flat direction may disappear and a definite value of $\langle M \rangle$ is selected. For the Wilsonian effective description in terms of the relevant degrees of freedom M, this is only possible if a superpotential $\mathcal{W}(M)$ is dynamically generated for M. By the perturbative non-renormalization theorem, this superpotential can only be generated by nonperturbative effects, since classically there was no superpotential for the massless gauge singlet M because of the masslessness of the quark multiplet.

If we turn on a bare mass for the quarks, m, the flat direction is lifted at classical level and a determined value of mass dependent function $\langle M \rangle$ is selected. But the advantage of the flat direction to carry $\langle M \rangle \to \infty$ to be at weak coupling is not completely lost. This limit can now be performed by sending the free parameter m to the appropriate limit, as far as we are able to know the mass dependence of the vacuum expectation value of the meson superfield M. Here holomorphy is very relevant.

10.2 Wilsonian effective actions and holomorphy

The concept of Wilsonian effective action is simple. Any physical process has a typical scale. The idea of the Wilsonian effective action is to give the Lagrangian of some physical processes at its corresponding characteristic scale μ:

$$\mathcal{L}^{(\mu)}(x) = \sum_i g^i(\mu)\mathcal{O}_i(x,\mu). \tag{108}$$

$\mathcal{O}_i(x, \mu)$ are relevant local composite operators of the effective fields $\varphi_a(p, \mu)$. These are the effective degrees of freedom at scale μ, with momentum modes p running from zero to μ. There could be some symmetries in the operators $\mathcal{O}_i$ that our physical system could realize in some way, broken or unbroken. The constants $g^i(\mu)$ measure the strength of the interaction $\mathcal{O}_i$ of φ_a at scale μ.

Behind some macroscopic physical processes, there is usually a microscopic theory, with a bare Lagrangian $\mathcal{L}^{(\mu_0)}(x)$ defined at scale μ_0. The microscopic theory has also its characteristic scale μ_0, much higher than the low energy scale μ. Also its corresponding microscopic degrees of freedom, $\phi_j(p, \mu_0)$, may be completely different than the macroscopic ones $\varphi_a(p, \mu)$. The bare Lagrangian encodes the dynamics at scales below the ultraviolet cut-off μ_0. The effective Lagrangian (108) is completely determined by the microscopic Lagrangian $\mathcal{L}^{(\mu_0)}(x)$. It is obtained by integrating out the momentum modes p between μ and μ_0. It gives the values of the effective couplings in terms of the bare couplings $g_0^i(\mu_0)$,

$$g^i(\mu) = g^i(\mu; \mu_0, g_0^i(\mu_0)). \tag{109}$$

In the macroscopic theory there is no reference to the scale μ_0. Physics is independent of the ultraviolet cut-off μ_0:

$$\frac{\partial g^i}{\partial \mu_0} = 0. \tag{110}$$

The μ_0-dependence on the bare couplings $g_0^i(\mu_0)$ cancel the explicit μ_0-dependence in (109). This is the action of the renormalization group. It allows to perform the continuum limit $\mu_0 \to \infty$ without changing the low energy physics.

In supersymmetric theories, there are some operators $\mathcal{O}_i(z)$, depending only on $z = (x, \theta)$, the chiral superspace coordinate, not on $\bar{\theta}$. Clearly, their field content can only be made of chiral superfields. Those of most relevant physical importance are the superpotential $\mathcal{W}(\Phi_i, \tau_0, m_f)$, and the gauge kinetic operator $\tau(\mu/\mu_0, \tau_0)W^\alpha W_\alpha$. We say that the superpotential $\mathcal{W}$ and the effective gauge coupling τ are holomorphic functions, with the chiral superfields Φ_i, the adimensional quotient μ/μ_0 and the bare parameters τ_0 and m_f playing the role of the complex variables. The Kähler potential $K(\Phi^\dagger, \Phi)$ is a real function of the variables Φ_i, but as far as supersymmetry is not broken and the theory is not on some Coulomb phase, the vacuum structure is determined by the superpotential in the limit $\mu \to 0$.

We know that complex analysis is substantially more powerful than just real analysis. For instance, there are a lot of real functions $f(x)$ that at $x \to 0$ and $x \to \infty$ goes like $f(x) \to x$. But there is only one holomorphic function $f(z)$ $(\partial_{\bar{z}} f(z) = 0)$ with those properties: $f(z) = z$. The holomorphic constraint is so strong that sometimes just the symmetries of the theory, together with some consistency conditions, are enough to determine the unique possible form of the functions $\mathcal{W}$ and τ (Seiberg 1994b).

An illustrative example is the saturation at one-loop of the holomorphic gauge coupling $\tau(\mu/\mu_0, \tau_0)$ at any order of perturbation theory. Since $\tau_0 = \theta_0/2\pi +$

$i4\pi/g_0^2$, physical periodicity in θ_0 implies

$$\tau(\frac{\mu}{\mu_0},\tau_0)=\tau_0+\sum_{n=0}^{\infty}c_n\left(\frac{\mu}{\mu_0}\right)e^{2\pi n i\tau_0}, \tag{111}$$

where the sum is restricted to $n\geq 0$ to ensure a well defined weak coupling limit $g_0\to 0$. The unique term compatible with perturbation theory is the $n=0$ term. Terms with $n>0$ correspond to instanton contributions. The function $c_0(t)$ must satisfy $c_0(t_1t_2)=c_0(t_1)+c_0(t_2)$ and hence it must be a logarithm. Hence

$$\tau_{\rm pert}\left(\frac{\mu}{\mu_0},\tau_0\right)=\tau_0+\frac{ib_0}{2\pi}\ln\frac{\mu}{\mu_0}, \tag{112}$$

with b_0 the one-loop coefficient of the beta function. We can use the definition (106) of the dynamically generated scale Λ to absorb the bare coupling constant inside the logarithm

$$\tau_{\rm pert}\left(\frac{\mu}{\Lambda}\right)=\frac{ib_0}{2\pi}\ln\frac{\mu}{\Lambda}, \tag{113}$$

showing explicitly the independence of the effective gauge coupling in the ultraviolet cut-off μ_0.

We would like to comment that the one-loop saturation of the perturbative beta function and the renormalization group invariance of the scale Λ can be lost by the effect of the Konishi anomaly (Konishi Shizuya 1985), (Arkani-Hamed and Murayama 1997). In general, after the integration of the modes $\mu<p<\mu_0$ the kinetic terms of the matter fields Φ_i are not canonically normalized,

$$\mathcal{L}^{(\mu)}=\sum_i Z_i(\frac{\mu}{\mu_0},g_0)\int d^4\theta\Phi_i^\dagger e^{-2V}\Phi_i+\cdots \tag{114}$$

These terms have an integral on the whole superspace $(\theta,\bar\theta)$ and hence are not protected by any non-renormalization theorem. For $N=1$ gauge theories, holomorphy is absent there, and the functions $Z_i(\frac{\mu}{\mu_0},g_0)$ are just real functions with perturbative multi-loop contributions. A canonical normalization of the matter fields in the effective action, defining the canonical fields $\Phi_i'=Z_i^{1/2}\Phi_i$ do not leave invariant the path integral measure $\Pi_i\mathcal{D}\Phi_i$. The anomaly is proportional to $(\sum_i\ln Z_i)\,W^\alpha W_\alpha$, giving a non-holomorphic contribution to the effective coupling τ. For $N=2$ theories, $Z_i=1$ and holomorphy is not lost for τ (Arkani-Hamed and Murayama 1997), (de Wit et al. 1985).

11 $N=1$ SQCD

11.1 Classical Lagrangian and symmetries

We now analyze $N=1$ SQCD with gauge group $SU(N_c)$ and N_f flavors. [3] The field content is the following: There is a spinor chiral superfield W_α in the

[3] Some reviews on exacts results in $N=1$ supersymmetric gauge theories are (Intriligator and Seiberg 1996).

adjoint of $SU(N_c)$, which contains the gluons A_μ and the gluinos λ. The matter content is given by $2N_f$ scalar chiral superfields Q_f and $\tilde{Q}_f$, $f, \tilde{f} = 1, ..., N_f$, in the $\mathbf{N_c}$ and $\overline{\mathbf{N}}_\mathbf{c}$ representations of $SU(N_c)$ respectively. The renormalizable bare Lagrangian is the following:

$$\begin{aligned}\mathcal{L}_{SQCD} &= \frac{1}{8\pi}\mathrm{Im}\left(\tau_0 \int d^2\theta\; W^\alpha W_\alpha\right) \\ &+ \int d^4\theta \left(Q_f^\dagger e^{2V} Q_f + \tilde{Q}_f e^{-2V} \tilde{Q}_f^\dagger\right) \\ &+ \left(\int d^2\theta\; m_f \tilde{Q}_f Q_f \;+\mathrm{h.c.}\right), \end{aligned} \tag{115}$$

with $\tau_0 = \theta_0/2\pi + i4\pi/g_0^2$ and m_f the bare couplings. In the massless limit the global symmetry of the classical Lagrangian is $SU(N_f)_L \times SU(N_f)_R \times U(1)_B \times U(1)_A \times U(1)_R$. For $N_c = 2$ the representations $\mathbf{2}$ and $\overline{\mathbf{2}}$ are equivalent, and the global symmetry group is enlarged. In general, we consider $N_c > 2$. The $U(1)_A$ and $U(1)_R$ symmetries are anomalous and are broken by instanton effects. But we can perform a linear combination of $U(1)_A$ and $U(1)_R$, call it $U(1)_{AF}$, that is anomaly free. We have the following table of representations for the global symmetries of SQCD:

	$SU(N_f)_L$	$SU(N_f)_R$	$U(1)_B$	$U(1)_{AF}$
W_α	$\mathbf{1}$	$\mathbf{1}$	0	1
Q_f	$\mathbf{N_c}$	$\mathbf{1}$	1	$\frac{(N_f-N_c)}{N_f}$
$\tilde{Q}_f$	$\mathbf{1}$	$\overline{\mathbf{N}}_\mathbf{c}$	-1	$\frac{(N_f-N_c)}{N_f}$

The anomaly free R-charges, R_{AF}, are derived by the following. The superfield W_α is neutral under $U(1)_A$ and its R-transformation is fixed to be

$$W_\alpha(x,\theta) \rightarrow e^{i\beta} W_\alpha(x, e^{-i\beta}\theta). \tag{116}$$

Consider now that the fermionic quarks ψ have charge R_ψ under an $U(1)_{AF}$ transformation. In the one-instanton sector, λ has $2N_c$ zero modes, and one for each Q_f and $\tilde{Q}_f$. In total we have $2N_c + 2N_f R_\psi = 0$ to avoid the anomalies. We derive that $R_\psi = -N_c/N_f$. Since this is the charge of the fermions, the superfields $(Q_f, \tilde{Q}_f)$ have R_{AF} charge $1 - N_c/N_f = (N_f - N_c)/N_f$.

11.2 The classical moduli space

The classical equations of motion of the auxiliary fields are

$$\begin{aligned}\overline{F}_{q_f} &= -m_f \tilde{q}_f = 0 \\ \overline{F}_{\tilde{q}_f} &= -m_f q_f = 0 \\ D^a &= \sum_f \left(q_f^\dagger T^a q_f - \tilde{q}_f T^a \tilde{q}_f^\dagger\right) = 0. \end{aligned} \tag{117}$$

If there is a massive flavor $m_f \neq 0$, then we must have $q_f = \tilde{q}_f = 0$. As we want to go to the infrared limit to analyze the vacuum structure, the interesting case is the situation of N_f massless flavors. If some quark has a non-zero mass m, its physical effects can be decoupled at very low energy, by taking into account the appropriate physical matching conditions at the decoupling scale m (see below). If all quarks are massive, in the infrared limit we only have a pure $SU(N_c)$ supersymmetric gauge theory. The Witten index of pure $SU(N_c)$ super Yang-Mills is $\mathrm{tr}(-1)^F = N_c$ (Witten 1982). We know that supersymmetry is not broken dynamically in this theory, and that there are N_c equivalent vacua. The $2N_c$ gaugino zero modes break the $U(1)_R$ symmetry to Z_{2N_c} by the instantons. Those N_c vacua corresponds to the spontaneously broken discrete symmetry Z_{2N_c} to Z_2 by the gaugino condensate $\langle\lambda\lambda\rangle \neq 0$.

If there are some massless super-quarks, they can have non-trivial physical effects on the vacuum structure. Consider that we have N_f massless flavors. We can look at the q_f and $\tilde{q}_f$ scalar quarks as $N_c \times N_f$ matrices. Using $SU(N_c) \times SU(N_f)$ transformations, the q_f matrix can be rotated into a simple form. There are two cases to be distinguished:

a) $N_f < N_c$:

In this case we have that the general solution of the classical vacuum equations (117) is:

$$q_f = \tilde{q}_f^\dagger = \begin{pmatrix} v_1 & 0 & \cdots & 0 \\ 0 & v_2 & & \\ & & \ddots & \\ 0 & \cdots & & v_{N_f} \\ \vdots & & & \vdots \\ 0 & \cdots & & 0 \end{pmatrix}, \tag{118}$$

with v_f arbitrary. These scalar quark's vacuum expectation values break spontaneously the gauge group to $SU(N_c - N_f)$. By the super-Higgs mechanism, $N_c^2 - (N_c - N_f)^2 = 2N_cN_f - N_f^2$ chiral superfields are eaten by the vector superfields. This leaves $2N_fN_c - (2N_fN_c - N_f^2) = N_f^2$ chiral superfields. They can be described by the meson operators

$$M_{fg} \equiv \tilde{Q}_f Q_g. \tag{119}$$

which provide a gauge invariant description of the classical moduli space.

b) $N_f \geq N_c$:

In this case the general solution of (117) is:

$$q_f = \begin{pmatrix} v_1 & 0 & \cdots & 0 & \cdots & 0 \\ 0 & v_2 & & \vdots & & \vdots \\ & & \ddots & & & \vdots \\ & & & v_{N_c} & \cdots & 0 \end{pmatrix}, \tag{120}$$

$$\tilde{q}_f^\dagger = \begin{pmatrix} \tilde{v}_1 & 0 & \cdots & 0 & \cdots & 0 \\ 0 & \tilde{v}_2 & & \vdots & & \vdots \\ & & \ddots & & & \vdots \\ & & & \tilde{v}_{N_c} & \cdots & 0 \end{pmatrix}, \tag{121}$$

with the parameters v_i, $\tilde{v}_i$ $(i = 1, ..., N_c)$ subject to the constraint

$$|v_i|^2 - |\tilde{v}_i|^2 = \text{constant independent of } i. \tag{122}$$

Now the gauge group is completely higgsed. The gauge invariant parametrization of the classical moduli space must be done by $2N_f N_c - (N_c^2 - 1)$ chiral superfields. For instance, if $N_f = N_c$, we need $N_c^2 + 1$ superfields. The meson operators M_{fg} provide N_c^2. The remaining degree of freedom comes from the baryon-like operators

$$\begin{aligned} B &= \epsilon^{f_1 \cdots f_{N_f}} Q_{f_1} \cdots Q_{f_{N_f}}, \\ \tilde{B} &= \epsilon^{f_1 \cdots f_{N_f}} \tilde{Q}_{f_1} \cdots \tilde{Q}_{f_{N_f}}, \end{aligned} \tag{123}$$

with the color indices also contracted by the ϵ-tensor. These are two superfields, but there is a holomorphic constraint:

$$\det M - \tilde{B}B = 0. \tag{124}$$

For $N_f = N_c+1$, we need $2N_c(N_c+1)-(N_c^2-1) = N_c^2+2N_c+1$ independent chiral superfields. We can construct the baryon operators:

$$\begin{aligned} B^f &= \epsilon^{f f_1 \cdots f_{N_c}} Q_{f_1} \cdots Q_{f_{N_c}}, \\ \tilde{B}^f &= \epsilon^{f f_1 \cdots f_{N_c}} \tilde{Q}_{f_1} \cdots \tilde{Q}_{f_{N_c}}. \end{aligned} \tag{125}$$

M_{fg}, B^f and $\tilde{B}^f$ have $(N_c + 1)^2 + 2(N_c + 1)$ components. The matrix M_{fg} has rank N_c, which can be expressed by the $2(N_c + 1)$ constraints:

$$M_{fg} B^g = M_{fg} \tilde{B}^g = 0. \tag{126}$$

And in total we get the needed $N_c^2 + 2N_c + 1$ independent chiral superfields.

As N_f increases, we get more and more constraints. Each case with $N_f \geq N_c$ is interesting by itself and we will have to look at them in different ways.

12 The vacuum structure of SQCD with $N_f < N_c$

12.1 The Affleck-Dine-Seiberg's superpotential

First we consider the case of massless flavors. At the classical level there are flat directions parametrized by the free vacuum expectation values of the meson fields M_{fg}. They belong to the representation $(\mathbf{N_f}, \overline{\mathbf{N}}_\mathbf{f}, 0, 2(N_f - N_c)/N_f)$ of the global symmetry group $SU(N_f)_L \times SU(N_f)_R \times U(1)_B \times U(1)_{AF}$. If nonperturbative effects generate a Wilsonian effective superpotential $\mathcal{W}$, it must depend in a holomorphic way on the light chiral superfields M_{fg} and the bare coupling constant τ_0. The renormalization group invariance of the Wilsonian effective action demands that the dependence on the bare coupling constant τ_0 of $\mathcal{W}$ enters through the dynamically generated scale Λ_{N_f,N_c}. The invariance of $\mathcal{W}$ under $SU(N_f)_L \times SU(N_f)_R$ rotations reduces the dependence in the mesons fields to the combination $\det M$. There is only one holomorphic function $\mathcal{W} = \mathcal{W}(\det M, \Lambda_{N_f,N_c})$, with R_{AF} charge two that can be built from the variables $\det M$ and Λ_{N_f,N_c}, which have R_{AF} charge $2(N_f - N_c)$ and zero, respectively. It is the Afleck-Dine-Seiberg's superpotential (Davis et al. 1983), (Affleck et al. 1984)

$$\mathcal{W} = c_{N_f,N_g} \left(\frac{\Lambda_{N_f,N_c}}{\det M} \right)^{\frac{1}{(N_c - N_f)}}, \tag{127}$$

where c_{N_f,N_c} are some undetermined adimensional constants. If $c_{N_f,N_c} \neq 0$, (127) corresponds to an exact nonperturbative dynamically generated Wilsonian superpotential. It has catastrophic consequences, the theory has no vacuum. If we try to minimize the energy derived from the superpotential (127) we find that $|\langle \det M \rangle| \to \infty$.

12.2 Massive flavors

When we add mass terms for all the flavors we expect to find some physical vacua. In fact, by Witten index, we should find N_c of them. To verify this, let us try to compute $\langle M_{fg} \rangle$ taking advantage of its holomorphy and symmetries.

A bare mass matrix $m_{fg} \neq 0$ breaks explicitly the $SU(N_f)_L \times SU(N_f)_R \times U(1)_{AF}$ global symmetry of the bare Lagrangian (115). In terms of the meson operator the mass term is

$$\mathcal{W}_{\text{tree}} = \operatorname{tr}(mM). \tag{128}$$

We see that, under an L and R rotation of $SU(N_f)_L$ and $SU(N_f)_R$ respectively, we can recover the $SU(N_f)_L \times SU(N_f)_R$ invariance if we require m to transform as $m \to L^{-1} m R$. In the same way, as the superpotential has R-charge two, the $U(1)_{AF}$ invariance is recovered if we assign the charge $2 - 2(N_f - N_c)/N_f = 2N_c/N_f$ to the mass matrix m. The vacuum expectation value of the matrix chiral superfield M is a holomorphic function of Λ_{N_f,N_c} and m. To implement the same action under $SU(N_f)_L \times SU(N_f)_R$ rotations, we must have

$$\langle M \rangle = f(\det m, \Lambda_{N_f,N_c}) m^{-1}. \tag{129}$$

The dependence in $\det m$ of the function f is determined by the R_{AF} charge. Then, the Λ_{N_f,N_c} dependence is worked out by dimensional analysis. The result is

$$\langle M \rangle = (\text{const}) \left(\Lambda_{N_f,N_c}^{3N_c - N_f} \det m \right)^{\frac{1}{N_c}} m^{-1}. \tag{130}$$

The N_c roots give N_c vacua. Observe that this is an exact result, and valid also for $N_f \geq N_c$. There is only an adimensional constant (in general N_f and N_c

dependent) to be determined. It would be nice to be able to carry its computation in the weak coupling limit, since holomorphy would allow to extend (130) also to the strong coupling region.

The result (130) suggests the existence of an effective superpotential out of which (130) can be obtained. Holomorphy and symmetries tell us that the possible superpotential would have to be

$$\mathcal{W}(M, \Lambda_{N_f,N_c}, m) = \left(\frac{\Lambda_{N_f,N_c}}{\det M}\right)^{\frac{1}{(N_c-N_f)}} f\left(t = \mathrm{tr}(mM)\left(\frac{\Lambda_{N_f,N_c}}{\det M}\right)^{\frac{-1}{(N_c-N_f)}}\right). \qquad (131)$$

In the limit of weak coupling, $\Lambda_{N_f,N_c} \to 0$, we know that $f(t) = c_{N_f,N_c} + t$. But we can play at the same time with the free values of m to reach any desired value of t. This fixes the function $f(t)$ and the superpotential $\mathcal{W}(M, \Lambda_{N_f,N_c}, m)$ to be

$$\mathcal{W}(M, \Lambda_{N_f,N_c}, m) = c_{N_f,N_c}\left(\frac{\Lambda_{N_f,N_c}}{\det M}\right)^{\frac{1}{(N_c-N_f)}} + tr\,(mM). \qquad (132)$$

As a consistency check, when we solve the equations $\partial\mathcal{W}/\partial M = 0$, we obtain the previously determined vacuum expectation values (130).

Finally, we have to check the non-vanishing of c_{N_f,N_c}. We take advantage of the decoupling theorem to obtain further information about the constants c_{N_f,N_c}. Let us add a mass term m only for the N_f flavor,

$$\mathcal{W}(M, \Lambda_{N_f,N_c}, m) = \left(\frac{\Lambda_{N_f,N_c}}{\det M}\right)^{\frac{1}{(N_c-N_f)}} + mM_{N_fN_f}. \qquad (133)$$

Solving for the equations:

$$\frac{\partial\mathcal{W}}{\partial M_{fN_f}}(M, \Lambda_{N_f,N_c}, m) = 0, \qquad \frac{\partial\mathcal{W}}{\partial M_{N_ff}}(M, \Lambda_{N_f,N_c}, m) = 0, \qquad (134)$$

for $f < N_f$ gives that $M_{fN_f} = M_{N_ff} = 0$. Hence $\det M = M_{N_fN_f} \cdot \det\hat{M}$, with $\hat{M}$ the $(N_f-1)\times(N_f-1)$ matrix meson operator of the $N_f - 1$ massless flavors. At scales below m, the N_f-th flavor decouples and its corresponding $M_{N_fN_f}$ meson operator is frozen to the value that satisfies:

$$\frac{\partial\mathcal{W}}{\partial M_{N_fN_f}}(M, \Lambda_{N_f,N_c}, m) = -\frac{c_{N_f,N_c}}{(N_f-N_c)} \cdot \Lambda_{N_f,N_c}^{(3N_f-N_c)/(N_f-N_c)}(\det M)^{\frac{1}{(N_c-N_f)}-1}\det\hat{M} + m = 0. \qquad (135)$$

Substituting the solution $\langle M_{N_f N_f}\rangle$ of the previous equation into the superpotential $\mathcal{W}(M, \Lambda_{N_f,N_c}, m)$, we should obtain the superpotential $\mathcal{W}(\hat{M}, \Lambda_{N_f-1,N_c}, 0)$ of $N_f - 1$ massless flavors with the dynamically generated scale Λ_{N_f-1,N_c}. The matching conditions at scale m between the theory with N_f flavors and the theory with $N_f - 1$ flavors gives the relation

$$m\Lambda_{N_f,N_c}^{3N_c-N_f} = \Lambda_{N_f-1,N_c}^{3N_c-N_f+1}\,, \tag{136}$$

thus,

$$\mathcal{W}(M, \Lambda_{N_f,N_c}, m)|_{\langle M_{N_f N_f}\rangle} = (N_c - N_f + 1)\,\cdot \\ \left(\frac{c_{N_f,N_c}}{N_c - N_f}\right)^{\frac{N_c-N_f}{N_c-N_f+1}} \left(\frac{\Lambda_{N_f-1,N_c}}{\det\hat{M}}\right)^{\frac{1}{(N_c-N_f+1)}}, \tag{137}$$

and we obtain the relation

$$\left(\frac{c_{N_f,N_c}}{N_c - N_f}\right)^{N_c-N_f} = \left(\frac{c_{N_f-1,N_c}}{N_c - N_f + 1}\right)^{N_c-N_f+1}. \tag{138}$$

Similarly, we can try to obtain another relation between the constants c_{N_f,N_c} for different numbers of colors. To this end we give a large expectation value to $M_{N_f N_f}$ with respect to the expectation values of $\hat{M}$. Then below the scale $\langle M_{N_f N_f}\rangle$ we have SQCD with $N_c - 1$ colors and N_f flavors. Following the same strategy as before we find that $c_{N_f-1,N_c-1} = c_{N_c,N_f}$. It means that $c_{N_c,N_f} = c_{N_f-N_c}$, which together with the relation (138) gives

$$c_{N_f,N_c} = (N_c - N_f)c_{1,2}\,. \tag{139}$$

We just have to compute the adimensional constant $c_{1,2}$ of the gauge group $SU(2)$ with one flavor. In this case, or for the general case of $N_f = N_c - 1$, the gauge group is completely higgsed and there are not infrared divergences in the instanton computation. In the weak coupling limit the unique surviving nonperturbative contributions come from the one-instanton sector. A direct instanton calculation reveals that the constant $c_{2,1} \neq 0$ (Affleck et al. 1984) [4].

For $N_f < N_c - 1$ there is an unbroken gauge group $SU(N_c - N_f)$. At scales below the smallest eigenvalue of the matrix $\langle M_{fg}\rangle$ we have a pure super Yang-Mills theory with $N_c - N_f$ colors. This theory is believed to confine with a mass gap given by the gaugino condensate $\langle\lambda\lambda\rangle \neq 0$. Consider the simplest case of $\langle M_{fg}\rangle = \mu^2 \mathbf{1}_{N_f}$. Matching the gauge couplings at scale μ gives $\Lambda_{N_f,N_c}^{3N_c-N_f} = (\det M)\,\Lambda_{0,N_c-N_f}^{3(N_c-N_f)}$, which implies for the effective superpotential

$$\mathcal{W} = (N_c - N_f)\Lambda_{0,N_c-N_f}^3\,. \tag{140}$$

[4] In the $\overline{\mathrm{DR}}$ scheme $c_{2,1} = 1$ (Finnell and Pouliot 1995). If we do not say the contrary, we will work on such a scheme.

On the other hand, the gaugino bilinear $\lambda\lambda$ is the lowest component of the chiral superfield $S = W^\alpha W_\alpha$, which represents the super-glueball operator. The bare gauge coupling τ_0 acts as the source of the operator S. If we differentiate (140) with respect to $\ln \Lambda^{3(N_c-N_f)}$ we obtain the gaugino condensate

$$\langle\lambda\lambda\rangle = \Lambda^3_{0,N_c-N_f}. \tag{141}$$

In fact, following the 'integrating in' procedure (Intriligator et al. 1994), (Intriligator 1994) we would obtain the Veneziano-Yankielowicz effective Lagrangian (Veneziano and Yankielowicz 1982).

It is not possible to extend the Afleck-Dine-Seiberg's superpotential to the case of $N_f \geq N_c$. For these values the quantum corrections do not lift the flat directions, and we still have a moduli space which may be different from the classical one. This is the case of $N_f = N_c$.

13 The vacuum structure of SQCD with $N_f = N_c$

13.1 A quantum modified moduli space

For $N_f = N_c$, the classical moduli space is spanned by the gauge singlet operators M_{fg}, B and $\tilde{B}$ subject to the constraint $\det M - \tilde{B}B = 0$. At quantum level, instanton effects could change the classical constraint to

$$\det M - \tilde{B}B = \Lambda^{2N_c}, \tag{142}$$

since $\Lambda^{2N_c} \sim e^{-8\pi/g^2+i\theta}$ corresponds to the one-instanton factor, it has the right dimensions, and the operators $(Q_f, \tilde{Q}_f)$ have R_{AF} charge zero.

To check if the quantum correction (142) really takes place, add a mass term for the quarks. The unique possible holomorphic term with R_{AF} charge two that can be generated with the variables $(M_{fg}, B, \tilde{B}, \Lambda, m)$ is

$$\mathcal{W} = \operatorname{tr} mM. \tag{143}$$

Imagine now that the N_c-flavor is much heavier, with bare mass m, than the $N_c - 1$ other ones, with bare mass matrix $\hat{m}$. The degree of freedom $M_{N_cN_c}$ is given by the constraint. Locate at $B = \tilde{B} = M_{fN_c} = 0$. By equation (130) we know that the $(N_c - 1) \times (N_c - 1)$ matrix $\hat{M}$ is determined to be

$$\hat{M} = \left(\Lambda^{2N_c+1}_{N_c-1,N_c} \det\hat{m}\right)^{\frac{1}{N_c}} \hat{m}^{-1}, \tag{144}$$

which has a non-zero determinant. It indicates that the constraint (142) is really generated at quantum level (Seiberg 1994a). As a final check, consider the simplest situation of $N_c - 1$ massless flavors. When we use the constraint (142) to express $M_{N_cN_c}$ as function of $\det\hat{M}$ we obtain

$$\mathcal{W} = \frac{m\Lambda^{2N_c}}{\det\hat{M}}, \tag{145}$$

the Afleck-Dine-Seiberg's superpotential for $N_f = N_c - 1$ massless flavors.

Far from the origin of the moduli field space we are at weak coupling and the quantum moduli space given by the constraint (142) looks like the classical moduli space (124). But far from the origin of order Λ, the one-instanton sector is sufficiently strong to change significantly the vacuum structure. Observe that the classically allowed point $M = B = \tilde{B} = 0$ is not a point of the quantum moduli space and the gluons never become massless.

13.2 Patterns of spontaneous symmetry breaking and 't Hooft's anomaly matching conditions

Our global symmetries are $SU(N_f)_L \times SU(N_f)_R \times U(1)_B \times U(1)_{AF}$. Since for $N_f = N_c$ the super-quarks are neutral with respect to the non-anomalous symmetry $U(1)_{AF}$, it is never spontaneously broken. The other symmetries present different patterns of symmetry breakings depending on which point of the moduli space the vacuum is located [5].

For instance, the point

$$M = \Lambda^2 \mathbf{1}_{N_f}, \quad B = \tilde{B} = 0, \tag{146}$$

suggests the spontaneous symmetry breaking

$$\begin{aligned} &SU(N_f)_L \times SU(N_f)_R \times U(1)_B \times U(1)_{AF} \\ &\longrightarrow SU(N_f)_V \times U(1)_B \times U(1)_{AF}, \end{aligned} \tag{147}$$

with $SU(N_f)_V$ the diagonal part of $SU(N_f)_L \times SU(N_f)_R$. To check it, the unbroken symmetries must satisfy the 't Hooft's anomaly matching conditions ('t Hooft 1994).

With respect to the unbroken symmetries the quantum numbers of the elementary and composite massless fermions, at high and low energy respectively, are

	$SU(N_f)_V$	$U(1)_B$	$U(1)_{AF}$
λ	$\mathbf{1}$	0	1
ψ_q	$\mathbf{N_f}$	1	-1
$\psi_{\tilde{q}}$	$\overline{\mathbf{N_f}}$	-1	-1
ψ_M	$\mathbf{N_f^2 - 1}$	0	-1
ψ_B	$\mathbf{1}$	N_f	-1
$\psi_{\tilde{B}}$	$\mathbf{1}$	$-N_f$	-1

[5] Different patterns of symmetry breakings have also been observed in softly broken $N = 2$ SQCD (Álvarez-Gaume et al. 1997).

Observe there are only $N_f^2 - 1$ independent meson fields, arranged in the adjoint of $SU(N_f)_V$, since the constraint (142) eliminates one of them. There are $N_f^2 - 1$ gluinos and N_f extra components for each quark ψ_q and anti-quark $\psi_{\tilde{q}}$ because of the gauge group $SU(N_c)$. The anomaly coefficients are:

triangles	high energy	low energy
$SU(N_f)^2 \times U(1)_{AF}$	$-2N_f T(\mathbf{N_f})$	$-T(\mathbf{N_f^2 - 1})$
$U(1)^3_{AF}$	$-2N_f^2 + (N_f^2 - 1)$	$-(N_f^2 - 1) - 2$
$U(1)^2_B \times U(1)_{AF}$	$-N_f^2 - N_f^2$	$-2N_f^2$
$\mathrm{tr}\, U(1)_{AF}$	$-2N_f^2 + N_f^2 - 1$	$-(N_f^2 - 1) - 2$

The constants $T(R)$ are defined by $\mathrm{tr}(T^a T^a) = T(R)\delta^{ab}$, with T^a in the representation R of the group $SU(N)$. For the fundamental representation, $T(\mathbf{N}) = 1/2$. For the adjoint representation, $T(\mathbf{N^2 - 1}) = N$. The coefficient of $\mathrm{tr}\, U(1)_{AF}$ corresponds to the gravitational anomaly. One can check that all the anomalies match perfectly, supporting the spontaneous symmetry breaking pattern of (147).

The quantum moduli space of $N_f = N_c$ allows another particular point with a quite different breaking pattern. It is:

$$M = 0, \quad B = -\tilde{B} = \Lambda^{N_c}. \tag{148}$$

At this point, only the vectorial baryon symmetry is broken, all the chiral symmetries $SU(N_f)_L \times SU(N_f)_R \times U(1)_{AF}$ remain unbroken. We check this pattern with the help of the 't Hooft's anomaly matching conditions again. In this case we have the quantum numbers:

	$SU(N_f)_L$	$SU(N_f)_R$	$U(1)_{AF}$
λ	$\mathbf{1}$	$\mathbf{1}$	1
ψ_q	$\mathbf{N_f}$	$\mathbf{1}$	-1
$\psi_{\tilde{q}}$	$\mathbf{1}$	$\overline{\mathbf{N_f}}$	-1
ψ_M	$\mathbf{N_f}$	$\overline{\mathbf{N_f}}$	-1
ψ_B	$\mathbf{1}$	$\mathbf{1}$	-1
$\psi_{\tilde{B}}$	$\mathbf{1}$	$\mathbf{1}$	-1

and the anomaly coefficients are:

triangles	high energy	low energy
$SU(N_f)^3_L$	$N_f C_3$	$N_f C_3$
$SU(N_f)^3_R$	$N_f C_3$	$N_f C_3$
$SU(N_f)^2 \times U(1)_{AF}$	$-N_f T(\mathbf{N_f})$	$-N_f T(\mathbf{N_f})$
$U(1)^3_{AF}$	$-2N_f^2 + N_f^2 - 1$	$-N_f^2 - 1$

where C_3 is defined by $\mathrm{tr}(T^a\{T^b,T^c\}) = C_3 d^{abc}$, with T^a in the fundamental representation of $SU(N_f)$. Because of the constraint (142) there is only one independent baryonic degree of freedom. The anomaly coefficients match perfectly.

14 The vacuum structure of SQCD with $N_f = N_c + 1$

14.1 The quantum moduli space

First we consider if the classical constraints:

$$M_{fg}B^g = M_{fg}\tilde{B}^f = 0, \tag{149}$$

$$\mathrm{det}M(M^{-1})^{fg} - B^f\tilde{B}^g = 0 \tag{150}$$

are modified quantum mechanically. For $N_f = N_c + 1$ the quark multiplets $(Q_f, \tilde{Q}_f)$ have R_{AF} charge equal to $1/N_f$. The mass matrix breaks the $U(1)_{AF}$ symmetry with a charge of $2 - 2/N_f = 2N_c/N_f$. It is exactly the charge $U(1)_{AF}$ of equation (150). On the other hand, the instanton factor Λ^{2N_c-1} supplies the right dimensionality. Then, there is the possibility that the classical constraint (150) is modified by nonperturbative contributions to

$$\mathrm{det}M(M^{-1})^{fg} - B^f\tilde{B}^g = \Lambda^{2N_c-1}m^{fg}. \tag{151}$$

On the other hand, one can see that the classical constraints (149) do not admit modification. Then if $M \neq 0$ we have $B^f = \tilde{B}^g = 0$. Using (130), we obtain

$$\mathrm{det}M(M^{-1})^{fg} = \Lambda^{2N_c-1}m^{fg}\,, \tag{152}$$

and the quantum modification (151) really takes place (Seiberg 1994a).

14.2 S-confinement

In the massless limit $m^{fg} \to 0$, (149) and (150) are satisfied at the quantum level. It means that the origin of field space, $M = B = \tilde{B} = 0$, is an allowed point of the quantum moduli space. On such a point, there is no spontaneous symmetry breaking at all. We use the 't Hooft's anomaly matching conditions to check it. The quantum numbers of the massless fermions at high and low energy are:

	$SU(N_f)_L$	$SU(N_f)_R$	$U(1)_B$	$U(1)_{AF}$
λ	$\mathbf{1}$	$\mathbf{1}$	0	1
ψ_q	$\mathbf{N_f}$	$\mathbf{1}$	1	$\frac{1}{N_f}-1$
$\psi_{\tilde{q}}$	$\mathbf{1}$	$\overline{\mathbf{N_f}}$	-1	$\frac{1}{N_f}-1$
ψ_M	$\mathbf{N_f}$	$\overline{\mathbf{N_f}}$	0	$\frac{2}{N_f}-1$
ψ_B	$\overline{\mathbf{N_f}}$	$\mathbf{1}$	N_f-1	$-\frac{1}{N_f}$
$\psi_{\tilde{B}}$	$\mathbf{1}$	$\mathbf{N_f}$	$1-N_f$	$-\frac{1}{N_f}$

and the anomaly coefficients are:

triangles	high energy	low energy
$SU(N_f)^3$	$N_c C_3$	$N_f C_3 + \overline{C}_3$
$SU(N_f)^2 \times U(1)_{AF}$	$N_c T(\mathbf{N_f})(-\frac{N_c}{N_f})$	$N_f T(\mathbf{N_f})(\frac{2}{N_f}-1) + T(\mathbf{N_f})(-\frac{1}{N_f})$
$U(1)_B^2 \times U(1)_{AF}$	$2N_c N_f(-\frac{N_c}{N_f})$	$2N_f N_c^2(-\frac{1}{N_f})$
$U(1)_{AF}^3$	$(N_c^2-1) +2N_f N_c(-\frac{N_c}{N_f})^3$	$N_f^2(\frac{2}{N_f}-1)^3 +2N_f(-\frac{1}{N_f})^3$
$\mathrm{tr}U(1)_{AF}$	$(N_c^2-1) +2N_f N_c(-\frac{N_c}{N_f})$	$N_f^2(\frac{2}{N_f}-1) +2N_f(-\frac{1}{N_f})$

with complete agreement. Hence, at the origin of field space we have massless mesons and baryons, and the full global symmetry is manifest. It is a singular point, with the number of massless degrees of freedom larger than the dimensionality of the space of vacua. As we move along the moduli space away from the origin, the 'extra' fields become massive and the massless fluctuations match with the dimensionality of the moduli space. As we are in a Higgs/confining phase, there should be a smooth connection of the dynamics at the origin of field space with the one away from it. This dynamics must be given by some nonperturbative superpotential of mesons and baryons. A theory with the previous characteristics is called s-confining.

There is a unique effective superpotential yielding all the constraints (Seiberg 1994a),

$$\mathcal{W} = \frac{1}{\Lambda^{2N_f-3}}(\tilde{B}^g M_{gf} B^f - \det M)\,, \tag{153}$$

it satisfies:

i) Invariance under all the symmetries.

ii) The equations of motion $\partial\mathcal{W}/\partial M = \partial\mathcal{W}/\partial B = \partial\mathcal{W}/\partial\tilde{B} = 0$ give the constraints (149, 150).
iii) At the origin all the fields are massless.
iv) Adding the bare term $\mathrm{tr}\,(mM) + b_f B^f + \tilde{b}_f \tilde{B}^f$ we recover the $N_f < N_c + 1$ results.

15 Seiberg's duality

15.1 The dual SQCD

If we try to extend the same view of $SU(N_c)$ SQCD for the case of $N_f > N_c+1$, *i.e.* as being in a Higgs/confining phase with the vacuum structure determined by meson and baryon operators satisfying the corresponding classical constraints, to the case of $N_f > N_c + 1$ (it is not possible to modify the classical constraints for $N_f > N_c + 1$), we obtain inconsistencies. It is not possible to generate a superpotential yielding to the constraints, and the 't Hooft's anomaly matching conditions are not satisfied. It indicates that for $N_f > N_c+1$ the Higgs/confining description of SQCD at large distances in terms of just M, B and $\tilde{B}$ is no longer valid.

For $N_f > N_c + 1$, Seiberg conjectured (Seiberg 1995) that the infrared limit of SQCD with N_f flavors admits a dual description in terms of an $N = 1$ super Yang-Mills gauge theory with $\tilde{N}_c = N_f - N_c$ number of colors, N_f flavors D^f and $\tilde{D}^f$ in the fundamental and anti-fundamental representations of $SU(N_f - N_c)$ respectively, and N_f^2 gauge singlet chiral superfields $M_{gf}^{(m)}$. The fields $M_{gf}^{(m)}$ couple to D_f and $\tilde{D}_f$ through the relevant bare superpotential

$$\mathcal{W} = M_{gf}^{(m)} \tilde{D}^g D^f . \tag{154}$$

If both theories are going to describe the same physics at large distances, we must be able to give a prescription of the gauge invariant operators M_{gf}, $B^{f_1 \cdots f_{\tilde{N}_c}}$ and $\tilde{B}^{f_1 \cdots f_{\tilde{N}_c}}$ in terms of the dual microscopic operators $(D^f, \tilde{D}^f)$ and $M_{gf}^{(m)}$. The simplest identification is:

$$\begin{aligned} M_{gf} &= \mu M_{gf}^{(m)} , \\ B^{f_1 \cdots f_{\tilde{N}_c}} &= D^{f_1} \dots D^{f_{\tilde{N}_c}} , \\ \tilde{B}^{f_1 \cdots f_{\tilde{N}_c}} &= \tilde{D}^{f_1} \dots \tilde{D}^{f_{\tilde{N}_c}} . \end{aligned} \tag{155}$$

In the baryon operators the $SU(\tilde{N}_c)$ color indices of $(D^f, \tilde{D}^f)$ are contracted with the $\tilde{N}_c$ antisymmetric tensor. The scale μ is introduced because the dimension of the bare operator $M_{gf}^{(m)}$, derived from (154), is one. This mass scale relates the intrinsic scales Λ and $\tilde{\Lambda}$ of the $SU(N_c)$ and $SU(\tilde{N}_c)$ gauge theories through the equation

$$\Lambda^{3N_c - N_f} \tilde{\Lambda}^{3\tilde{N}_c - N_f} = (-1)^{N_f - N_c} \mu^{N_f} . \tag{156}$$

We see that a strongly coupled $SU(N_c)$ gauge theory corresponds to a weakly coupled $SU(\tilde{N}_c)$ gauge theory, in analogy with the electric-magnetic duality. From this analogy, we call the $SU(N_c)$ gauge theory the electric one, and the $SU(\tilde{N}_c)$ gauge theory the magnetic one.

Both theories must have the same global symmetries. The mapping (155) give the quantum numbers of the magnetic degrees of freedom. Once more, 't Hooft's anomaly matching conditions for the electric and magnetic theories give a non-trivial check of (155). In the following table we write the quantum numbers for the fermions of the magnetic theory:

	$SU(N_f)_L$	$SU(N_f)_R$	$U(1)_B$	$U(1)_{AF}$
$\tilde{\lambda}$	$\mathbf{1}$	$\mathbf{1}$	0	1
ψ_d	$\overline{\mathbf{N_f}}$	$\mathbf{1}$	$\frac{N_c}{\tilde{N}_c}$	$\frac{\tilde{N}_c}{N_f}$
$\psi_{\bar{d}}$	$\mathbf{1}$	$\mathbf{N_f}$	$-\frac{N_c}{\tilde{N}_c}$	$\frac{\tilde{N}_c}{N_f}$
ψ_m	$\mathbf{N_f}$	$\overline{\mathbf{N_f}}$	0	$1-2\frac{N_c}{N_f}$

with $\tilde{\lambda}$ the magnetic gluinos. One can check that both theories give the same anomalies.

It can be verified that applying duality again we obtain the original theory.

15.2 $N_c + 1 < N_f \leq 3N_c/2$. An infrared free non-abelian Coulomb phase

In this range of N_f the magnetic theory is not asymptotically free and has a trivial infrared fixed point. At large distances the physical effective degrees of freedom are the fields D^f, D^f, M_{gf} and the massless super-gluons of the gauge group $SU(N_f - N_c)$. At the origin of field space we are in an infrared free non-abelian Coulomb phase, with a complete screening of its charges in the infrared limit. Observe that the strongly coupled electric theory is weakly coupled in terms of the magnetic degrees of freedom, according to the philosophy of the electric-magnetic duality.

15.3 $3N_c/2 < N_f < 3N_c$. An interacting non-abelian Coulomb phase

As in QCD, the $N = 1$ SQCD has a Banks-Zaks fixed point (Banks and Zaks 1982) for $N_c, N_f \to \infty$, when $N_f/N_c = 3-\epsilon$ with $\epsilon \ll 1$. We still have asymptotic freedom and under the renormalization group transformations the theory flows from the ultraviolet free fixed point to an infrared fixed point with a non-zero finite value of the gauge coupling constant. If there is an interacting superconformal gauge theory the scaling dimensions of some gauge invariant operators should be non-trivial.

The superconformal invariance includes an R-symmetry, from which the scaling dimensions of the operators satisfy the lower bound

$$D \geq \frac{3}{2}|R| \tag{157}$$

with equality for chiral and anti-chiral operators. The R-current is in the same supermultiplet as the energy-momentum tensor, whose trace anomaly is zero on the fixed point. It implies that there the R-symmetry must be the anomaly-free $U(1)_{AF}$ symmetry. It gives the scaling dimensions of the following chiral operators:

$$D(M) = \frac{3}{2} R_{AF}(M) = 3\frac{N_f - N_c}{N_f}, \tag{158}$$

$$D(B) = D(\tilde{B}) = \frac{3}{2}\frac{N_c(N_f - N_c)}{N_f}. \tag{159}$$

Unitarity restricts the scaling dimensions of the gauge invariant operators to be $D \geq 1$. If $D = 1$, the corresponding operator $\mathcal{O}$ satisfies the free equation of motion $\partial^2 \mathcal{O} = 0$. If $D > 1$, there are non-trivial interactions between the operators.

For the range $3N_c/2 < N_f < 3N_c$, the gauge invariant chiral operators M, B and $\tilde{B}$ satisfy the unitarity constraint with $D > 1$. Seiberg conjectured the existence of such a non-trivial fixed point for any value of $3N_c/2 < N_f < 3N_c$, at least for large N_c.

As $\frac{3}{2}(N_f - N_c) < N_f < 3(N_f - N_c)$, there is also a non-trivial fixed point in the magnetic theory. Seiberg's claim is that both theories flow to the same infrared fixed point (Seiberg 1995).

16 $N = 2$ supersymmetry

16.1 The supersymmetry algebra and its massless representations

The $N = 2$ supersymmetry algebra, without central charge, is

$$\begin{aligned} \{Q_\alpha^{(I)}, \bar{Q}_{\dot{\beta}(J)}\} &= 2(\sigma^\mu)_{\alpha\dot{\beta}} P_\mu \delta^I_J\,, \\ \{Q_\alpha^{(I)}, Q_\beta^{(J)}\} &= 0 \end{aligned} \tag{160}$$

with $I, J = 1, 2$. The algebra (160) has a new symmetry. We can perform unitary rotations of the two supercharges $Q_\alpha^{(I)}$ that do leave the anticommutator relations (160) invariant. We have an $U(2)_R = U(1)_R \times SU(2)_R$ symmetry. The abelian factor $U(1)_R$ corresponds to the familiar R-symmetry of supersymmetric theories that rotate the global phase of the supercharges $Q_\alpha^{(I)}$. With respect to the $SU(2)_R$ group, the supercharges $Q_\alpha^{(I)}$ are in the doublet representation **2**.

As in massless $N = 1$ supersymmetric representations, half of the supercharges are realized as vanishing operators: $Q_2^{(I)} = 0$. We normalize the other two supercharges,

$$a_1^{(I)} = \frac{1}{2\sqrt{E}} Q_1^{(I)} , \tag{161}$$

which are an $SU(2)_R$ doublet. The massless $N = 2$ vector multiplet is a representation constructed from the Clifford vacuum $|1>$, which has helicity $\lambda = 1$ and is an $SU(2)_R$ singlet. From it we obtain two fermionic states, $|1/2>^{(I)} = (a^{(I)})^\dagger |1>$, and a scalar boson $|0> = (a^{(1)})^\dagger (a^{(2)})^\dagger |1>$. After CPT doubling we obtain the $N = 2$ vector multiplet:

$$\begin{pmatrix} & \{\ |1>,\ |-1>_{CPT}\ \} & \\ \{\ |\tfrac{1}{2}>^{(1)},\ |-\tfrac{1}{2}>^{(1)}_{CPT}\ \} & & \{\ |\tfrac{1}{2}>^{(2)},\ |-\tfrac{1}{2}>^{(2)}_{CPT}\ \} \\ & \{\ |0>,\ |0>_{CPT}\ \} & \end{pmatrix} \tag{162}$$

In terms of local fields we have: a vector A_μ (the gauge bosons of some gauge group G, since we consider massless representations), which is $SU(2)_R$ singlet; two Weyl spinors $\lambda^{(I)}$, the gauginos, arranged in an $SU(2)_R$ doublet, and a complex scalar ϕ, playing the role of the Higgs, a singlet of $SU(2)_R$ but in the adjoint of the gauge group G. These fields arrange as

$$\begin{pmatrix} & A_\mu & \\ & \swarrow & \\ \lambda^{(1)} & & \lambda^{(2)} \\ & \swarrow & \\ & \phi & \end{pmatrix} \tag{163}$$

where the arrows indicate the action of the supercharge $\overline{Q}^{(1)}_{\dot\alpha}$. We can use a manifest $N = 1$ supersymmetry representation taking into account that the $N = 2$ vector multiplet is composed of an $N = 1$ vector multiplet $W_\alpha = (A_\mu, \lambda^{(1)})$ and an $N = 1$ chiral multiplet $\Phi = (\phi, \lambda^{(2)})$.

The massless $N = 2$ hypermultiplet is a representation constructed from a Clifford vacuum $|1/2>$, which is an $SU(2)_R$ singlet. The action of the two grassmanian operators a^I_α seems to produce the same particle content as the $N = 1$ chiral multiplet, but $|1/2> = |1/2, \mathbf{R}>$ is usually in some non-trivial representation $\mathbf{R}$ of a gauge group G. As $\mathbf{R} \to \overline{\mathbf{R}}$ under a CPT transformation, it forces to make the CPT doubling, and the $N = 2$ hypermultiplet is built from two $N = 1$ chiral multiplets in complex conjugate gauge group representations:

$$\begin{pmatrix} & \{\ |\tfrac{1}{2}, \mathbf{R}>,\ |-\tfrac{1}{2}, \overline{\mathbf{R}}>_{CPT}\ \} & \\ \{\ |0, \mathbf{R}>^{(1)},\ |0, \overline{\mathbf{R}}>^{(1)}_{CPT}\} & & \{\ |0, \mathbf{R}>^{(2)},\ |0, \overline{\mathbf{R}}>^{(2)}_{CPT}\} \\ & \{\ |-\tfrac{1}{2}, \mathbf{R}>,\ |\tfrac{1}{2}, \overline{\mathbf{R}}>_{CPT}\ \} & \end{pmatrix} \tag{164}$$

which represents the local fields

$$\begin{pmatrix} & \psi_q & \\ & \swarrow & \\ q & & \tilde{q}^\dagger \\ & \swarrow & \\ & \overline{\psi}_{\tilde{q}} & \end{pmatrix} \tag{165}$$

with the complex scalar fields $(q, \tilde{q}^\dagger)$ in a doublet representation of $SU(2)_R$. In terms of $N = 1$ superfields we have one chiral superfield $Q = (q, \psi_q)$ in gauge representation $\mathbf{R}$ and another chiral superfield $\tilde{Q} = (\tilde{q}, \tilde{\psi}_{\tilde{q}})$ in gauge representation $\overline{\mathbf{R}}$. All the fields in the hypermultiplet have spin $\leq 1/2$. Because of the CPT doubling, the matter content of extended supersymmetry $(N > 1)$ is always in vectorial representations of the gauge group.

16.2 The central charge and massive short representations

As shown by Haag, Lapuszanski and Sohnius (Haag et al. 1975), the $N = 2$ supersymmetry algebra admits a central extension:

$$\begin{aligned} \{Q^a_\alpha, Q^b_\beta\} &= 2\sqrt{2}\epsilon_{\alpha\beta}\epsilon^{ab} Z \,, \\ \{\bar{Q}_{\dot{\alpha}a}, \bar{Q}_{\dot{\beta}b}\} &= 2\sqrt{2}\epsilon_{\dot{\alpha}\dot{\beta}}\epsilon_{ab}\bar{Z} \,. \end{aligned} \tag{166}$$

Since Z commutes with all the generators, we can fix it to be the eigenvalue for the given representation. Now, let us define:

$$a_\alpha = \frac{1}{2}\{Q^1_\alpha + \epsilon_{\alpha\beta}(Q^2_\beta)^\dagger\} \,, \tag{167}$$

$$b_\alpha = \frac{1}{2}\{Q^1_\alpha - \epsilon_{\alpha\beta}(Q^2_\beta)^\dagger\} \,. \tag{168}$$

Then, in the rest frame, the $N = 2$ supersymmetry algebra reduces to

$$\{a_\alpha, a^\dagger_\beta\} = \delta_{\alpha\beta}(\mathcal{M} + \sqrt{2}Z) \,, \tag{169}$$

$$\{b_\alpha, b^\dagger_\beta\} = \delta_{\alpha\beta}(\mathcal{M} - \sqrt{2}Z) \,, \tag{170}$$

with all other anticommutators vanishing. Since all physical states have positive definite norm, it follows that for massless states, the central charge is trivially realized ($i.e., Z = 0$), as we stated before. For massive states, this leads to a bound on the mass $\mathcal{M} \geq \sqrt{2}|Z|$. When $\mathcal{M} = \sqrt{2}|Z|$, the operators in (170) are trivially realized and the algebra resembles the massless case. The dimension of the representation is greatly reduced. For example, a reduced massive $N = 2$ multiplet has the same number of states as a massless $N = 2$ multiplet. Thus the representations of the $N = 2$ algebra with a central charge can be classified as either long multiplets (when $\mathcal{M} > \sqrt{2}|Z|$) or short multiplets (when $\mathcal{M} = \sqrt{2}|Z|$).

From (170) it is clear that the BPS states (Bogomol'nyi 1976), (Prasad and Sommerfield 1975) (which saturate the bound) are annihilated by half of the supersymmetry generators and thus belong to reduced representations of the supersymmetry algebra. An important consequence of this is that, for BPS states, the relationship between their charges and masses is dictated by supersymmetry and does not receive perturbative or non-perturbative corrections in the quantum theory. This is so because a modification of this relation implies that the states no longer belong to a short multiplet. On the other hand, quantum corrections are not expected to generate the extra degrees of freedom needed to convert a short multiplet into a long multiplet. Since there is no other possibility, we conclude that for short multiplets the relation $\mathcal{M} = \sqrt{2}|Z|$ is not modified either perturbatively or nonperturbatively.

17 $N = 2$ $SU(2)$ super Yang-Mills theory in perturbation theory

17.1 The $N = 2$ Lagrangian

The $N = 2$ superspace has two independent chiral spinors $\theta^{(I)}$, $I = 1, 2$. The $N = 2$ vector multiplet can be written in terms of $N = 2$ superspace by the $N = 2$ superfield $\Psi(x, \theta^{(I)})$ subject to the superspace constraints (Gates 1984):

$$\overline{\nabla}^{(I)}_{\dot{\alpha}} \Psi = 0\,, \\ \nabla_{(I)}\nabla_{(J)}\Psi = \epsilon_{IK}\epsilon_{JL}\overline{\nabla}^{(K)}\overline{\nabla}^{(L)}\overline{\Psi}\,, \tag{171}$$

where $\nabla_{(I)\alpha} = D_{(I)\alpha} + \Gamma_{(I)\alpha}$ is the generalized supercovariant derivative of the variable $\theta^{(I)}$, with $\Gamma_{(I)\alpha}$ the superconnection. The $N = 1$ superfields are connected to the $N = 2$ vector superfield through the equations:

$$\Psi|_{\theta^{(2)}=\overline{\theta}^{(2)}=0} = \Phi(x, \theta^{(1)}, \overline{\theta}^{(1)})\,, \\ D_{2\alpha}\Psi|_{\theta^{(2)}=\overline{\theta}^{(2)}=0} = i\sqrt{2}W_\alpha(x, \theta^{(1)}, \overline{\theta}^{1})\,. \tag{172}$$

It results that the renormalizable $N = 2$ super Yang-Mills Lagrangian is

$$\mathcal{L} = \frac{1}{8\pi}\mathrm{Im}\left(\tau \int d^2\theta^{(1)} d^2\theta^{(2)}\; \Psi^a\Psi^a\right) \tag{173}$$

with our old friend $\tau = \theta/2\pi + i4\pi/g^2$. In terms of $N = 1$ superspace, using (171) and (172), with $\theta \equiv \theta^{(1)}$, the Lagrangian is

$$\mathcal{L} = \frac{1}{8\pi}\mathrm{Im}\left(\tau \int d^2\theta\; W^\alpha W_\alpha\right) + \frac{1}{g^2}\int d^2\theta d^2\bar{\theta}\; \Phi^\dagger e^{-2V}\Phi\,. \tag{174}$$

It looks like $N = 1$ $SU(2)$ gauge theory with an adjoint chiral superfield Φ. The point is that the $1/g^2$ normalization in front of the kinetic term of Φ gives $N = 2$

supersymmetry. In fact, when we perform the remaining superspace integral in (174), we obtain a Lagrangian that looks like a Georgi-Glashow model with a complex Higgs triplet and the addition of a Dirac spinor $(\lambda^{(1)}, \bar{\lambda}^{(2)})$ in the adjoint also. This Lagrangian does not have all the gauge invariant renormalizable terms. $N = 2$ supersymmetry restricts the possible terms and gives relations between their couplings, such that at the end there are only the parameters g^2 and θ.

If we apply perturbation theory to the Lagrangian (173) we only have to perform a one loop renormalization. This is an indication that in $N = 2$ supersymmetry, holomorphy is not lost by radiative corrections. The reason is the following: We explained that the multi-loop renormalization of the coupling τ came from the generation of non-holomorphic factors $Z(\mu/\mu_0, g)$ in front of the complete $N = 1$ superspace integrals. At the level of the Lagrangian (174), consider the bare coupling τ_0 at scale μ_0 and integrate out the modes between μ_0 and μ. If we consider only the renormalizable terms, $N = 1$ supersymmetry gives us

$$\mathcal{L}_{ren} = \frac{1}{8\pi}\mathrm{Im}\left(\tau(\mu/\Lambda)\int d^2\theta W^\alpha W_\alpha\right) + Z\left(\frac{\mu}{\mu_0}, g_0\right)\frac{1}{g^2(\frac{\mu}{\Lambda})}\int d^2\theta d^2\bar{\theta} \quad \Phi^\dagger e^{-2V}\Phi, \tag{175}$$

where

$$\tau(\frac{\mu}{\Lambda}) = \frac{2i}{\pi}\ln\frac{\mu}{\Lambda} + \sum_{n=0}^{\infty} c_n\left(\frac{\Lambda}{\mu}\right)^{4n} \tag{176}$$

is the renormalized coupling constant at scale μ. We used the one-loop beta function of $N = 2$ $SU(2)$ gauge theory $b_0 = 4$ and the renormalization group invariant scale $\Lambda \equiv \mu_0 \exp(i\pi\tau_0/2)$. The adimensional constants c_n are the coefficients of the n-instanton contribution $(\Lambda/\mu)^{4n} = \exp(-8\pi n/g^2(\mu) + i\theta(\mu)n)$.

If we compare with the $N = 2$ renormalizable Lagrangian (174) we derive that $Z(\mu/\mu_0, g_0) = 1$. Then, there is no Konishi anomaly and the one-loop renormalization of τ is all there is in perturbation theory.

17.2 The flat direction

Unlike $N = 1$ super Yang-Mills, $N = 2$ super Yang-Mills theory includes a complex scalar ϕ in the adjoint of the gauge group. This scalar plays the role of a Higgs field through the potential derived from the Lagrangian (174),

$$V(\phi, \phi^\dagger) = \frac{1}{2g^2}[\phi^\dagger, \phi]^2 . \tag{177}$$

The supersymmetric minimum is obtained by the solution of

$$[\phi^\dagger, \phi] = 0 , \tag{178}$$

whose solution, up to gauge transformations, is $\phi = a\sigma^3$, with a an arbitrary complex number. This is our flat direction. Along it, the $SU(2)$ gauge group is spontaneously broken to the $U(1)$ subgroup. The $\Psi^{\pm} = \frac{1}{\sqrt{2}}(\Psi^1 \pm i\Psi^2)$ superfield components have $U(1)$ electric charge $Q_e = \pm g$, respectively, and they have the classical squared mass

$$\mathcal{M}_W^2 = 2|a|^2 . \tag{179}$$

The $\mathcal{A} \equiv \Psi^3$ superfield component remains massless. We know that the Lagrangian (173) admits semi-classical dyons with electric charge $Q_e = n_e g + \theta/2\pi$ and magnetic charge $Q_m = (4\pi/g)$, *i.e.*, the points $(1, n_e)$ in the charge lattice. They have the classical squared mass

$$\mathcal{M}^2(1, n_e) = 2|a|^2 |n_e + \tau|^2 . \tag{180}$$

Physical masses are gauge invariant. We can use the gauge invariant parametrization of the moduli space in terms of the chiral superfield

$$U = \mathrm{tr}\Phi^2 , \tag{181}$$

and translate the a-dependence in previous formulae by an u-dependence through the relation $u = \mathrm{tr}\langle\phi^2\rangle$. The classical relation is just $u = a^2/2$.

Then, semi-classical analysis gives $\mathcal{A}$ as the unique light degree of freedom. Only at $u = 0$ the full $SU(2)$ gauge symmetry is restored. How is this picture modified by the nonperturbative corrections? The Seiberg-Witten solution answers this question (Seiberg and Witten 1994a) [6].

18 The low energy effective Lagrangian

The $N = 2$ vector superfield $\mathcal{A}$ is invariant under the unbroken $U(1)$ gauge transformations. At a scale of the order of the $\mathcal{M}_W$ mass, *i.e.*, of the order of $|u|^{1/2}$, the most general $N = 2$ Wilsonian Lagrangian, with two derivatives and four fermions terms, that can be constructed from the light degrees of freedom in $\mathcal{A}$ is

$$\mathcal{L}_{eff} = \frac{1}{4\pi}\mathrm{Im}\left(\int d^2\theta^{(1)} d^2\theta^{(2)} \; \mathcal{F}(\mathcal{A})\right) \tag{182}$$

with $\mathcal{F}$ a holomorphic function of $\mathcal{A}$, called the prepotential. We stress that the unique inputs to equation (182) are $N = 2$ supersymmetry and that $\mathcal{A}$ is a vector multiplet. We derive an immediate consequence of the general form of the effective Lagrangian (182): $N = 2$ supersymmetry prevents the generation of a superpotential for the $N = 1$ chiral superfield of $\mathcal{A}$. It means that the previously derived flat direction, parametrized by the arbitrary value $u = \mathrm{tr}\langle\phi^2\rangle$, is not lifted by nonperturbative corrections.

[6] Some additional reviews on the Seiberg-Witten solution are (Álvarez-Gaumé and Hassan 1997).

In terms of $N = 1$ superspace we have

$$\mathcal{L}_{eff} = \frac{1}{4\pi}\mathrm{Im}\left(\int d^2\theta \frac{1}{2}\tau(A) W^\alpha W_\alpha\right) + \int d^2\theta d^2\bar{\theta}\; K(A,\bar{A}), \tag{183}$$

where

$$\tau(A) = \frac{\partial^2 \mathcal{F}}{\partial A^2}(A), \tag{184}$$

$$K(A,\bar{\mathcal{A}}) = \mathrm{Im}\left(\frac{\partial \mathcal{F}}{\partial A}\overline{\mathcal{A}}\right), \tag{185}$$

and A is the $N = 1$ chiral multiplet of $\mathcal{A}$.

The Wilsonian Lagrangian (183) is an abelian gauge theory defined at some scale of order $\mathcal{M}_W \sim |u|^{1/2}$. Interaction terms come out after the expansion $A = a + \hat{A}$, with a the vacuum expectation value of the Higgs field, and $\hat{A}$ the quantum fluctuations of the chiral superfield. The matching at scale $|u|^{1/2}$ with the high energy $SU(2)$ theory is performed by the renormalization group:

$$\tau(u) = \frac{i}{\pi}\mathrm{ln}\frac{u}{\Lambda^2} + \sum_{n=0}^{\infty} c_n \left(\frac{\Lambda^2}{u}\right)^{2n}. \tag{186}$$

Observe that the phase of the adimensional quotient u/Λ^2 plays the role of the bare θ_0 angle. If we are able to know the relation between the u and a variables, *i.e.*, the function $u(a)$, we can insert it into (186) to obtain $\tau(a)$. Integrating twice in the variable a we obtain the prepotential

$$\mathcal{F}(a) = \frac{i}{2\pi}a^2\mathrm{ln}\frac{a^2}{\Lambda^2} + a^2\sum_{n=1}^{\infty}\mathcal{F}_k\left(\frac{\Lambda}{a}\right)^{4k}. \tag{187}$$

If we look at the terms of the Lagrangian (183) proportional to the adimensional constant $\mathcal{F}_n$, they correspond to the effective interaction terms created by the n-instanton contribution, as expected. For $a \to \infty$, the instanton contributions go to zero. This is an expected result, since at $a \to \infty$ the matching takes place at weak coupling due to asymptotic freedom. In this region perturbation theory is applicable and we can believe the semi-classical relation, $u \sim a^2/2$.

19 BPS bound and duality

The $N = 2$ supersymmetry algebra gives the mass bound

$$\mathcal{M} \geq \sqrt{2}|Z|, \tag{188}$$

with Z the central charge. The origin of the central charge is easy to understand: the supersymmetry charges Q and $\bar{Q}$ are space integrals of local expressions in the fields (the time component of the super-currents). In calculating their anticommutators, one encounters surface terms which are normally neglected. However, in the presence of electric and magnetic charges, these surface terms are non-zero and give rise to a central charge. When one calculates the central charge that arises from the classical Lagrangian (173) one obtains (Witten and Olive 1978)

$$Z = ae(n + m\tau)\,, \tag{189}$$

so that $\mathcal{M} \geq \sqrt{2}|Z|$ coincides with the Bogomol'nyi bound (57).

But the equation (189) is a classical result. The effective Lagrangian (182) includes all the nonperturbative quantum corrections of the higher modes. To get their contribution to the BPS bound, we just have to compute the central charge that is derived from the effective Lagrangian (182). The result is

$$Z(n_m, n_e) = n_e a + n_m a_D\,, \tag{190}$$

for a supermultiplet located in the charge lattice at (n_m, n_e). We have defined the a_D function

$$a_D \equiv \frac{\partial \mathcal{F}}{\partial a}(a)\,. \tag{191}$$

This function plays a crucial role in duality. Observe that under the $SL(2, \mathbf{Z})$ transformation $M = \begin{pmatrix} \alpha & \beta \\ \gamma & \delta \end{pmatrix}$ of the charge lattice,

$$(n_m, n_e) \to (n_m, n_e)M^{-1}\,, \tag{192}$$

the invariance of the central charge demands

$$\begin{pmatrix} a_D \\ a \end{pmatrix} \to M \begin{pmatrix} a_D \\ a \end{pmatrix}\,. \tag{193}$$

Its action on the effective gauge coupling $\tau = \partial a_D / \partial a$ is

$$\tau \to \frac{\alpha\tau + \beta}{\gamma\tau + \delta}\,. \tag{194}$$

The S-transformation, that interchanges electric with magnetic charges, makes

$$\begin{aligned} a_D &\to a \\ a &\to -a_D\,. \end{aligned} \tag{195}$$

Then, a_D is the dual scalar photon, that couples locally with the monopole $(1, 0)$ through the dual gauge coupling $\tau_D = -1/\tau$.

From (184) and (185), we see that $\mathrm{Im}\tau(a)$ is the Kähler metric of the Kähler potential $K(a, \overline{a})$,

$$d^2 s = [\mathrm{Im}\tau(a)]da d\overline{a}\,. \tag{196}$$

Physical constraints demands the metric be positive definite, $\mathrm{Im}\tau > 0$. However, if $\tau(a)$ is globally defined the metric cannot be positive definite as the harmonic function $\mathrm{Im}\tau(a)$ cannot have a minimum. This indicates that the above description of the metric in terms of the variable a must be valid only locally. In the weak coupling region, $|u| \gg |\Lambda|$, where $\tau(a) \sim (2i/\pi)\ln(a/\Lambda)$, we have that $\mathrm{Im}\tau(a) > 0$, but for $a \sim \Lambda$, when the theory is at strong coupling and the non-perturbative effects become important, the perturbative result does not give the correct physical answer. Two things should happen: the instanton corrections must secure the positivity of the metric and physics must be described in terms of a new local variable a'. Which is this new local variable? If we do not want to change the physics, the change of variables must be an isometry of the Kähler metric (196). In terms of the variables (a_D, a) the Kähler metric is

$$d^2s = \mathrm{Im}(da_D d\bar{a}) = -\frac{i}{2}(da_D d\bar{a} - da d\bar{a}_D)\,. \tag{197}$$

The complete isometry group of (197) is $\begin{pmatrix} a_D \\ a \end{pmatrix} \to M \begin{pmatrix} a_D \\ a \end{pmatrix} + \begin{pmatrix} p \\ q \end{pmatrix}$ with $M \in SL(2,\mathbf{R})$ and $p, q \in \mathbf{R}$. But the invariance of the central charge puts $p = q = 0$ [7] and the Dirac quantization condition restricts $M \in SL(2,\mathbf{Z})$. We arrive at an important result: in some region of the moduli space we have to perform an electric-magnetic duality transformation.

20 Singularities in the moduli space

As $\mathrm{Im}\tau$ cannot be globally defined on the u plane, there must be some singularities u_i indicating the multivaluedness of $\tau(u)$. If we perform a loop around a singularity u_i, there is a non-trivial monodromy action M_i on $\tau(u)$. This action should be an isometry of the Kähler metric, if we do not want to change the physics. It implies that the monodromies M_i are elements of the $SL(2,\mathbf{Z})$ group.

In fact, we have found already one non-trivial monodromy because of the perturbative contributions. The multivalued logarithmic dependence of τ gives the monodromy. For $u \sim \infty$, $\tau \sim (i/\pi)\ln(u/\Lambda^2)$. In that region, the loop $u \to e^{2\pi i}u$ applied on $\tau(u)$ gives

$$\tau \to \tau - 2\,. \tag{198}$$

Its associated monodromy is

$$M_\infty = \begin{pmatrix} -1 & 2 \\ 0 & -1 \end{pmatrix} = PT^{-2}\,, \tag{199}$$

which acts on the variables (a_D, a) as

$$a_D \to -a_D + 2a \tag{200}$$

$$a \to -a\,. \tag{201}$$

[7] In $N = 2$ SQCD with massive matter, the central charge allows to have $p, q \neq 0$ (Seiberg and Witten 1994b).

As it should be, the monodromy is a symmetry of the theory. T^{-2} just shifts the θ parameter by -4π, and P is the action of the Weyl subgroup of the $SU(2)$ gauge group. Then, the monodromy at infinity, M_∞, leaves the a variable invariant (up to a gauge transformation).

The monodromy at infinity means that there must be some singularity in the u plane. How many singularities? We know that the anomalous $U(1)_R$ symmetry is broken by instantons, and that there is an unbroken $\mathbf{Z}_8$ subgroup because the one-instanton sector has eight fermionic zero modes. The $U = \mathrm{tr}\,\Phi^2$ operator has R-charge four. It means that the $u \to -u$ symmetry is spontaneously broken, leading to equivalent physical vacua. Then, if u_0 is a singular point, $-u_0$ must be also another singular point.

Let us assume that there is only one singularity. If this were the situation, the monodromy group would be abelian, generated only by the monodromy at infinity. From the monodromy invariance of the variable a under M_∞, we would have that a is a good variable to describe the physics of the whole moduli space. This is in contradiction with the holomorphy of $\tau(a)$.

Seiberg and Witten made the assumption that there are only two singularities, which they normalized to be $u_1 = \Lambda^2$ and $u_2 = -\Lambda^2$. This assumption leads to a unique and elegant solution that passes many tests.

21 The physical interpretation of the singularities

The most natural physical interpretation of singularities in the u plane is that some additional massless particles appear at the singular point $u = u_0$.

The particles will arrange in some $N = 2$ supermultiplet and will be labeled by some quantum numbers (n_m, n_e). If the massless particle is purely electric, the Bogomolńyi bound implies $a(u_0) = 0$. It would mean that the W-bosons become massless at u_0 and the whole $SU(2)$ gauge symmetry is restored there. It would imply the existence of a non-abelian infrared fixed point with $\langle \mathrm{tr}\phi^2 \rangle \neq 0$. By conformal invariance, the scaling dimension of the operator $\mathrm{tr}\phi^2$ at this infrared fixed point would have to be zero, *i.e.*, it would have to be the identity operator. It is not possible since $\mathrm{tr}\phi^2$ is odd under a global symmetry.

Then, the particles that become massless at the singular point u_0 are arranged in an $N = 2$ supermultiplet of spin $\leq 1/2$. The possibilities are severely restricted by the structure of $N = 2$ supersymmetry: the multiplet must be an hypermultiplet that saturates the BPS bound. As we have derived that we should have $a \neq 0$ for all the points of the moduli space, the singular BPS state must have a non-zero magnetic charge.

Near its associated singularity, the light $N = 2$ hypermultiplet is a relevant degree of freedom to be considered in the low energy Lagrangian. The coupling to the massless photon of the unbroken $U(1)$ gauge symmetry has to be local. Therefore, we apply a duality transformation to describe the relevant degree of freedom (n_m, n_e) as a purely electric state $(0, 1)$,

$$(0,1) = (n_m, n_e) N^{-1} \,, \tag{202}$$

with N the appropriate $SL(2,\mathbf{Z})$ transformation. The dual variables are the good local variables near the u_0 singularity. It implies that the monodromy matrix must leave invariant the singular state (n_m, n_e). This constraint plus the $U(1)$ β-function give the monodromy matrix

$$M(n_m, n_e) = \begin{pmatrix} 1+2n_m n_e & 2n_e^2 \\ -2n_m^2 & 1-2n_e n_m \end{pmatrix} . \tag{203}$$

In fact, in terms of the local variables,

$$\begin{pmatrix} a'_D \\ a' \end{pmatrix} = N \begin{pmatrix} a_D \\ a \end{pmatrix} , \tag{204}$$

the monodromy matrix is just T^2. This result can be understood as follows: the renormalizable part of the low energy Lagrangian is just $N=2$ QED with one light hypermultiplet with mass $\sqrt{2}|a'| = \sqrt{2}|n_m a_D + n_e a|$. It has a trivial infrared fixed point, and the theory is weakly coupled at large distances. Perturbation theory gives

$$\tau' \simeq -\frac{i}{\pi} \ln a' . \tag{205}$$

On the other hand, by the monodromy invariance of a', we have $a'(u) \simeq c_0(u - u_0)$, this gives the monodromy matrix T^2: $\tau' \to \tau' + 2$.

With all the monodromies taken in the counter-clockwise direction, and the monodromy base point chosen in the negative imaginary part of the complex u plane, we have the topological constraint

$$M_{-\Lambda^2} M_{\Lambda^2} = M_\infty . \tag{206}$$

If we use the expression (203) for the monodromies $M_{\pm\Lambda^2}$ and that $M_\infty = PT^{-2}$, (206) implies that the magnetic charge of the singular states must be ± 1. Then, they exist semi-classically and are continuously connected with the weak coupling region. Moreover, if the state $(1, n_e)$ becomes massless at $u = \Lambda^2$, then (206) gives the massless state $(1, n_e - 1)$ at $u = -\Lambda^2$. It is consistent with the action of the spontaneously broken symmetry $u \to -u$, since by the expression of $\tau(u)$ in (186) we have that $\theta_{eff}(-\Lambda^2) = 2\pi \mathrm{Re}(\tau(-\Lambda^2)) = 2\pi$, and by the Witten effect gives the same physical electric charge to the massless states at $u = \pm\Lambda^2$.

Seiberg and Witten took the simplest solution: a purely magnetic monopole $(1,0)$ [8] becomes massless at $u = \Lambda^2$. With our chosen monodromy base point, the state with quantum numbers $(1,-1)$ has vanishing mass at $u = -\Lambda^2$.

[8] Observe that by Witten's effect, the shift $\theta \to \theta + 2\pi n$ transforms $(1,0) \to (1,n)$. There is a complete democracy between the semi-classical stable dyons.

22 The Seiberg-Witten solution

22.1 The inputs

After this long preparation, we can present the solution of the model. The moduli space is the compactified u-plane punctured at $u = \Lambda^2, -\Lambda^2, \infty$. These singular points generate the monodromies:

$$\begin{aligned} M_{\Lambda^2} &= \begin{pmatrix} 1 & 0 \\ -2 & 1 \end{pmatrix}, \\ M_{-\Lambda^2} &= \begin{pmatrix} -1 & 2 \\ -2 & 3 \end{pmatrix}, \\ M_{\infty} &= \begin{pmatrix} -1 & 2 \\ 0 & -1 \end{pmatrix}, \end{aligned} \tag{207}$$

which act on the holomorphic function $\tau(u)$ by the corresponding modular transformations. Physically, the function $\tau(u)$ is the effective coupling at the vacuum u and its asymptotic behaviour near the punctured points $u = \Lambda^2, -\Lambda^2, \infty$, is known.

22.2 The geometric picture

A torus is a two-dimensional compact Riemann surface of genus one. Topologically it can be described by a two-dimensional lattice with complex periods ω and ω_D. The construction is the following: a point z in the complex plane is identified with the points $z+\omega$ and $z+\omega_D$ (with the convention $\mathrm{Im}(\omega_D/\omega) > 0$), to get the topology of a torus. Then, the $SL(2, \mathbf{Z})$ transformations

$$\begin{pmatrix} \omega_D \\ \omega \end{pmatrix} \to M \begin{pmatrix} \omega_D \\ \omega \end{pmatrix} \tag{208}$$

leave invariant the torus. If we rescale the lattice with $1/\omega$, the torus is characterized just by the modulus

$$\tau \equiv \frac{\omega_D}{\omega},$$

up to $SL(2, \mathbf{Z})$ transformations,

$$\tau \sim \frac{\alpha\tau + \beta}{\gamma\tau + \delta}.$$

Algebraically the torus can be described by a complex elliptic curve

$$y^2 = 4(x - e_1)(x - e_2)(x - e_3). \tag{209}$$

The toric structure arises because of the two Riemman sheets in the x plane joined through the two branch cuts going from e_1 to e_2 and e_3 to infinity:

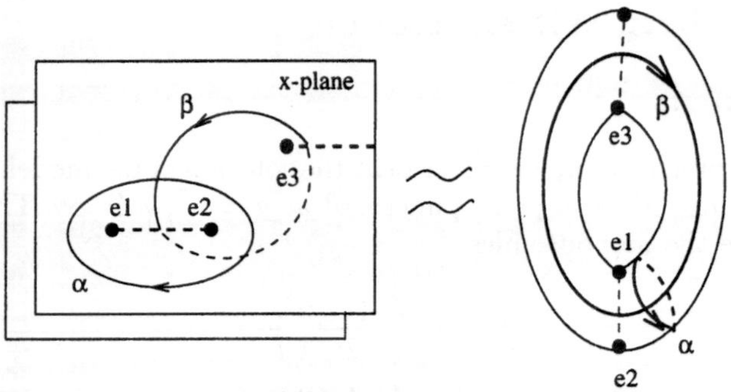

The lattice periods are obtained by integrating the abelian differential of first kind dx/y along the two homologically non-trivial one-cycles α and β, with intersection number $\beta \cdot \alpha = 1$,

$$\omega_D = \oint_\beta \frac{dx}{y}, \qquad \omega = \oint_\alpha \frac{dx}{y}, \tag{210}$$

They have the property that $\mathrm{Im}\tau > 0$.

22.3 The physical connection with $N = 2$ super Yang-Mills

The breakthrough of Seiberg and Witten for the solution of the model was the identification of the complex effective coupling $\tau(u)$ at a given vacuum u with the modulus of a u-dependent torus. At any point u of the moduli space, they associated an elliptic curve

$$y^2 = 4\prod_{i=1}^{3}(x - e_i(u)), \tag{211}$$

with its lattice periods given by (210).

The identification of the physical coupling $\tau(u) = \partial a_D/\partial a$ with the modulus $\tau_u = \omega_D(u)/\omega(u)$ of the elliptic curve (211),

$$\tau(u) = \frac{\partial a_D/\partial u}{\partial a/\partial u} = \frac{\oint_\beta dx/y}{\oint_\alpha dx/y} = \tau_u, \tag{212}$$

leads to the formulae:

$$a_D = \oint_\beta \lambda(u), \tag{213}$$

$$a = \oint_\alpha \lambda(u), \tag{214}$$

where $\lambda(u)$ is an abelian differential with the property that

$$\frac{\partial\lambda}{\partial u} = f(u)\frac{dx}{y} + dg\,. \tag{215}$$

Then, the solution of the problem is reduced to finding the family of elliptic curves (211) and the holomorphic function $f(u)$. The conditions at the beginning of this section fix a unique solution. The family of elliptic curves is determined by the monodromy group generated by the monodromy matrices. The matrices (207) generate the group $\Gamma(2)$, the subgroup of $SL(2,\mathbf{Z})$ consisting of matrices congruent to the identity modulo 2. It gives the elliptic curves

$$y^2 = (x^2 - \Lambda^4)(x-u)\,. \tag{216}$$

Finally, the function $f(u)$ is determined by the asymptotic behaviour of (a_D, a) at the singular points. The answer is $f = -\sqrt{2}/4\pi$.

23 Breaking $N = 2$ to $N = 1$. Monopole condensation and confinement

In this section we will exhibit an explicit realization of the confinement mechanism envisaged by Mandelstam (Mandelstam 1976) and 't Hooft through the condensation of light monopoles.

In the $N = 2$ model, we have found points in the moduli space where the relevant light degrees of freedom are magnetic particles. Since we have the exact solution of the low energy $N = 2$ model, it would be nice to answer in which phase the dynamics of the model, or controllable deformations of it, locates the vacuum.

For the $N = 2$ model we already know from section 18 that $N = 2$ supersymmetry does not allow the generation of a superpotential just for the $N = 1$ chiral multiplet of the $N = 2$ vector multiplet. It means that the theory is always in an abelian Coulomb phase. The exact solution of the model allowed us to know which are all the instanton corrections to the low energy Lagrangian. Remarkably enough, the instanton series admits a resummation in terms of magnetic variables.

To go out of the Coulomb branch, we need a superpotential for the chiral superfield Φ. In (Seiberg and Witten 1994a) an explicit mass term for the chiral superfield was added in the bare Lagrangian,

$$\mathcal{W}_{tree} = m\,\mathrm{tr}\,\Phi^2\,. \tag{217}$$

It breaks $N = 2$ to $N = 1$ supersymmetry. At low energy, we will have an effective superpotential $\mathcal{W}(m, M, \tilde{M}, A_D)$. Once again, holomorphy of the superpotential and selection rules from the symmetries will fix the exact form of $\mathcal{W}$. In terms of $N = 1$ superspace, only the subgroup $U(1)_J \subset SU(2)_R$ is manifestly a symmetry. It is a non-anomalous R-symmetry (rotates the complex phases of $\theta^{(I)}$, $I = 1, 2$, in opposite directions.). The corresponding charge of Φ is zero. As superpotentials should have charge two, from (217) we derive that the parameter $m \neq 0$ breaks the $U(1)_J$ symmetry by two units. On the other hand, the $N = 1$ chiral superfields M and $\tilde{M}$ are in an $N = 2$ hypermultiplet and therefore, both have charge one. Imposing that $\mathcal{W}$ is a regular function at $m = \tilde{M}M = 0$, we find that it is of the form $\mathcal{W} = mf_1(A_D) + \tilde{M}Mf_2(A_D)$. For $m \to 0$, the effective superpotential flows to the tree level superpotential (217) plus the term $\sqrt{2}A_D\tilde{M}M$. As the functions f_1 and f_2 are independent of m, we obtain the exact result

$$\mathcal{W} = \sqrt{2}A_D\tilde{M}M + mU(A_D)\,. \tag{218}$$

We found what we were looking for: an exact effective superpotential with a term which depends only on the $N = 1$ chiral composite operator U. It presumably will remove the flat direction. The $N = 2$ to $N = 1$ breaking makes no longer valid the hidden $N = 2$ holomorphy in the Kähler potential $K(A, \bar{A})$. But as long as there is an unbroken supersymmetry, the vacuum configuration corresponds to the solution of the equations

$$d\mathcal{W} = 0\,, \tag{219}$$

$$D = |M|^2 - |\tilde{M}|^2 = 0\,. \tag{220}$$

From the exact solution we know that $du/da_D \neq 0$ at $a_D = 0$. Thus (up to gauge transformations)

$$\begin{aligned} M = \tilde{M} &= \left(-\, mu'(0)/\sqrt{2}\right)^{1/2} \\ a_D &= 0\,. \end{aligned} \tag{221}$$

Expanding around this vacuum we find:

i) There is a mass gap of the order $(m\Lambda)^{1/2}$.

ii) The objects that condense are magnetic monopoles. There are electric flux tubes with a non-zero string tension of the order of the mass gap, that confine the electric charges of the $U(1)$ gauge group.

The spontaneously broken symmetry $u \to -u$ carries the theory to the 'dyon region', with the local variable $a_D - a$. The perturbing superpotential there, $mU(a_D - a)$, also produces the condensation of the 'dyon' with physical electric charge zero at the point $a_D - a = 0$. Then, we have two physically equivalent vacua, related by an spontaneously broken symmetry, in agreement with the Witten index of $N = 1$ $SU(2)$ gauge theory.

24 Breaking $N = 2$ to $N = 0$

When the $N = 2$ theory is broken to the $N = 1$ theory through the decoupling of the chiral superfield Φ in the adjoint, we have seen that the mechanism of confinement takes place because of the condensation of a magnetic monopole.

The natural question is if this results can be extended to non supersymmetric gauge theories.

The $N = 1,2$ results were based on the use of holomorphy; the question is whether the properties connected with holomorphy can be extended to the $N = 0$ case. The answer is positive provided supersymmetry is broken via soft breaking terms.

The method is to promote some couplings in the supersymmetric Lagrangian to the quality of frozen superfields, called spurion superfields. We could think they correspond to some heavy degrees of freedom which at low energies have been decoupled. Their trace is only through their vacuum expectation values appearing in the Lagrangian and are parametrized by the spurion superfields (Girardello and Grisaru 1982).

In the $N = 2$ theory we will promote some couplings to the status of spurion superfields. The property of holomorphy in the prepotential will be secured if the introduced spurions are $N = 2$ vector superfields (Álvarez-Gaumé et al. 1996), (Álvarez-Gaume et al. 1997) [9].

In the bare Lagrangian of the $N = 2$ $SU(2)$ gauge theory (173), there is only one parameter: τ_0. The $N = 2$ softly broken theory is obtained by the bare prepotential

$$\mathcal{F}_0 = \frac{1}{\pi}\mathcal{S}\mathcal{A}^a\mathcal{A}^a\,, \tag{222}$$

where S is an adimensional $N = 2$ vector multiplet whose scalar component gives the bare coupling constant, $s = \frac{\pi}{2}\tau_0$. The factor of proportionality is related with the one loop coefficient of the beta function, such that $\Lambda = \mu_0\exp(is)$. Inspired by String Theory, we call S the dilaton spurion. The source of soft breaking comes from the non vanishing auxiliary fields, F_0 and D_0, in the dilaton spurion S.

The tree level mass terms arising from the softly broken bare Lagrangian (222) are the following: the W-bosons get a mass term by the usual Higgs mechanism, with the mass square equal to $2|a|^2$; the photon of the unbroken $U(1)$ remains massless; the gauginos get a mass square $\mathcal{M}_{1/2}^2 = (|F_0|^2 + D_0^2/2)(4\mathrm{Im}S)^{-1}$; all the scalar components, except the real part of ϕ^3 which do not have a bare mass term, get a square mass $\mathcal{M}_0^2 = 4\mathcal{M}_{1/2}^2$.

At low energy, *i.e.*, at scales of the order $|u|^{1/2} \sim \Lambda$, the Wilsonian effective Lagrangian up to two derivatives and four fermion terms is given by the effective prepotential $\mathcal{F}(a,\Lambda)$ found in the $N = 2$ model, but with the difference that the bare coupling constant is replaced by the dilaton spurion, *i.e.*, $\Lambda \to \mu_0\exp(i\mathcal{S})$. Then, the prepotential depends on two vector multiplets and the effective Lagrangian becomes

$$\begin{aligned}\mathcal{L} = \frac{1}{4\pi}\mathrm{Im}\left(\int d^4\theta \frac{\partial\mathcal{F}}{\partial A^i}\overline{A}^i + \int d^2\theta \frac{1}{2}\frac{\partial^2\mathcal{F}}{\partial A^i\partial A^j}W^iW^j\right) \\ + \mathcal{L}_{HM}\,.\end{aligned} \tag{223}$$

[9] Soft breaking of $N = 1$ SQCD has been studied in (Evans et al. 1995).

with $A^i = (S, A)$ and $\mathcal{L}_{HM}$ the $N = 2$ Lagrangian that includes the monopole hypermultiplet. Observe that the dilaton spurion does not enter in the Lagrangian of the hypermultiplets, in agreement with the $N = 2$ non-renormalization theorem of (de Wit et al. 1985). The low energy couplings are determined by the 2×2 matrix

$$\tau_{ij}(a, s) = \frac{\partial^2 \mathcal{F}}{\partial a^i \partial a^j} . \qquad (224)$$

The supersymmetry breaking generates a non-trivial effective potential for the scalar fields,

$$\begin{aligned} V_{eff} = & \left(b_{00} - \frac{b_{01}^2}{b_{11}} \right) \left(|F_0|^2 + \frac{1}{2} D_0^2 \right) \\ & + \frac{b_{01}}{b_{11}} \left[\sqrt{2} (F_0 m \tilde{m} + \overline{F}_0 \overline{m \tilde{m}}) + D_0 (|m|^2 - |\tilde{m}|^2) \right] \\ & + \frac{1}{2 b_{11}} (|m|^2 + |\tilde{m}|^2)^2 + 2|a|^2 (|m|^2 + |\tilde{m}|^2) , \end{aligned} \qquad (225)$$

where we have defined $b_{ij} = (4\pi)^{-1} \mathrm{Im} \tau_{ij}$. m and $\tilde{m}$ are the scalar components of the chiral superfields M and $\tilde{M}$ of the monopole hypermultiplet, respectively. Observe that the first line of equation (225) is independent of the monopole degrees of freedom. To be sure that such quantity gives the right amount of energy at any point of the moduli space, where different local descriptions of the physics are necessary, it must be duality invariant. This is the case for any $SL(2, \mathbf{Z})$ transformation.

The auxiliary fields of the dilaton spurion are in the adjoint representation of the group $SU(2)_R$ and have $U(1)_R$ charge two. We can consider the situation of $D_0 = 0, F_0 = f_0 > 0$ without any loss of generality, since it is related with the case of $D_0 \neq 0$ and complex F_0 just by the appropriate $SU(2)_R$ rotation.

We have to be careful with the validity of our approximations. Because of supersymmetry, the expansion in derivatives is linked with the expansion in fermions and the expansion in auxiliary fields. The exact solution of Seiberg and Witten is only for the first terms in the derivative expansion of the effective Lagrangian, in particular up to two derivatives. At the level of the softly broken effective Lagrangian, the exact solution of Seiberg and Witten only gives us the terms at most quadratic in the supersymmetry breaking parameter f_0. The expansion is performed in the dimensionless parameter f_0/Λ. Our ignorance on the higher derivative terms of the effective Lagrangian is translated into our ignorance of the terms of $\mathcal{O}((f_0/\Lambda)^4)$. Hence our results are reliable for small values of f_0/Λ, and this is far from the supersymmetry decoupling limit $f_0/\Lambda \to \infty$.

But for moderate values of the supersymmetry breaking parameter, the effective Lagrangian (223) gives the large distance physics of a non-supersymmetric gauge theory at strong coupling. If we minimize the effective potential (225) with respect to the monopoles, we obtain the energy of the vacuum u

$$V_{eff}(u) = \left(b_{00}(u) - \frac{b_{01}^2(u)}{b_{11}(u)} \right) |F_0|^2 - \frac{2}{b_{11}(u)} \rho^4(u) , \qquad (226)$$

where $\rho(u)$ is a positive function that gives the monopole condensate at u

$$|m|^2 = |\tilde{m}|^2 = \rho^2(u) = \frac{|b_{01}|f_0}{\sqrt{2}} - b_{11}|a|^2 > 0 \tag{227}$$

or $m = \tilde{m} = \rho(u) = 0$ if $|b_{01}|f_0 < \sqrt{2}b_{11}|a|^2$. ρ is depicted below, and the effective potential in the next figure.

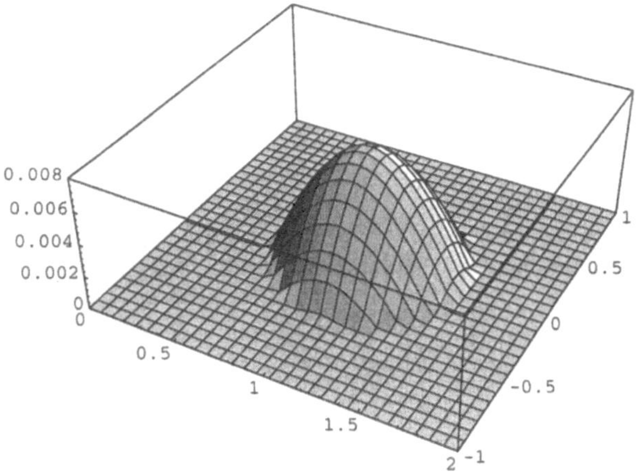

Notice that b_{11} diverges logarithmically at the singularities $u = \pm\Lambda^2$, but the corresponding local variable a vanishes linearly at $u = \pm\Lambda^2$. It implies that $b_{11}|a|^2 \to 0$ for $u \to \pm\Lambda^2$. It can be shown that the Seiberg-Witten solution gives $b_{01} \sim \Lambda/8\pi$ for $u \sim \Lambda$. It means that the monopole condenses at the monopole region, since from the expression of the effective potential (226), such condensation is energetically favoured. If we look at the dyon region, we find that $b_{01} \to 0$ for $u \to -\Lambda^2$. Numerically, there is a very small dyon condensate without any associated minimum in the effective potential in that region. On the other hand, there is a clear absolute minimum in the monopole region. The different behaviours of the broken theory under the transformation $u \to -u$ is an expected result if we take into account that $f_0 \neq 0$ breaks explicitly the $U(1)_R$ symmetry.

The softly broken theory selects a unique minimum at the monopole region, with a non-vanishing expectation value for the monopole. The theory confines and has a mass gap of order $(f_0\Lambda)^{1/2}$.

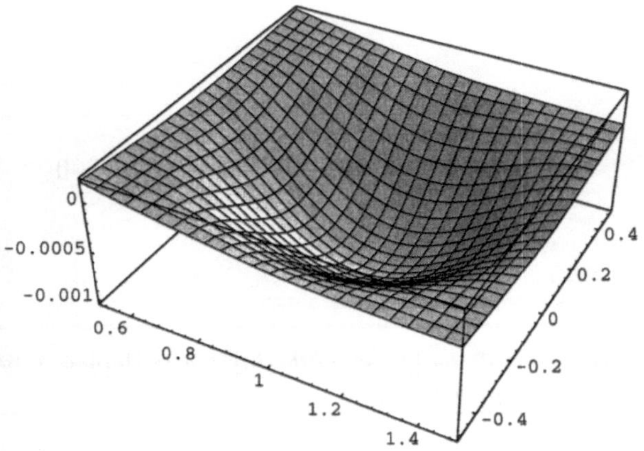

25 String Theory in perturbation theory

String Theory is a multifaceted subject. In the sixties, strings were first introduced to model the dynamics of hadron dynamics. In section 7 we described the confining phase as the dual Higgs phase, where magnetic degrees of freedom condense. The topology of the gauge group allows the existence of electric vortex tubes, ending on quark-antiquark bound states. The transverse size of the electric tubes is of the order of the compton wave length of the 'massive' W-bosons. At large distances, these electric tubes can be considered as open strings with a quark and an anti-quark at their end points. This is the QCD string, with a string tension of the order of the characteristic length square of the hadròns, $\alpha' \sim (1\text{GeV})^{-2}$.

But the major interest in String Theory comes from being a good candidate for quantum gravity (Green et al. 1987). The macroscopic gravitational force includes an intrinsic constant, G_N, with dimensions of length square

$$G_N = l_p^2 = (1.6 \times 10^{-33}\text{cm})^2 \,. \tag{228}$$

In a physical process with an energy scale E for the fundamental constituents of matter, the strength of the gravitational interaction is given by the dimensionless coupling $G_N E^2$ to the graviton. This interaction can be neglected when the graviton probes length scales much larger than the Planck's size, $G_N E^2 \ll 1$. The interaction is also non-renormalizable. From the point of view of Quantum Field Theory, it corresponds to an effective low energy interaction, with l_p the natural length scale at which the effects of quantum gravity become important. The natural suspicion is that there is new physics at such short distances, which smears out the interaction. The idea of String Theory is to replace the point particle description of the interactions by one-dimensional objects, strings with size of the order of the Planck's length $l_p \sim 10^{-33}$ cm (see the figure below). Such a simple change has profound consequences on the physical behaviour of the theory, as we will briefly review below. It is still not clear whether the stringy solution to quantum gravity should work. Because Planck's length scale is so small, up to now String Theory is only constructed from internal consistency. But it is at the moment the best candidate we have. Let us quickly review some of the major implications of String Theory, derived already at perturbative level.

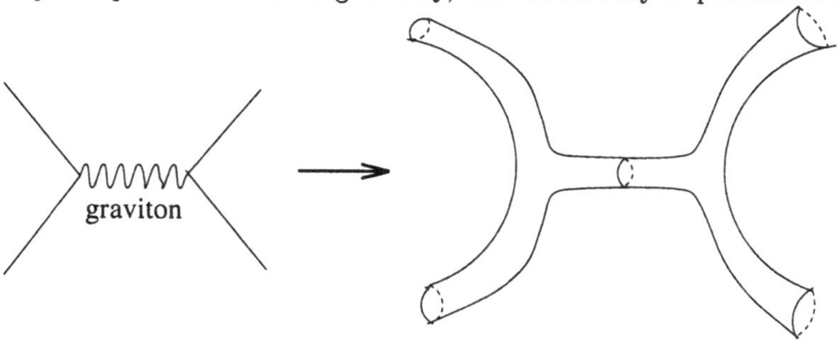

The first important consequence of String Theory is the existence of vibrating modes of the string. They correspond to the physical particle spectrum. For phenomenology the relevant part comes from the massless modes, since the massive modes are excited at energies of the order of the Planck's mass l_p^{-1}. At low energies all the massive modes decouple and we end with an effective Quantum Field Theory for the massless modes. In the massless spectrum of the closed string, there is a particle of spin two. It is the graviton. Then String Theory includes gravity. If we know how to make a consistent and phenomenologically satisfactory quantum theory of strings, we have quantized gravity.

Up to now, String Theory is only well understood at the perturbative level. The field theory diagrams are replaced by two-dimensional Riemann surfaces, with the loop expansion being performed by an expansion in the genus of the surfaces. It is a formulation of first quantization, where the path integral is weighed by the area of the Riemann surface and the external states are included by the insertion of the appropriate vertex operators. The perturbative string coupling constant is determined by the vacuum expectation value of a massless real scalar field, called the dilaton, through the relation $g_s = \exp\langle s \rangle$. The thickening of Feynman diagrams into 'surface' diagrams improves considerably the ultraviolet behaviour of the theory. String Theory is ultraviolet finite.

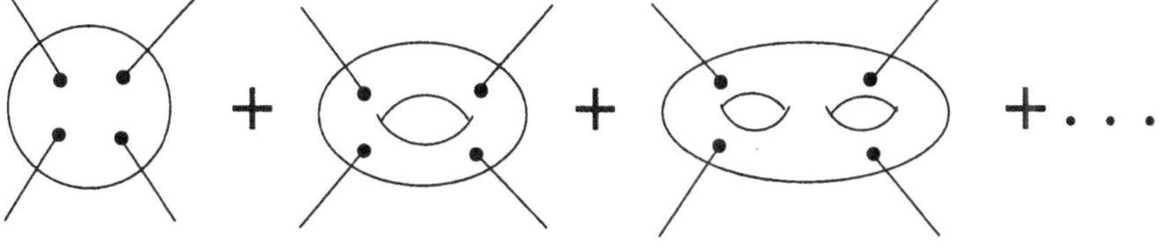

The third important consequence is the introduction of supersymmetry. For the bosonic string, the lowest vibrating mode corresponds to a tachyon. It indicates that we are performing perturbation theory around an unstable minimum.

imum. Supersymmetry gives a very economical solution to this problem. In a supersymmetric theory the hamiltonian operator is positive semi-definite and the ground state has always zero energy. It is also very appealing from the point of view of the cosmological constant problem. Furthermore, supersymmetry also introduces fermionic degrees of freedom in the physical spectrum. If nature really chooses to be supersymmetric at short distances, the big question is: How is supersymmetry dynamically broken? The satisfactory answer must include the observed low energy phenomena of the standard model and the vanishing of the cosmological constant. As a last comment on supersymmetry we will say that the Green-Schwarz formulation of the superstring action demands invariance under a world-sheet local fermionic symmetry, called κ-symmetry. It is only possible to construct κ-symmetric world-sheet actions if the number of spacetime symmetries is $N \leq 2$ (in ten spacetime dimensions).

The fourth important consequence is the prediction on the number of dimensions of the target space where the perturbative string propagates. Lorentz invariance on the target space or conformal invariance on the world-sheet fixes the number of spacetime dimensions (twenty-six for bosonic strings and ten for superstrings). As our low energy world is four dimensional, String Theory incorporates the Kaluza-Klein idea in a natural way. But again the one-dimensional nature of the string gives a quite different behaviour of String Theory with respect to field theory. The dimensional reduction of a field theory in D spacetime dimensions is another field theory in $D-1$ dimensions. The effect of a non-zero finite radius R for the compactified dimension is just a tower of Kaluza-Klein states with masses n/R. But in String Theory, the string can wind m times around the compact dimension. This process gives a contribution to the momentum of the string proportional to the compact radius, mR/α'. These quantum states become light for $R \to 0$. The dimensional reduction of a String Theory in D dimensions is another String Theory in D dimensions. This is T duality (Giveon et al. 1994).

The fifth important consequence comes from the cancellation of spacetime anomalies (gauge, gravitational and mixed anomalies). It gives only the following five anomaly-free superstring theories in ten spacetime dimensions.

25.1 The type IIA and type IIB string theories

A type II string theory is constructed from closed superstrings with $N = 2$ spacetime supersymmetries. The spectrum is obtained as a tensor product of left- and right-moving world-sheet sectors of the closed string. Working in the light-cone gauge, the massless states of each sector are in the representation $\mathbf{8}_v \oplus \mathbf{8}_\pm$ of the little group $SO(8)$. The representations $\mathbf{8}_v$ and $\mathbf{8}_\pm$ are the vector representation and the irreducible chiral spinor representations of $SO(8)$, respectively.

The type IIA string theory corresponds to the choice of opposite chiralities for the spinorial representations in the left- and right-moving sectors,

$$\text{Type IIA}: \quad (\mathbf{8}_v \oplus \mathbf{8}_+) \otimes (\mathbf{8}_v \oplus \mathbf{8}_-) \,. \qquad (229)$$

The bosonic massless spectrum is divided between the NS-NS fields:

$$\mathbf{8}_v \otimes \mathbf{8}_v = \mathbf{1} \oplus \mathbf{28} \oplus \mathbf{35}\,, \tag{230}$$

which corresponds to the dilaton s, the antisymmetric tensor $B_{\mu\nu}$ and the gravitation field $g_{\mu\nu}$, respectively, and the R-R fields:

$$\mathbf{8}_+ \otimes \mathbf{8}_- = \mathbf{8}_v \oplus \mathbf{56}, \tag{231}$$

which correspond to the light-cone degrees of freedom of the antisymmetric tensors A_μ and $A_{\mu\nu\rho}$, respectively. As the chiral spinors have opposite chiralities, in the vertex operators of the R-R fields only even forms appear, F_2 and F_4. The physical state conditions on the massless states give the following equations on these even forms:

$$dF = 0 \quad d \star F = 0\,, \tag{232}$$

with $\star F$ the Poincare dual $(10-n)$-form of the n-form F_n. These are the Bianchi identity and the equation of motion for a field strength. Their relation with the R-R fields is then $F_n = dA_{n-1}$. The abelian field strengths F_n are gauge invariant, and since these are the fields that appear in the vertex operators, the fundamental strings do not carry RR charges.

The fermionic massless spectrum is given by the $NS-R$ and $R-NS$ fields:

$$\begin{aligned} \mathbf{8}_v \otimes \mathbf{8}_- &= \mathbf{8}_+ \oplus \mathbf{56}_-\,, \\ \mathbf{8}_+ \otimes \mathbf{8}_v &= \mathbf{8}_- \oplus \mathbf{56}_+\,. \end{aligned} \tag{233}$$

The $\mathbf{8}_\pm$ states are the two dilatini. The $\mathbf{56}_\pm$ states are the two gravitini, with a spinor and a vector index. Observe that the fermions have opposite chiralities, which prevent the type IIA theory from gravitational anomalies.

The Type IIB String Theory corresponds to the choice of the same chirality for the spinor representations of the left- and right-moving sector,

$$\text{Type IIB}: \quad (\mathbf{8}_v \oplus \mathbf{8}_+) \otimes (\mathbf{8}_v \oplus \mathbf{8}_+)\,. \tag{234}$$

The NS-NS fields are the same as for the type IIA string. The difference comes from the R-R fields:

$$\mathbf{8}_+ \otimes \mathbf{8}_+ = \mathbf{1}_+ \oplus \mathbf{28} \oplus \mathbf{35}_+\,. \tag{235}$$

They correspond, respectively, to the forms A_0, A_2 and A_4 (self-dual).

For the massless fermions there are two dilatini and two gravitini, but now all of them have the same chirality. In spite of it, the theory does not have gravitational anomalies (Álvarez-Gaumé and Witten 1984).

Under spacetime compactifications, the type IIA and the type IIB string theories are unified by the T-duality symmetry. It is an exact symmetry of the theory already at the perturbative level and maps a type IIA string with a compact dimension of radius R to a type IIB string with radius α'/R.

25.2 The Type I string theory

It is constructed from unoriented open and closed superstrings, leading only $N = 1$ spacetime supersymmetry. The massless states are:

$$\text{Open}: \quad \mathbf{8}_v \otimes \mathbf{8}_+ \tag{236}$$

$$\begin{aligned}\text{Closed sym.}: \quad & [(\mathbf{8}_v \oplus \mathbf{8}_+) \otimes (\mathbf{8}_v \oplus \mathbf{8}_+)]_{\text{sym}} = \\ & = [\mathbf{1} \oplus \mathbf{28} \oplus \mathbf{35}]_{\text{bosonic}} \oplus [\mathbf{8}_- \oplus \mathbf{56}_-]_{\text{fermionic}} \,.\end{aligned} \tag{237}$$

The massless sector of the spectrum that comes from the unoriented open superstring (236) gives $N = 1$ super Yang-Mills theory, with a gauge group $SO(N_c)$ or $USp(N_c)$ introduced by Chan-Paton factors at the ends of the open superstring. The sector coming from the unoriented closed string (237) gives $N = 1$ supergravity. Cancellation of spacetime anomalies restricts the gauge group to $SO(32)$.

25.3 The $SO(32)$ and $E_8 \times E_8$ heterotic strings

The heterotic string is constructed from a right-moving closed superstring and a left-moving closed bosonic string. Conformal anomaly cancellation demands twenty-six bosonic target space coordinates in the left-moving sector. The additional sixteen left-moving coordinates X_L^I, $I = 1, ..., 16$, are compactified on a T^{16} torus, defined by a sixteen-dimensional lattice, Λ_{16}, with some basis vectors $\{e_i^I\}$, $i = 1, ..., 16$. The left-moving momenta p_L^I live on the dual lattice $\tilde{\Lambda}_{16}$. The mass operator gives an even lattice ($\sum_{I=1}^{16} e_i^I e_i^I = 2$ for any i). The modular invariance of the one-loop diagrams restricts the lattice to be self-dual ($\tilde{\Lambda}_{16} = \Lambda_{16}$). There are only two even self-dual sixteen-dimensional lattices. They correspond to the root lattices of the Lie groups $SO(32)/Z_2$ and $E_8 \times E_8$.

For the physical massless states, the supersymmetric right-moving sector gives the factor $\mathbf{8}_v \otimes \mathbf{8}_+$, which together with the lattice points of length squared two of the left-moving sector, give an $N = 1$ vector multiplet in the adjoint representation of the gauge group $SO(32)$ or $E_8 \times E_8$.

There is also a T-duality symmetry relating the two heterotic strings.

26 D-branes

Perturbation theory is not the whole history. In the field theory sections we have learned how much the nonperturbative effects could change the perturbative picture of a theory. In particular, there are nonperturbative stable field configurations (solitons) that can become the relevant degrees of freedom in some regime. In that situation it is convenient to perform a duality transformation to have an effective description of the theory in terms of these solitonic degrees of freedom as the fundamental objects.

What about the nonperturbative effects in String Theory? Does String Theory incorporate nonperturbative excitations (string solitons)? Are there also strong-weak coupling duality transformations in String Theory? Before the role of D-branes in String Theory were appreciated, the answers to these three questions were not clear.

For instance, it was known, by the study of large orders of string perturbation theory, that the nonperturbative effects in string theory had to be stronger than in field theory, in the sense of being of the order of $\exp(-1/g_s)$ instead of order $\exp(-1/g_s^2)$ (Shenker 1991), but it was not known which was the nature of such nonperturbative effects.

With respect to the existence of nonperturbative objects, the unique evidence came form solitonic solutions of the supergravity equations of motion which are the low energy limits of string theories. These objects were in general extended membranes in $p+1$ dimensions, called p-branes (Duff et al. 1995).

In relation to the utility of the duality transformation in String Theory, there is strong evidence of some string dualities (Hull and Townsend 1995). There is for instance the $SL(2,\mathbf{Z})$ self-duality conjecture of the type IIB theory (Font et al. 1990). Under an S-transformation, the string coupling value g_s is mapped to the value $1/g_s$, and the NS-NS field $B_{\mu\nu}$ is mapped to the R-R field $A_{\mu\nu}$. Then, self-duality of type IIB demands the existence of a string with a tension scaling as g_s^{-1} and non-zero RR charge.

26.1 Dirichlet boundary conditions

In open string theory, it is possible to impose two different boundary conditions at the ends of the open string:

$$\text{Neuman}: \quad \partial_\perp X^\mu = 0 \tag{238}$$

$$\text{Dirichlet}: \quad \partial_t X^\mu = 0\,. \tag{239}$$

An extended topological defect with $p+1$ dimensions is described by the following boundary conditions on the open strings:

$$\partial_\perp X^{0,1,\cdots p} = \partial_t X^{p+1,\cdots 9} = 0\,. \tag{240}$$

We call it a D p-brane (for Dirichlet (Polchinski 1996)), an extended (p+1)-dimensional object (located at $X^{p+1,\cdots 9} = 0$) with the end points of open strings attached to it.

The Dirichlet boundary conditions are not Lorentz invariant. There is a momentum flux going from the ends of open strings to the D-branes to which they are attached. In fact, the quantum fluctuations of the open string endpoints in the longitudinal directions of the D-brane live on the world-volume of the D-brane. The quantum fluctuations of the open string endpoints in the transverse directions of the D-brane make the D-brane fluctuate locally. It is a dynamical object, characterized by a tension T_p and a RR charge μ_p. If $\mu_p \neq 0$, the world-volume of a p-brane will couple to the R-R $(p+1)$-form A_{p+1}.

Far from the D-brane, we have closed superstrings, but the world-sheet boundaries (240) relate the right-moving supercharges to the left-moving ones, and only a linear combination of both is a good symmetry of the given configuration. In presence of the D-brane, half of the supersymmetries are broken. The D-brane is a BPS state. In fact, in (Dai et al. 1989) it was shown that the D-brane tension arises from the disk and therefore scales as g_s^{-1}. This is the same coupling constant dependence as for BPS solitonic branes carrying RR charges (Duff et al. 1995).

The Dirichlet boundary condition becomes the Neuman boundary condition in terms of the T-dual coordinates, and vice versa. It implies that if we T-dualize a direction longitudinal to the world-volume of the D p-brane, it becomes a $(p-1)$-brane. Equally, if the T-dualized direction is transverse to the D p-brane, we obtain a D $(p+1)$-brane. Consider a 9-brane in a type IIB background. The 9-brane fills the spacetime and the endpoints of the open strings attached to it are free to move in all the directions. It is a type I theory, with only $N = 1$ supersymmetry. Now T-dualize one direction of the target space. We obtain an 8-brane in a type IIA background. If we proceed further, we obtain that a type IIA background can hold $p = 9, 7, 5, 3, 1, -1$ p-branes. A D (-1)-brane is a D-instanton, a localized spacetime point. For a type IIB background we obtain $p = 8, 6, 4, 2, 0$ p-branes.

26.2 BPS states with RR charges

To check if really the D-branes are the nonperturbative string solitons required by string duality, Polchinski computed explicitly the tension and RR charge of a D p-brane (Polchinski 1995). He first computed the one-loop amplitude of an open string attached to two parallel D p-branes. The resulting Casimir force between the D-branes was zero, supporting its BPS nature. By modular invariance, it can also be interpreted as the amplitude for the interchange of a closed string between the D-branes.

In the large separation limit, only the massless closed modes contribute. These are the NS-NS fields (graviton and dilaton) and the R-R $(p+1)$ form. On

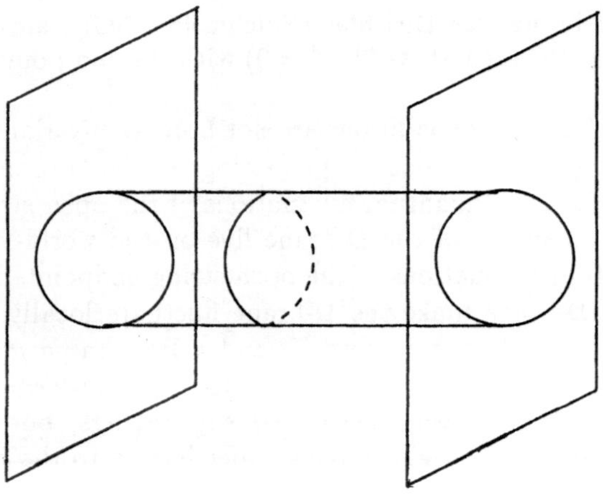

the space between the D-branes these fields follow the low energy type II action (type IIA for p even and type IIB for p odd). On the D p-branes, the coupling to the NS-NS and R-R fields is

$$S_p = T_p \int d^{p+1}\xi \; e^{-s} \, |\mathrm{det} G_{ab}|^{1/2} + \mu_p \int_{p-\mathrm{brane}} A_{p+1} \, . \tag{241}$$

From (241) we see that the actual D-brane action includes a dilaton factor $\tau_p = T_p/g_s$, with g_s the coupling constant of the closed string theory. Comparing the field theory calculation with the contribution of the massless closed modes in the string theory computation, one can obtain the values of T_p and μ_p. The result is (Polchinski 1995)

$$\mu_p^2 = 2T_p^2 = (4\pi^2\alpha')^{3-p} \, . \tag{242}$$

Observe that the R-R charge is really non-zero. In fact, if one checks (the generalization of) the Dirac's quantization condition for the charge μ_p and its dual charge $\mu_{(6-p)}$, one obtains that $\mu_p\mu_{(6-p)} = 2\pi$. They satisfy the minimal quantization condition. It means that the D-branes carry the minimal allowed RR charges.

27 Some final comments on nonperturbative String Theory

27.1 D-instantons and S-duality

The answers to the three questions at the beginning of the previous section can now be more concrete, since some nonperturbative objects in String Theory has been identified: the D-branes.

Consider a D p-brane wrapped around a non-trivial $(p+1)$-cycle. This configuration is topologically stable. Its action is $T_p V_{p+1}/g_s$, with V_{p+1} the volume of the non-trivial $(p+1)$ cycle. It contributes in amplitudes with factors $e^{-T_p V_{p+1}/g_s}$, a generalized instanton effect. Now we understand why the nonperturbative effects in String Theory are stronger than in field theory, it is related to the peculiar nature of the string solitons.

The D-branes also give the necessary ingredient for the $SL(2,\mathbf{Z})$ self-duality of the type IIB string theory. This theory allows D 1-branes, with a mass $\tau_1 \sim (2\pi\alpha' g_s)^{-1}$ in the string metric and non-zero RR charge. Also, one can see that on the D 1-brane there are the same fluctuations of a fundamental IIB string (Witten 1996). Then, it is the required object for the S-duality transformation of the type IIB string. In fact, at strong coupling the D 1-string becomes light and it is natural to formulate the type IIB theory in terms of weakly coupled D 1-branes.

There is another S-duality relation in String Theory. Observe that the type I theory and the $SO(32)$ heterotic theory have the same low energy limit. It could be that they correspond to the same theory but for different values of the string coupling constant. Again D-branes help to make this picture clearer. Consider a D 1-brane in a type I background with open strings attached to it, but also with open strings with one end point attached to a 9-brane. We call them $1-9$ strings. The 9-brane fills the spacetime, and the $1-9$ strings, having one Chan-Paton index, are vectors of $SO(32)$. One can see that the world-sheet theory of the D 1-brane is precisely that of the $SO(32)$ heterotic string (Polchinski and Witten 1996). Having a tension that scales as g_s^{-1}, one can argue that this D heterotic string sets the lightest scale in the theory when $g_s \gg 1$. The strong coupling behaviour of the type I string can be modeled by the weak coupling behaviour of the heterotic string.

27.2 An eleventh dimension

Type IIA allows the existence of 0-branes that couple to the R-R one-form A_1. The 0-brane mass is $\tau_0 \sim (\alpha')^{-1/2}/g_s$ in the string metric. At strong coupling in the type IIA theory, $g_s \gg 1$, this mass is the lightest scale of the theory. In fact, n 0-branes can form a BPS bound state with mass $n\tau_0$. This tower of states becoming a continuum of light states at strong coupling is characteristic of the appearance of an additional dimension. Type IIA theory at strong coupling feels an eleventh dimension of some size $2\pi R$, with the 0-branes playing the role of the Kaluza-Klein states (Townsen 1995).

If we compactify 11D supergravity (Cremmer et al.) on a circle of radius R and compare its action with the 10D type IIA supergravity action, we obtain the relation

$$R \sim g_s^{2/3} . \tag{243}$$

This eleventh dimension is invisible in perturbation theory, where we perform an expansion near $g_s = 0$.

This has been a lightning review of some aspects of duality in String Theory. We hope it will serve to whet the appetite of the reader and encourage her/him to learn more about the subject and to eventually contribute to some of the outstanding open problems. More information can be found from the references (Aspinwall 1996).

Acknowledgements

We have benefited from valuable conversations with many colleagues. We would like to thank in particular E. Álvarez, J.M. F. Barbón, J. Distler, D. Espriu, C. Gómez, J. Gomis, K. Kounnas, J. Labastida, W. Lerche, M. Mariño and E. Verlinde for discussions. F. Z. would like to thank the Theory Division at CERN for its hospitality.

References

Affleck, I., Dine, M., Seiberg, N. (1984): Nucl. Phys. **B241**, 493

Álvarez-Gaumé, L., Distler, J., Kounnas, C., Mariño, M. (1996): Int. J. Mod. Phys. **A11**, 4745; Álvarez-Gaumé, L., Mariño, M. (1997): Int. J. Mod. Phys. **A12**, 975

Álvarez-Gaumé, L., Hassan, S.F. (1997): Fortschr. Phys. **45**, 159; Bilal, A. (1996): hep-th/9601077; Gomez, C., Hernandez, R. (1995): Advanced School on Effective Theories, Almuñecar, hep-th/9510023; Lerche, W. (1997): Nucl. Phys. Proc. Suppl. **55B**, 83

Álvarez-Gaumé, L., Mariño, M., Zamora, F. (1997): hep-th/9703072, hep-th/9707017, to appear in Int. J. Mod. Phys. **A**

Álvarez-Gaumé, L., Witten, E. (1984): Nucl. Phys. **B234**, 269

Arkani-Hamed, N., Murayama, H. (1997): hep-th/9705189; hep-th/9707133

Aspinwall, P. (1996): TASI 96 lectures, hep-th/9611137; Schwarz, J. (1996): TASI 96 lectures, hep-th/9607202; Townsend, P.K. (1995): Trieste 95 lectures, hep-th/9612121; Witten, E. (1996): Nucl. Phys. **B460**, 335; Horava, P., Witten, E. (1996): Nucl. Phys. **B460**, 506; Banks, T., Fischler, W., Shenker, S.H., Susskind, L. (1997): Phys. Rev. **D55**, 5112; Hanany, A., Witten, E. (1997): Nucl. Phys. **B492**, 152; Witten, E. (1997): hep-th/9703166; Dijkgraaf, R., Verlinde, E., Verlinde, H. (1997): hep-th/9709107

Banks, T., Zaks, A. (1982): Nucl. Phys. **B196**, 189

Bogomol'nyi, E.B. (1976): Sov. J. Nucl. Phys. **24**, 449

Cremmer, E., Julia, B., Scherk, J. (1978): Phys. Lett. **76B**, 409

Dai, J., Leigh, R.G., Polchinski, J. (1989): Mod. Phys. Lett. **A4**, 2073; Leigh, R.G. (1989): Mod. Phys. Lett. **A4**, 2767

Davis, A.C., Dine, M., Seiberg, N. (1983): Phys. Lett. **125B**, 487; Affleck, I., Dine, M., Seiberg, N. (1983): Phys. Rev. Lett **51**, 1026

de Wit, B., Lauwers, P.G., van Proeyen, A. (1985): Nucl. Phys. **B255**, 569

Deser, S., Teitelboim, C. (1976): Phys. Rev. **D13**, 1592; Deser, S., Gomberoff, A., Henneaux, M., Teitelboim, C. (1997): Phys. Lett. **B400**, 80

Dirac, P.A.M (1931): Proc. Roy. Soc. **A133**, 60

Duff, M.J., Khuri, R.R., Lu, J.X. (1995): Phys. Rep. **259**, 213; Duff, M.J. (1996): 'Supermembranes', TASI 96 lectures, hep-th/9611203; and references therein

Evans, N., Hsu, S.D.H., Schwetz, M. (1995): Phys. Lett **355B**, 475; Nucl. Phys. **B 456**, 205; Aharony, O., Sonnenschein, J., Peskin, M.E., Yankielowicz, S. (1995): Phys. Rev. **D52**, 6157; D'Hoker, E., Mimura, Y., Sakai, N. (1996): Phys. Rev. **D54**, 7724

Evans N., Hsu, S.D.H., Schwetz, M. (1997): Nucl. Phys. **B484**, 124

Finnell, D., Pouliot, P. (1995): Nucl. Phys. **B453**, 225

Font, A., Ibañez, L., Lust, D., Quevedo, F. (1990): Phys. Lett. **249**, 35

Fradkin, E., Shenker, S.H. (1979): Phys. Rev. **D12**, 3682; Banks, T., Rabinovici, E. (1979): Nucl. Phys. **B160**, 349

Gates, S.J., Jr. (1984): Nucl. Phys. **B238**, 349

Girardello, L., Grisaru, M.T. (1982): Nucl. Phys. **B194**, 65

Giveon, A., Porrati, M., Rabinovici, E. (1994): Phys. Rep. **244**, 77

Green, M.B., Schwarz, J.H., Witten, E. (1987): *Superstring Theory* Cambridge University Press; and references therein

Grisaru, M.T., Siegel, W., Rocek, M. (1979): Nucl. Phys. **B159**, 429

Haag, R., Lopuszanski, J.T., Sohnius, M. (1975): Nucl. Phys. **B88**, 257

Hull, C.M., Townsend, P.K. (1995): Nucl. Phys. **B438**, 109; Witten, E. (1995): Nucl. Phys. **B443**, 83; Strominger, A. (1995): Nucl. Phys. **B451**, 96
Intriligator, K. (1994): Phys. Lett. **336B**, 409
Intriligator, K., Leigh, R.G., Seiberg, N. (1994): Phys. Rev. **D50**, 1092
Intriligator, K., Seiberg, N. (1996): Nucl. Phys. Proc. Suppl. **45B**, 1, hep-th/9509066; Peskin, M.E. (1997): TASI 96 lectures, hep-th/9702094; Shifman, M. (1997): hep-th/9704114
Julia, B., Zee, A. (1975): Phys. Rev. **D11**, 2227
Konishi, K., Shizuya, K. (1985): Nuovo Cim. **90A**, 111
Mandelstam, S. (1976): Phys. Rep. **C23**, 245
Nielsen, H.B., Olesen, P. (1973): Nucl. Phys. **B61**, 45-61
Polchinski, J. (1995): Phys. Rev. Lett. **75**, 4724
Polchinski, J. (1996): TASI 96 lectures, hep-th/9611050;
Bachas, C. (1997): 'Half a Lecture on D-branes', hep-th/9701019; and references therein
Polchinski, J., Witten, E. (1996): Nucl. Phys. **B460**, 525
Polyakov, A.M. (1974): JETP Lett. **20**, 194
Prasad, M.K., Sommerfield, C.M. (1975): Phys. Rev. Lett. **35**, 760
Salam, A., Strathdee, J. (1974): Nucl. Phys. **B76**, 477
Schwinger, J. (1966): Phys. Rev. **144**, 1087; **173**, (1968) 1536; Zwanziger, D. (1968): Phys. Rev. **176**, 1480
Seiberg, N. (1994): The Power of Holomorphy: Exact results in 4-D SUSY Gauge Theories. in PASCOS 94, p. 357, hep-th/9408013
Seiberg, N. (1994): Phys. Rev. **D49**, 6857
Seiberg, N. (1995): Nucl. Phys. **B435**, 129
Seiberg, N., Witten, E. (1994a): Nucl. Phys. **B426**, 19
Seiberg, N., Witten, E. (1994b): Nucl. Phys. **B431**, 484
Shenker, S.H. (1991): 'The Strength of Nonperturbative Effects in String Theory', in *Random Surfaces and Quantum Gravity* eds. O. Alvarez et al.
't Hooft, G. (1994): Nucl. Phys. **B190**, [FS3] (1981), 455, and *1981 Cargese Summer School Lecture Notes on Fundamental Interactions* in *Under the Spell of the Gauge Principle* World Scientific
't Hooft, G. (1974): Nucl. Phys. **B79**, 276
't Hooft, G. (1994): 'Naturalness, chiral symmetry, and spontaneous chiral symmetry breaking', Cargese 1979, in *Under the Spell of the Gauge Principle*, World Scientific
Townsend, P.K. (1995): Phys. Lett. **B350**, 184;
Witten, E. (1995): Nucl. Phys. **B443**, 85
Vafa, C., Witten, E. (1994): Nucl. Phys. **B431**, 3; Ferrari, F. (1997): hep-th/9702166
Veneziano, G., Yankielowicz, S. (1982): Phys. Lett. **113B**, 231; Taylor, T., Veneziano, G., Yankielowicz, S. (1983): Nucl. Phys. **B218**, 439
Weinberg, S. (1996): *The Quantum Theory of Fields I* Cambridge University Press
Wess, J., Bagger, J. (1992): *Supersymmetry and Supergravity* Princeton University Press, 2nd ed. and references therein
Wilson, K.G. (1974): Phys. Rev. **D10**, 2445
Witten, E. (1979): Phys. Lett. **86B**, 283
Witten, E. (1982): Nucl. Phys. **B202**, 253
Witten, E. (1996): Nucl. Phys. **B460**, 335
Witten, E., Olive, D. (1978): Phys. Lett. **78B**, 97
Zumino, B. (1979): Phys. Lett. **87B**, 203

Supersymmetry, Strings and Unification

Hans Peter Nilles

Physikalisches Institut, Universität Bonn,
Nussallee 12, D-53115 Bonn, Germany

Abstract. We review the construction of supersymmetric extensions of the standard model as well as grand unified models. Unification of fundamental interactions is considered in the framework of string theories. The low-energy effective action of heterotic string theory is discussed in detail, both in the framework of $d = 10$ and the $d = 11$ M-theory. Especially for the discussion of unification and supersymmetry breakdown the M-theoretic view seems to be particularly promising.

1 Introduction

The consideration of symmetries has played a major role in particle physics. Strong, weak and electromagnetic interactions are believed to be mediated by gauge particles. The underlying gauge symmetries seem to be required for a consistent field theoretic description of spin-1 particles. Unbroken gauge symmetries are the reasons why some of the gauge bosons like photons and gluons are massless. Chiral symmetries can play this role of mass protection in the case of spin-1/2 particles. Nonvanishing masses can then only appear when some of these symmetries are broken. In the case of the mediators of weak interactions, masses of the W and Z bosons of order of 100 GeV can be understood through the breakdown of the $SU(2) \times U(1)$ gauge group. This breakdown mechanism then also controls the size of the masses, they cannot become arbitrarily heavy, because they are related to the size of the symmetry breakdown. If the symmetries are only broken by a small amount, this can lead to small mass parameters.

In the standard $SU(3) \times SU(2) \times U(1)$ model of strong and electroweak interactions, all the mass parameters are governed by the mass and the vacuum expectation value (vev) of a scalar field, the so-called Higgs-field. The vev breaks gauge and chiral symmetries and therefore all gauge boson, quark and lepton masses are proportional to this vev. If we then understand the size of this vev (e.g. why it is small compared to the Planck scale of 10^{19}GeV), then we also understand why masses of gauge bosons and fermions are small compared to that large scale.

Unfortunately the mass of this scalar field is not protected by either gauge nor chiral symmetries. Thus we have a problem, as we do not understand why the mass and vev of this particle are small compared to larger scales in physics, like e.g. the Planck scale. This is called the hierarchy problem. Technically the lack of a symmetry to protect the mass of a scalar particle reflects itself in the fact that this mass is unstable in quantum field theory, it is quadratically divergent in perturbation theory.

A theory with fundamental scalar particles should thus contain a symmetry that protects the mass of scalar particles. The only known symmetry that can play this role is supersymmetry. It is for this reason that supersymmetry has been discussed as a serious candidate for the physics beyond the standard model. The above mentioned instability of the scalar masses disappears in the supersymmetric case. In fact, supersymmetric quantum field theories have many astonishing properties, such as various nonrenormalization theorems in perturbation theory. Especially N extended supersymmetries could even lead to finite theories. The supersymmetric extension of the standard model, however, cannot tolerate too much supersymmetry. The presence of parity violation in the weak interactions limits ourselves to simple $N = 1$ supersymmetry.

Supersymmetry generators have fermionic character leading to multiplets with an equal number of bosons and fermions. A supersymmetric generalization of the standard model requires the introduction of supersymmetric partners of known particles. The explicit constructions show, that even in models with minimal particle content (MSSM), the number of particles has been more than doubled: each particle of the standard model needs a new supersymmetric partner and, in addition, one has to introduce a second Higgs supermultiplet. We have not seen these supersymmetric partners yet. In unbroken supersymmetry they would be degenerate in mass with their partners. Since we know that there is no spin 0 particle with mass and quantum numbers of e.g. the electron, supersymmetry cannot be an exact symmetry of nature. The breakdown of supersymmetry should be responsible for the non-degeneracy of the supermultiplets, pushing up the masses of the newly introduced superpartners. But where should they be? Remember the reason for the introduction of supersymmetry, the instability of the mass and vev of the scalar Higgs particle. This vev should be stabilized in the mass range of several hundred GeV. Thus the breakdown scale of supersymmetry and the splitting of the supermultiplets cannot be arbitrarily large, it is expected in the region of several hundreds of GeV to a TeV. Again we have seen that a broken symmetry could be an important ingredient in the discussion of the generalization of the standard model.

Experimental searches for supersymmetry have not yet been successful, but the new supersymmetric particles could be around the corner. So far we can only hope that they exist. They could manifest themselves indirectly in various physical processes through radiative corrections. Still no positive evidence is known. A study of the evolution of gauge coupling constants to energies far beyond the weak scale of 100 GeV has revealed a surprise in the context of the supersymmetric models. The outcome of such an evolution might be of interest in the search for a unification of strong and electroweak interactions, such as grand unified theories (GUTs). According to this hypothesis, all three coupling constants of $SU(3) \times SU(2) \times U(1)$ approach each other at a large scale to a unique coupling of a GUT group, like e.g. SU(5). The evolution of coupling constants to higher energies depends on the matter content of the low energy theory. For the standard model it is known that, with its particle content and the present precision determination of the coupling constants, such a unification does not happen. On the other hand the supersymmetric model yields a unification at a scale $M_{GUT} \sim 10^{16}$ GeV, provided that the supersymmetric partners are in the TeV range, exactly where they are supposed to be. Future experiments have to tell whether they are really there.

Grand unification is a very attractive scheme to accommodate the known three gauge symmetries. What about the fourth fundamental force: gravity. After all the Planck scale, the mass scale attached to the gravitational coupling constant, is not that far away from the GUT-scale. The theoretical description of gravitational interactions up to now is less satisfactory than the one for the gauge interactions. Gravity is supposed to be mediated by the exchange of a spin 2 particle, the graviton. A description in terms of a renormalizable quantum field theory is not possible. Many problems remain when trying to formulate a consistent quantum theoretical description of general relativity.

Supersymmetry can lead to progress in that direction. Up to now, we had considered supersymmetry as a global symmetry, where the parameter of supersymmetry transformations is constant and does not depend on space time coordinates. Trying to consider a gauged version of supersymmetry reveals another surprise. The gauge particle of supersymmetry is a spinor and carries in addition a world index: it is a spin-3/2 particle. Supersymmetry requires a partner for this fermion and this partner can be shown to be a spin-2 particle with all the properties of the graviton. For this reason the gauge particle is called gravitino. But one should not forget that in this construction one started from gauge supersymmetry and arrived at gravitation for free. In this sense we can also consider gravity as a kind of gauge symmetry in view of a unification of all fundamental interactions.

Nowadays the supersymmetric generalization of the standard model is discussed in the framework of supergravity. Supersymmetry is broken in a hidden sector that couples only gravitationally to our visible world, the so-called observable sector. The low-energy effective theory feels supersymmetry breakdown through soft breaking terms that push the mass of the supersymmetric partners in the TeV range. Still a consistent quantum theoretical treatment of general relativity cannot be achieved in the framework of supersymmetric field theories. The theories are still nonrenormalizable. They can be considered as an effective low-energy approximation for a more complete theory, that might solve the problems of a quantum theory of gravity.

Such a candidate might be string theory or, more general, a theory of extended objects. Instead of working with the notion of a pointlike particle, one here considers extended objects, strings, membranes etc. String theory again is a theory that contains gravity and gauge interactions and could serve as a candidate for a unification of all (gravitational and gauge) interactions. String theories are more difficult to construct than ordinary quantum field theories. They cannot be constructed in all space time dimensions. Typically there is a maximal (so-called critical) dimension for such theories. In supersymmetric string theories this critical dimension is $d = 10$. Theories in less than $d = 10$ can be obtained via a generalized Kaluza-Klein mechanism with some compactified dimensions. Two dimensional extended objects (2-branes or membranes) have critical dimension $d = 11$.

Superstring theories in $d = 10$ come in different versions. There are two theories with maximal supersymmetry ($N = 2$) in $d = 10$, type IIA and IIB that contain no gauge group in $d = 10$. Three different theories with $N = 1$ come as a theory of open and closed strings with gauge group $SO(32)$ and two theories of oriented closed strings: heterotic $E_8 \times E_8$ or SO(32). It is believed that all these theories represent different manifestations of one fundamental theory: $d = 11$ dimensional M-theory. For practical applications such as the generalization of the standard model up to now one has mostly considered the heterotic $E_8 \times E_8$ theory. The question of unification of fundamental interactions as well as the search for an effective low energy $d = 4$ supergravity action can be addressed in this framework. This is what we try to learn in these lectures.

We shall start with a short introduction to supersymmetry, followed by the construction of the minimal supersymmetric extension to the standard model and a discussion of its phenomenological properties. The next step will deal with supersymmetric grand unified models. We then move to supergravity and the question of supersymmetry breakdown. Time does not permit us to give a thorough introduction to string theory. We shall content ourselves with the investigation of the corresponding properties of the $d = 10$ supersymmetric field theories and try to work out the possible consequences for the $d = 4$ supergravity theories. Although this is a simple approximation to full string theory, it has turned out that many of the basic questions can already be addressed in that framework. Symmetry arguments and the consideration of the absence of gauge and gravitational anomalies prove to be useful tools in that framework. This will all be discussed in the framework of the (weakly coupled) heterotic string theory. We shall then move on to more recent developments in the field initiated through the discovery of string dualities. This will eventually lead us to the $d = 11$ $E_8 \times E_8$ theory and its consequences for unification, supersymmetry breakdown and the properties of the low-energy effective supergravity action.

In this written up version of the lectures, I shall not include the introduction to supersymmetry and supergravity, the construction of the supersymmetric standard model and grand unified models. The lectures follow quite closely to lectures I have given earlier. For the written versions the interested reader might consult [Nilles (1984)][Nilles (1990a)][Nilles (1993)]. We shall start here immediately from the $d = 10$ supergravity theory motivated by the $E_8 \times E_8$ heterotic string.

2 Low-energy limit of superstring theories in $d = 10$

We shall discuss here the effective action of superstring theories in the supergravity field theory framework. Superstring theories typically require $d = 10$ and we will mostly concentrate on the heterotic theory with gauge group $E_8 \times E_8$. The underlying theory is a candidate for a finite theory, including gravity, and therefore unifies all known interactions. Being defined in $d = 10$, some compactifications of the six extra dimensions would be required to make contact with phenomenology. This process is at the moment not understood very well; one has to make crude approximations and then check for consistency a posteriori. One well-defined starting point for such an approach is the theory in the so-called zero slope limit, i.e., the $d = 10$ field theory of the massless string states. For the known superstring theories this is $N = 1$ supergravity in $d = 10$ coupled to pure $E_8 \times E_8$ or 0(32) gauge multiplets. The spectrum of this theory is given by the supergravity multiplet $(g_{MN}, \psi_{M\alpha}, B_{MN}, \lambda_\alpha, \phi)$, where $M, N = 0, \ldots, 9$ are world indices and α is a Majorana-Weyl spinor index, as well as the gauge multiplet (A_M^A, χ_α^A), where $A = 1, \ldots, 496$ lables the adjoint representation of $E_8 \times E_8$ or 0(32). In the Type I theory, these correspond to the massless closed or open string states, respectively. The action of such a theory, including terms up to two derivatives, is unique and given by [Chamseddine (1981)]:

$$
\begin{aligned}
e_{10}^{-1}\mathcal{L} = &-\frac{1}{2}R - \frac{i}{2}\bar{\psi}^M \Gamma^{MNP} D_N(\omega)\psi_P + \frac{9}{16}\left(\frac{\partial_M \varphi}{\varphi}\right)^2 + \\
&+\frac{3}{4}\varphi^{-3/2} H_{MNP}H^{MNP} + \frac{i}{2}\bar{\lambda}\Gamma^M D_M(\omega)\lambda + \frac{3\sqrt{2}}{8}\bar{\psi}_M\left(\frac{\Gamma^P \partial_P \varphi}{\varphi}\right)\Gamma^M \lambda \\
&-\frac{\sqrt{2}}{16}\varphi^{-3/4} H_{MNP}\Big(i\bar{\psi}_Q \Gamma^{QMNPR}\psi_R + 6i\bar{\psi}^M \Gamma^N \psi^P + \\
&+\sqrt{2}\bar{\psi}_Q \Gamma^{MNP}\Gamma^Q \lambda - i\bar{\chi}\Gamma^{MNP}\chi\Big) - \\
&-\frac{1}{4}\varphi^{-3/4} F_{MN}F^{MN} + \frac{i}{2}\bar{\chi}\Gamma^M D_M(\omega)\chi - \\
&-\frac{i}{4}\varphi^{-3/8}\left(\bar{\chi}\Gamma^M \Gamma^{NP} F_{NP}\right)\left(\psi_M + \frac{i\sqrt{2}}{12}\Gamma_M \lambda\right) \\
&+ \text{four fermion interactions},
\end{aligned}
\tag{1}
$$

where Γ denote Dirac matrices in $d = 10$ and

$$
F_{MN}^A = \frac{1}{2}\partial_{[M}A_{N]}^A + f^{ABC}A_M^B A_N^C \tag{2}
$$

(written for short as $F = dA + A^2$) denotes the gauge field strength. Supersymmetry requires the field strength H_{MNP} of the antisymmetric tensor field B_{MN} not just to be the curl of B, but

$$
H_{MNP}^A = \partial_{[M}B_{NP]} + \omega_{MNP}^{YM}, \tag{3}
$$

where the Chern-Simons term is given by

$$
\omega^{YM} = Tr\left(AF - \frac{2}{3}A^3\right), \tag{4}
$$

i.e., B_{NP} has to transform non-trivially under the $E_8 \times E_8$ [or 0(32)] gauge transformations. This theory as it stands has gravitational anomalies and is too naive an approximation to the anomaly-free superstring theory. The absence of anomalies can be simulated by adding an additional term to (3)[Green and Schwarz (1984)]:

$$H = dB + \omega^{YM} - \omega^{L} \tag{5}$$

with

$$\omega^{L} = Tr(\omega R - \frac{2}{3} w^{3}), \tag{6}$$

where ω_M^{ab} is the spin connection. ω contains a derivative, thus ω^L contains three and appears squared in the action. This term is purely bosonic and for a supersymmetric action requires additional terms which up to now are only partially known. The action in (1) thus requires further terms in order to be an adequate low-energy limit of string theory. The action (1) was derived by truncating all heavy string states. For a better approximation they should be integrated out, leaving a low-energy theory with higher derivatives and terms of higher order in α' (the slope parameter). These terms appear in what is usually called "σ -model perturbation theory", not to be confused with the string loop expansion, which, at least in the heterotic case, is an expansion in g, the gauge coupling constant. This expansion in powers of α' is classical at the string level. There might also be world-sheet non-perturbative effects that play a role at this classical level. Looking at (1), one might wonder what g (the gauge coupling constant) is. Observe that the gauge fields have non-minimal gauge kinetic terms. Here g is not an input parameter, but g will be determined dynamically.

$$\frac{1}{g^2} = < \varphi^{-3/4} > \tag{7}$$

consistent with the expectations in string theory. We have to be aware of the fact that the coupling constant as determined by this naive approximation might be different from that determined by the string theory. Also, the problem that we have only one naive field theory but [at least in the case of 0(32)] two different string theories cannot be resolved in this context. This approximation is probably only useful in defining the important interactions at low energies. In order to ask more fundamental questions, like the determination of the fundamental coupling constants, the approximation probably has to be improved. This can already be seen when we discuss compactification. One possible way is to compactify on a six-torus T^6 , leading to $N = 4$ supergravity in $d = 4$, which does not resemble known $d = 4$ phenomenology. One might therefore ask the question for more non-trivial compactifications (still postponing the question of why these should be more likely than the trivial ones). Defining $\phi = (3/4)\log\varphi$ and neglecting fermionic terms, the equation of motion for ϕ is:

$$\Box\phi = \exp(-\phi)\left[F_{MN}^2 + \exp(-\phi) H_{MNP}^2\right] . \tag{8}$$

Integrating $\Box\phi$ over a compact manifold without boundary leads to a vanishing result. The right-hand side is positive definite and therefore has to vanish. This implies trivial compactification unless $\phi \to \infty$, which is outside the validity of our approximation. The addition of ω^L in H does not change the situation, but this term requires supersymmetric completion which necessitates the presence of R^2 terms. They actually appear in the Euler combination

$$-\exp(-\phi)\left[R^2_{MNPQ} - 4R^2_{MN} + R^2\right] \tag{9}$$

on the right-hand side of (8), ensuring the absence of ghosts. With these terms from the α' expansion, non-trivial compactification is possible: R^2 can be compensated by F^2, and this implies a breakdown of gauge symmetries in the presence of compactification [Candelas et al. (1985)]. Notice, however, that the scale of compactification is not yet fixed. There exists an independent argument confirming this result. For the H field to be well defined, the integral of the curl of H over a compact manifold without boundary should vanish:

$$\int_{C_4} dH = \int_{C_4} [TrF \wedge F - TrR \wedge R] = 0 \tag{10}$$

leading to a compensation of F and R in extra dimensions. These results are very encouraging. If $E_8 \times E_8$ or 0(32) were to remain unbroken in $d = 4$, they would not be able to lead to chiral fermions. The discussed constraints involve integrated quantities and could have various solutions. Only the simplest possibility – a vanishing integrand – can be studied easily [Witten (1985)]. It implies a direct identification of F and R. The spin connection $\omega^{ab}_m (m = 4, \ldots, 9\ a, b = 1, \ldots, 6)$ can be viewed as a gauge field of an 0(6) subgroup of the Lorentz group 0(9, 1), identified with A^A_m in an 0(6) subgroup of $E_8 \times E_8$ or 0(32) in order to fulfill the constraints. The question of a remaining supersymmetry in $d = 4$ is related to the holonomy group of the compact manifold, which in turn is a subgroup of 0(6). I shall not explain this relation here in detail, but just give a heuristic argument. The gravitino ψ^α_M transforms like a 4 of 0(6). $N = 1$ supersymmetry will be present in $d = 4$ if the decomposition of the 4 with respect to the holonomy group contains exactly one singlet. If there are more singlets, one will have extended supersymmetries, e.g., in the case of the torus the holonomy group is trivial and $4 = 1 + 1 + 1 + 1$, resulting in $N = 4$ supersymmetry. The simplest choice for $N = 1$ is to have $SU(3)$ holonomy, which leads to $4 = 1 + 3$ and $6 = 3 + \bar{3}$, and is used in the Calabi-Yau approach. But there are certainly more possibilities, even with discrete subgroups of $SU(3)$ corresponding to certain orbifolds. For simplicity, I shall assume here $SU(3)$ holonomy. With this identification of ω and A at least one $SU(3)$ subgroup of 0(32) or $E_8 \times E_8$ will break down during compactification. In the case of 0(32), this will lead to $0(26) \times U(1)$ with possible zero modes in the decomposition of the adjoint of 0(32), giving exclusively real representations of 0(32). Based on this argument, one usually concludes that 0(32) will not lead to a phenomenologically successful model, although not all possibilities have yet been studied. The situation in the case of $E_8 \times E_8$ looks better. A decomposition of the adjoint of E_8 with respect to $E_6 \times SU(3)$ leads to $248 = (78, 1) + (27, 3) + (\overline{27}, \bar{3}) + (1, 8)$ and contains chiral representations. Moreover, E_6 is one of the more successful candidates for a grand unified gauge group with a family of quarks and leptons in 27, the number of these zero modes being defined by topolocigal properties of the compact manifold. Here is then a common starting point for "superstring-inspired models" involving a further breakdown of E_6, renormalization group analysis of coupling constants, intermediate scale breaking, possibilities of additional $U(1)$'s at low energies and the question of Yukawa couplings, where one seems to need more input to explain neutrino masses and the absence of proton decay. I have not the time to discuss this here, but will concentrate on questions which are less model-dependent.

3 A first look at the possible theory in $d = 4$

We have first to discuss the possible zero modes. Let us define indices $M = (\mu, m)$ ($\mu = 0, \ldots, 3$; $m = 4, \ldots, 9$) and start with the metric

$$g_{MN} = \left(\begin{array}{c|c} g_6^{-1/2}\hat{g}_{\mu\nu} & \\ \hline & g_{mn} \end{array}\right), \tag{11}$$

where $g_6 = \det g_{mn}$ is used to redefine $g_{\mu\nu}$ in order to have usual kinetic terms for the graviton. The integral over extra dimensions

$$\int d^6y\sqrt{-g_6} = R_c^6 \sim \frac{1}{M_c^6} \tag{12}$$

defines the average radius of compactification. Defining $g_{mn} = \exp(\sigma)\hat{g}_{mn}$, one can then normalize $\int d^6y\sqrt{-\hat{g}_6} = M_P^{-6}$ and $\exp(\sigma)$ defines the radius of compactification in units of the Planck length. Depending on the topological properties of the manifold, g_{mn} gives rise to zero modes that are scalars in $d = 4$ (we will not discuss off-diagonal terms in g_{MN} like $g_{\mu m}$ that give rise to gauge bosons depending on the isometries of the manifold). g_{mn} corresponds to a symmetric tensor of 0(6) with respect to the $SU(3)$ subgroup discussed earlier; we have $21 = 1 + 8 + 6 + \bar{6}$. With the notation $m = (i, \bar{j})$, the latter correspond to modes of $g_{i\bar{j}}, g_{ij}, g_{\bar{i}\bar{j}}$, while σ is the singlet.

Turning to the gravitino ψ_M^α, we can view α as an eight-dimensional index which transforms as a 4 of 0(6) and a Weyl spinor of 0(3, 1). ψ_μ^α corresponds to spin-3/2 particles in $d = 4$ with $N_{max} = 4$ as already discussed. ψ_m^α can give rise to spin-$\frac{1}{2}$ zero modes. To obtain canonical kinetic terms for the gravitino, as in the case of the metric, a rescaling

$$\tilde{\psi}_\mu = \exp(-3\sigma/4)\psi_\mu \tag{13}$$

is required.

The antisymmetric tensor field B_{MN} could give rise to $B_{\mu\nu}, B_{m\nu}$ and B_{mn} (corresponding to the Betti numbers b_0, b_1 and b_2). A zero mode from $B_{\mu\nu}$ corresponds to one pseudoscalar degree of freedom θ defined through a duality transformation

$$H_{\mu\nu\varrho}\epsilon^{\mu\nu\varrho\sigma} = \varphi^{3/2} exp(-6\sigma)\partial^\sigma\theta + \ldots \tag{14}$$

$B_{m\nu}$ could give rise to extra gauge bosons which (although possibly interesting) we shall not discuss here. B_{mn} will again correspond to pseudoscalars in $d = 4$. A decomposition with respect to $SU(3)$ gives $15 = 1 + 3 + \bar{3} + 8$ with the singlet corresponding to the "trace" $\eta = \epsilon^{mn} B_{mn}$ and $B_{\bar{i}\bar{j}}$, and B_{ij} and $B_{i\bar{j}}$ corresponding to $3, \bar{3}$, and 8 respectively. All these modes appear in the action only through the field strength H implying derivative couplings, i.e., they show axion-like behaviour. From the λ, ϕ members of the supergravity multiplet, we expect additional spin $-\frac{1}{2}(0)$ particles in $d = 4$.

The discussion of the zero modes of A_M^A involves some complication because of the identification of ω_m^{ab} and A_m^A in an $SU(3)$ subgroup. A_μ^A will, of course, give rise to gauge bosons in the adjoint representations of the unbroken gauge group, e.g., $A = 1, \ldots, 78$ for E_6. A_m^A will give rise to scalars in $d = 4$, and we are mostly interested in those transforming as 27 (or $\overline{27}$) under E_6. Let us therefore write $A = (a, i)$ or $(\bar{a}, \bar{i})$ $a = 1, \cdots, 27$. The states $C^b = A_{\bar{i}}^{b,i}$ and $B^{\bar{b}} = A_i^{\bar{b},\bar{i}}$ then transform as $27, \overline{27}$ with respect to E_6 and are singlets under the diagonal subgroup $SU(3)$ of the product of $SU(3) \subset 0(6)$ and $SU(3) \subset E_8$. These bosons will have supersymmetric partners from the zero modes of χ_α^A. The number of the possible zero models is of course entirely defined by the topological properties of the manifold under consideration.

We can now have a first look at the possible interactions of these zero models in $d = 4$ starting from the $d = 10$ action given in (1). Of course, in general we expect here not only the influence of topological properties, but also the explicit form of the metric of the compact manifold will become important. Nonetheless we will be able to obtain some non-trivial results that are rather independent of the special form of the metric. We will do that exclusively in the framework of $N = 1$ supergravity in $d = 4$, firstly because of the reasons given in Section 2, and secondly because this theory is simpler than the non-supersymmetric case.

$N = 1$ supergravity in $d = 4$ (with action including terms up to two derivatives [Cremmer et al. (1983)]) is defined through two functions of the chiral superfields ϕ_i. The first is an analytic function $f(\phi_i)$ defining the gauge kinetic terms $f(\phi_i) W^\alpha W_\alpha$. In a component language, f appears in many places, but it can be extracted most efficiently from

$$\mathrm{Ref}(\varphi_i) F_{\mu\nu} F^{\mu\nu} + \mathrm{Imf}(\varphi_i) \epsilon_{\mu\nu\varrho\sigma} F^{\mu\nu} F^{\varrho\sigma}, \tag{15}$$

where φ_i denotes the (complex) scalar component of ϕ_i. The second is the so-called Kähler potential

$$G(\phi_i, \phi_i^*) = K(\phi_i, \phi_i^*) + \log|W(\phi_i)|^2. \tag{16}$$

Unlike f, G is not analytic and contains the left-handed chiral superfields along with their complex conjugates. The second term in (16) contains the analytic function $W(\phi_i)$: the superpotential. The action in component form usually contains G in complicated form; the scalar kinetic terms, e.g., are

$$G_i^j (\partial_\mu \varphi^i)(\partial^\mu \varphi_j^*); \quad G_i^j \equiv \frac{\partial^2 G}{\partial \varphi^i \partial \varphi_j^*}, \tag{17}$$

whereas the scalar potential is given by

$$V = \exp(G)[G_k(G^{-1})^k_l G^l - 3] \tag{18}$$

which makes it difficult to extract G once an action is given in component form. There is only one term which allows a rather simple identification of G, and this is a term involving the gravitino

$$e_4 \exp(G/2)\, \bar{\psi}_\mu \gamma^{\mu\nu} \gamma_5 \psi_\nu \tag{19}$$

which will later be used extensively after the correct redefinitions of the gravitino in $d = 4$ have been performed. Let us now consider the action in $d = 10$ in order to learn something about the possible action in $d = 4$. We start with the gauge kinetic term

$$e_{10} \varphi^{-3/4} F_{MN} F^{MN}. \tag{20}$$

Since we are interested in the $F^2_{\mu\nu}$ part, we write

$$e_4 e_6 \varphi^{-3/4} F_{\mu\nu} F_{\varrho\sigma} g^{\mu\varrho} g^{\nu\sigma}, \tag{21}$$

where, with the definitions given earlier, we would like to extract f from

$$\hat{e}_4 \mathrm{Re} f\, F_{\mu\nu} F^{\mu\nu} \tag{22}$$

with $\hat{e}_4 = (det\ \hat{g}_{\mu\nu})^{\frac{1}{2}} = \exp(6\sigma) e_4$, and indices are contracted with the "hatted" metric. Integrating the extra six dimensions with the normalization given in (13) using $M_P \equiv 1$, we obtain

$$ReS \equiv Ref = \varphi^{-3/4} \exp(3\sigma) \tag{23}$$

as the real part of the scalar component of a chiral superfield denoted by S. This is a rather amazing result. Remember that at no point in the derivation did we have to know something about the metric of the compact six-dimensional space, so this constitutes a rather model-independent result. Observe that f is usually non-trivial, that its vacuum expectation value (vev) will determine the gauge coupling constant, and that the couplings of E_8 (or E_6) and E'_8 coincide.

Let us now discuss the imaginary part of f, to be extracted from $F_{\mu\nu} F_{\rho\sigma} \varepsilon^{\mu\nu\rho\sigma}$. The relevant degree of freedom comes from $B_{\mu\nu}$ as discussed earlier. $B_{\mu\nu}$ couples only through its field strength $H_{\mu\nu\rho}$ and has therefore only derivative couplings. Taking the relevant terms in the $d = 10$ action and integrating the extra dimensions, we obtain

$$\varphi^{-3/2} \exp(6\sigma) H_{\mu\nu\varrho} H^{\mu\nu\varrho} + H_{\mu\nu\varrho} O^{\mu\nu\varrho}, \tag{24}$$

where $O^{\mu\nu\varrho}$ contains fermion bilinears. H has to satisfy a constraint (neglecting R^2-terms for the moment)

$$\partial_{[\mu} H_{\nu\varrho\sigma]} = -Tr F_{[\mu\nu} F_{\varrho\sigma]} \tag{25}$$

which we take into account by adding a Lagrange multiplier

$$\theta \epsilon^{\mu\nu\varrho\sigma} \left(\partial_\mu H_{\nu\varrho\sigma} + Tr F_{\mu\nu} F_{\varrho\sigma}\right) \tag{26}$$

Next we eliminate H via the equations of motion and arrive at an action containing the terms

$$\varphi^{3/2}\exp(-6\sigma)(\partial_\mu\theta)^2 + \theta\epsilon^{\mu\nu\varrho\sigma}Tr(F_{\mu\nu}F_{\varrho\sigma}) \tag{27}$$

which tells us that $Imf = \theta$, and for the scalar component of S we obtain

$$S = \varphi^{-3/4}\exp(+3\sigma) + i\theta \tag{28}$$

as a mixture of g_{MN} and B_{MN} zero modes. The partner is a combination of ψ_m and λ zero modes which we will not discuss here in detail. Observe that θ couples only with derivatives except for the last term in (27), and that the $d = 4$ action has a Peccei-Quinn-like symmetry under shifts of θ by a real constant, thus θ couples like an axion. Let me stress again that all these statements about the action and the form of (28) are model-independent and could be derived without explicit knowledge of the metric.

Unfortunately, the situation changes once we try to extract the Kähler potential. As already indicated, the term to investigate is the $d = 4$ "gravitino mass term" (19). The extraction of this term is rather complicated due to several redefinitions of the gravitino field. A general form has been given in [Derendinger et al. (1986)]), and we will not repeat the derivation here. Many of the terms appearing there depend explicitly on the metric and spin-connection of the six-dimensional compact space. A model-independent statement can only be made about the structure of the superpotential, because it is an analytic function in the chiral superfields. Symbolically the "gravitino mass term" is obtained as

$$\exp(G/2) = \varphi^{-3/4}\exp(-3\sigma)\Gamma^{mnp}H_{mnp} \tag{29}$$

and from (16) we can try to read off the superpotential. $W(\phi_i)$ is defined to be an analytic function in the chiral superfields and should not contain derivatives. A first inspection of (29) therefore suggests that a possible candidate for a superpotential is the A^3 term contained in the Yang-Mills Chern-Simons term (4) included in H. This then gives rise to a trilinear superpotential involving the C and B fields defined earlier. At the moment it is not clear whether these are the only possible terms in the superpotential, although at the classical level this seems to be the complete expression. Observe that, for example, the superfield S as defined in (28) cannot appear in the superpotential, since its pseudoscalar component has only derivative couplings. We will come back to these points later. In any case, a more detailed discussion of the Kähler potential requires more information (or approximations) about the $d = 6$ metric. Before we tackle this topic, let me first present a discussion about supersymmetry breakdown in $d = 4$.

4 Gaugino condensation and supersymmetry breakdown

$N = 1$ supergravity in $d = 4$ still needs the incorporation of supersymmetry breakdown at a scale small compared to the Planck mass. For the phenomenological reasons mentioned earlier, this should appear in a hidden sector only coupled gravitationally to the observable sector. Some superstring models now miraculously contain such a hidden sector, the sector that contains the particles transforming non-trivially under the second E_8. Notice that the observable sector (for definiteness called the E_6 sector) only couples gravitationally to the E'_8 sector (there are no particles that transform non-trivially both under E_6 and E'_8). Moreover, the E'_8 sector contains a $d = 10$ pure super-Yang-Mills multiplet, suggesting a possible breakdown of supersymmetry via gaugino condensates. This breakdown has already been discussed in the framework of supergravity models, both at the level of an effective Lagrangian [Nilles (1982)] and at the level of the complete classical action [Ferrara et al. (1983)]. Assume asymptotically-free gauge interactions (here E'_8 or a subgroup thereof) with a scale

$$\Lambda = \mu \exp\left(-1/b_0 g^2(\mu)\right) \tag{30}$$

which is renormalization-group invariant at the level of the one-loop β-function. In analogy to QCD, which leads to $q\bar{q}$ condensates, we will assume here that the gauge fermions condense at a scale

$$< \chi\chi >= \Lambda^3 \tag{31}$$

As long as Λ is small compared to M_p, we assume that gravity will not qualitatively disturb this dynamic mechanism. The question whether such a condensate breaks supersymmetry can be studied by investigating the supersymmetry transformation laws of the fermionic fields of the theory. The non-derivative terms in these transformations will give us the auxiliary fields that serve as order parameters for supersymmetry breakdown. The relevant objects here are the auxiliary fields of the chiral superfields

$$F_k = \exp(G/2)G_k - \frac{1}{4} f_k(\chi\chi) + \ldots, \tag{32}$$

where f is the gauge kinetic function discussed earlier and f_k is its derivative with respect to ϕ_k. A necessary condition for the breakdown of supersymmetry via gaugino condensates is therefore a non-trivial f-function. This condition is fulfilled in the framework of superstring-inspired models [Derendinger et al. (1985)] [Dine et al. (1985)], since we have seen in the last section that $f = S$ in a rather model-independent way. Whether this is also sufficient for the breakdown of supersymmetry can only be checked by minimizing the potential

$$V = F_k \left(G^{-1}\right)^k_\ell F^\ell - 3\exp(G), \tag{33}$$

since the different terms in (32) might cancel at the minimum. But let us for the moment assume that only the second term in (32) receives a vev. Since $f_s = 1$ in units of M_p, we find a supersymmetry breakdown scale

$$< F_s >= M_s^2 \approx \Lambda^3/M_p \tag{34}$$

and a scale of $\Lambda \sim 10^{13}$ GeV would lead to a gravitino mass in the TeV range. Once we understand why Λ is five orders of magnitude smaller than M_p, we shall understand why $m_{3/2}/M_p \sim 10^{-15}$. Λ now depends on the E_8' gauge coupling and the spectrum of low-energy modes. Identifying g_6 with g_8 would in many circumstances lead to too large a value for Λ, and one might speculate that E_8' should break during compactification. We shall, however, see later that the equality of g_6 and g_8 seems to be only an artifact of the classical approximation, which is not true in the full theory. Thus the shadow E_8' (or a subgroup thereof) sector of the superstring takes the role of the hidden sector of supergravity models and might explain the smallness of $m_{3/2}$ compared to M_p. But how does this breakdown of SUSY in the hidden sector influence the observable sector? In general, we would expect gaugino masses (m_0), scalar masses ($\tilde{m}$) and the trilinear couplings (Am) to be of the order of magnitude of $m_{3/2}$. A naive inspection shows that this might also be true here. Gaugino masses in the observable sector are in general given by

$$m_0 = f_k \left(G^{-1}\right)^k_\ell F^\ell, \tag{35}$$

where f is the gauge kinetic function of the observable sector. With $F^\ell = (1/4)f^\ell < \chi\chi >$ we would therefore obtain $m_0 = m_{3/2}$. In the same way we would obtain under these circumstances trilinear couplings $A = 1$ and scalar masses of order $m_{3/2}$. To make a definite statement we have to watch out for possible cancellations, which can only be studied once we have a better knowledge of the Kähler potential, a question which we want to discuss now.

5 Reduction and Truncation

A first approximation for G (that might simulate an orbifold approximation of interest in this context) is obtained through reduction and truncation. One first compactifies the $d = 10$ theory on a six-torus T^6. The resulting theory is $N = 4$ supersymmetric in $d = 4$. From this theory one truncates unwanted states, to obtain an $N = 1$ theory. From the gauge singlet sector one keeps only those states that transform as singlets under an $SU(3) \subset 0(6)$ of the Lorentz group. Since ψ^α_μ transforms as a 4 of 0(6) and thus as 1 + 3 under $SU(3)$, we remain with one gravitino. As already explained in Section 3, there are only a few gauge singlets that survive this truncation. For the bosonic modes we have φ, σ from the metric as well as θ and η from the antisymmetric tensor. For the gauge non-singlet fields one has to remember the identification of spin-connection and gauge fields. Here one keeps those states which are singlets under the diagonal subgroup of the product of $SU(3) \subset 0(6)$ and $SU(3) \subset E_8$. This leaves us with one 27 of E_6 in this case, corresponding to $C^b = A^{b,i}_{\bar{i}}$; ($b = 1, \ldots, 27$, cf. Section 3). With this well-defined procedure based on simple reduction on T^6, the component Lagrangian in $d = 4$ can be deduced. From this we can immediately read off $f = S$ and $W = d_{abc}C^aC^bC^c$, which should not be surprising. Moreover, from the "gravitino mass term" formula (29) one obtains

$$G = \log\left(e^{-6\sigma}\varphi^{-3/2}\right) + \log|W|^2. \tag{36}$$

The components φ and σ should correspond to lowest components of chiral superfields. One combination $S = \varphi^{-3/4}\exp(3\sigma) + i\theta$ has already been defined earlier. To define the other combination, the information from (36) is not enough. The charged fields C do not yet appear in the first term of G in (36) and the correct definition of the superfields has yet to be found. This can be done, for example, by using the scalar kinetic terms. It leads to a second superfield in which φ, σ and the C-modes mix

$$T = \exp(\sigma)\varphi^{3/4} + |C_a|^2 + i\eta, \tag{37}$$

where η is the mode from $\epsilon^{mn}B_{mn}$ as discussed earlier, and the Kähler potential from (36) thus reads

$$G = -\log(S + S^*) - 3\log\left(T + T^* - 2|C|^2\right) + \log|W|^2, \tag{38}$$

a form already previously mentioned in the framework of supergravity models. The scalar potential derived from this G-function has some remarkable properties

$$V = \frac{1}{16st_c^3}\left[|W|^2 + \frac{t_c}{3}|W'|^2\right] + D^2 - \text{terms}, \tag{39}$$

where $s = ReS$ and $t_c = ReT - |C_a|^2 = t - |C_a|^2$ and W' is the derivative of W with respect to the C-field. The potential is positive definite ($t_c > 0$ is required by the kinetic terms) and has a minimum with vanishing vacuum energy $V = 0$. This minimum is obtained at $W = W' = 0$ independent of the values of s and t. This implies that at this level the gauge coupling constant and the radius of compactification is not yet fixed. The theory has classical symmetries which allow shifts of the values of s and t, as well as Peccei-Quinn symmetries corresponding to shifts in θ and η. This, of course, makes the use of this approximation as an effective low-energy limit of the superstring very problematic. Certain crucial parameters, like the value of the gauge coupling constant and the scale of compactification, which we believe to be dynamically determined in the full string theory, are not yet fixed. To determine these quantities in the truncated theory does not necessarily lead to the same results that would be obtained in the full theory. Actual calculations of radiative corrections in the truncated theory have been performed at the one-loop level. It was found that the resulting potential is unbounded from below with some vev running to infinity. This result should actually not be too surprising. The model was obtained by truncating states with mass of the order of the compactification scale M_c, which in a certain way corresponds to the limit $M_c \to \infty$. In the truncated theory this scale is determined by the vev of $\exp(-\sigma)$ and, if the approach is consistent, the only logical possibilities are either M_c undetermined or $M_c \to \infty$. These questions can only be solved once we know more about the theory of the massive states or the full string dynamics. The model as it stands should be regarded as an approximation for the possible interactions of the zero modes, rather than a tool to determine the fundamental dynamic quantities of string theories.

This becomes even more apparent when we include the concept of gaugino condensation within this framework. Since the gauge coupling constant is not determined, Λ in (30) is also unknown. Using (33) and (38), we get for the potential

$$V = \frac{1}{16st_c^3}\left[|W - 2(st_c)^{3/2}(\chi\chi)|^2 + \frac{t_c}{3}|W'|^2\right], \tag{40}$$

where $(\chi\chi)$ depends on g^2 through $\exp(-S/b_0)$. The potential is still positive definite and has a minimum at $V = 0$ which is still degenerate. Now the minimum needs not necessarily to imply $W = W' = 0$, but we could have a non-trivial vev of W. Given fixed $< W > \neq 0$ by some unknown mechanism (not to be understood in the truncated theory), such as a vev of dB in H or a slight mismatch of the vevs of the Chern-Simons terms, the value of the gauge coupling constant would be fixed. In order to minimize the potential, the theory slides to a coupling constant which, through (30), gives a value of the condensate that exactly cancels the contribution of W. In other words, this means that the dilaton S slides to a value that cancels the vacuum energy in the same way as an axion slides to cancel a possible θ-parameter of a gauge theory [observe that $\exp(-S/b_0)$ contains both s and θ]. Although we do not yet understand the magnitude of supersymmetry breakdown, this mechanism to ensure $E_{\text{vacuum}} = 0$ after $SUSY$ breakdown appears very attractive. We shall still need to convince ourselves that supersymmetry is actually broken, since in (40) a certain cancellation of $< W >$ and $< \chi\chi >$ appears. In fact it tells us that the auxiliary field F_S of the S-superfield vanishes in the vacuum. Nonetheless, here F_T requires a non-vanishing vev once $< W > \neq 0$, and supersymmetry is broken

$$F_T = exp(G/2)G_T \neq 0. \tag{41}$$

Another question is the breakdown of $SUSY$ as it is felt in the observable sector, and here things look different. For example, the gaugino masses are given by $m_0 = f_k(G^{-1})^k_\ell F^\ell$, and only f_S is different from zero. In this case we therefore obtain $m_0 = 0$, and the same is true for the scalar masses. That they remain zero in this case has brought people to the idea of constructing models in which $m_{3/2}$ is as big as M_P but still having small $SUSY$ breakdown parameters in the observable sector, a situation which is not likely to be meaningful once radiative corrections are fully considered. As already mentioned, the full corrections as naively computed at one loop in the truncated theory have to be interpreted carefully. Nonetheless, the naive approximation discussed in this section has some attractive properties as well as these obvious shortcomings. The question remains whether these shortcomings are a result of the special approximation or a more general property of classical considerations.

6 Classical Symmetries

A more complete picture of the classical approximation can be obtained by a study of the symmetries of the field theory Lagrangian. In this chapter, we will present such a discussion in the spirit of [Nilles (1986)].

We turn first to the Peccei-Quinn-like symmetries originating from the B_{MN} modes which couple only with derivatives. This implies that the superfields S and T in $d = 4$ which contain these modes as pseudoscalars cannot appear in the superpotential, at least in any order of perturbation theory. In a more general framework than that discussed previously, not only one T-field but several could appear. For our purpose it is, however, sufficient to discuss only one mode; the generalization to more than one is trivial. The symmetries of the classical action are

$$S \to S + i\alpha \quad T \to T + i\beta \tag{42}$$

with α, β being real constants. This implies that the exact value of the imaginary parts of S and T must be unphysical. As a consequence, the functions G and the real part of f should not depend on the imaginary part of S and T. Observe that the imaginary part of f which multiplies $F\tilde{F}$ can very well depend on these fields.

Inspecting the $d = 10$ action given in (1), we can identify two more classical symmetries. They are

$$\begin{array}{llll} g_{MN} & \to t g_{MN} & \lambda & \to t^{-1/4}\lambda \\ \varphi & \to t^{-4/3}\varphi & \chi & \to t^{-1/4}\chi \\ \psi_a & \to t^{-1/4}\psi_a & \mathcal{L}_{10} & \to t^4\mathcal{L}_{10} \end{array} \tag{43}$$

where a is a flat index, and

$$\begin{array}{llll} \varphi & \to r^{2/3}\varphi & \chi & \to r^{-1}\chi \\ \psi_M & \to r^{-1}\psi_M & \kappa_{10} & \to r\kappa_{10} \\ \lambda & \to r^{-1}\lambda & \mathcal{L}_{10} & \to r^{-2}\mathcal{L}_{10} \end{array} \tag{44}$$

where κ_{10} is the gravitational coupling in the $d = 10$ theory. Observe that under both transformations the action is multiplied by an overall factor, implying that these symmetries are just classical and expected to be broken in string perturbation theory. These are symmetries of the field-theory Lagrangian; is there any connection to string theory? There is. The classical symmetries can be understood as a rescaling of κ_{10} into the string tension (keeping the relation $g_{10} = \sqrt{T_s}\kappa_{10}$ in the heterotic string theory), and it is understood that in string theory g_{10} is not an input parameter but the choice of a vacuum. For the $d = 4$ theory, this implies that classically the gauge coupling constant and the scale of compactification are not yet fixed. We have therefore (as already previously explained) a theory of the zero modes which in itself cannot solve these fundamental questions. A solution has to come from higher modes or non-classical or non-perturbative effects. In the $d = 4$ action, the symmetries correspond to

$$\begin{array}{llll} \hat{g}_{\mu\nu} & \to t^4\hat{g}_{\mu\nu} & \tilde{\psi}_\mu & \to t\tilde{\psi}_\mu \\ S & \to t^4 S & \chi & \to t^{-1}\chi \\ T & \to T & \lambda & \to t^{-1}\lambda \end{array} \tag{45}$$

with $\mathcal{L}_4 \to t^4 \mathcal{L}_4$; $\tilde{\psi}_\mu$ is the rescaled gravitino and

$$\begin{aligned} S &\to r^{-1/2} S \\ C^a &\to r^{1/4} C^a \\ T &\to r^{1/2} T \end{aligned} \tag{46}$$

with $\mathcal{L}_4 \to r^{-1/2} \mathcal{L}_4$. These classical symmetries imply the existence of flat directions in the potential leading to degenerate vacua in which T and S are not determined. Sometimes these symmetries are discussed as a part of $SU(1,1) \times SU(n,1)$ symmetries present at the level of the scalar kinetic terms. These bigger symmetries seem to be as relevant as the $SU(n)$ symmetries of the kinetic Lagrangian in a usual theory of complex scalar fields. They are broken if the string theory provides a superpotential, and only (42), (45) and (46) remain.

We can now study the implications of these symmetries for the classical form of f and G. To discuss f, we consider the term $\hat{e}^4 Ref F^2_{\mu\nu}$ and deduce that

$$Ref \to t^4 r^{-1/2} Ref \tag{47}$$

which is the transformation behaviour of S. Including the restrictions from the Peccei-Quinn symmetries (independence of Re f on Im S,T), we deduce

$$f = S \tag{48}$$

which should not be too big a surprise, since we had already deduced this result earlier in a model-independent way.

Again, the discussion of G is more complicated because G is not an analytic function of the chiral superfields. Investigating the gravitino mass term, we obtain

$$\exp(G/2) \to t^{-2} r^{1/4} \exp(G/2). \tag{49}$$

Writing according to (16)

$$\exp(G) = |W|^2 \exp(K), \tag{50}$$

we observe that W has to be homogeneous of degree n. There are no restrictions on n from the symmetries, but we have earlier discussed an argument that the most likely choice is $n = 3$. To present the general form of G, we restrict ourselves to the modes discussed in the last chapter. Generalizations are conceptually trivial but technically more complicated. We assume that W is just a function of the C^a 's, and arrive at

$$\begin{aligned} G = &-\log(S + S^*) - n\log(T + T^* + g(C, C^*)) \\ &+ \log|W|^2 + \bar{K}\left(\frac{C}{\sqrt{T+T^*}}, \frac{C^*}{\sqrt{T+T^*}}\right), \end{aligned} \tag{51}$$

where $g(C, C^*)$ should scale like T, i.e., CC^*. Actually, the functions g and $\bar{K}$ can be transformed into each other, giving the general form of a supergravity potential with flat directions. We also see that (51) is quite close to (38) found in the simple approximation, and the classical symmetries restrict the form of f and G drastically. But it should again be stressed that these symmetries have to be broken before we can understand the magnitude of the gauge coupling constant and the scale of compactification. There are many sources of such a possible breakdown: higher terms in σ-model perturbation theory including non-perturbative effects such as world-sheet instantons, even at the classical level. These effects might also break the non-renormalization of the superpotential usually expected in perturbation theory (S, T might appear exponentially in the superpotential). These results seem to depend strongly, however, on the chosen compactification scheme. Another source is given by new terms in the string-loop expansion which we will discuss in the next section.

7 Beyond the classical level

The Peccei-Quinn-like symmetries (42) are believed to survive this loop expansion in any finite order of perturbation theory and can only be broken by non-perturbative effects like gauge or world-sheet instantons. (Actually, the symmetry related to S will not be disturbed by world-sheet non-perturbative effects.) For the other classical symmetries the situation is different. Classically the action is scaled and the symmetries are broken by quantum effects. In the heterotic string this loop expansion is governed by the coupling constant g, which in turn is defined through a vev of the dilation field. This will allow us to construct a definite loop expansion in the dilation field and still give us restrictions on how the classical symmetries are broken by loop effects. But before we discuss the loop expansion in more general terms, let us examine some aspects at the one-loop level. We can do that because of the mechanism of anomaly cancellation in the $d = 10$ field theory. Green and Schwarz have observed that the cancellation of anomalies [Green and Schwarz (1984)] requires certain new local counterterms with definite finite coefficients in the one-loop effective action to cancel the gauge non-invariance of present non-local terms. In general, such terms appear with infinite coefficients, but the possible symmetry of the effective action forces us to renormalize the theory in such a way that these gauge-variant local counterterms have a well-defined finite coefficient. An example of such a term is

$$\epsilon\epsilon^{MNPQRSTUVW} B_{MN} Tr(F_{PQ}F_{RS}) Tr(F_{TU}F_{VW}), \tag{52}$$

where $\epsilon = 1/720(2\pi)^5$. While this gives rise to many new interaction terms in the $d = 4$ theory, one possible manifestation seems to be of particular importance. Replacing one of the TrF^2 terms by their vev in extra dimensions, one arrives at

$$\eta\epsilon^{\mu\nu\varrho\sigma} Tr(F_{\mu\nu}F_{\varrho\sigma}). \tag{53}$$

η is the imaginary part of T, and unlike in the classical case it now (in addition to θ) couples to $F\tilde{F}$. Observe that (53) is gauge-invariant, while (52) is not, but is required by the absence of anomalies in $d = 10$. This shows that the remnants of such terms originate in ten dimensions, and are one of the few places where we could in principle observe whether we live in higher dimensions. (53) suggests that not only θ, but also η, couples like an axion. To make sure that this does not lead just to a redefinition of θ at the one-loop level, all anomaly cancellation terms have to be considered. Doing this and satisfying $TrF^2 = TrR^2$ in extra dimensions, one arrives at the result that η couples differently to E_6 and E_8':

$$\epsilon\eta[(F\tilde{F})_8 - (F\tilde{F})_6] \tag{54}$$

while

$$\theta[(F\tilde{F})_8 + (F\tilde{F})_6]. \tag{55}$$

This fact has interesting consequences, some of which we will list now.

a) The second axion could be a candidate to solve the strong CP problem of QCD in the observable sector. One axion (like θ alone) would not be sufficient, because it is used to adjust the θ-angle of E_8' and becomes massive, of order $m_{3/2}$. Things are not simple. For a relatively recent discussion see [Georgi et al. (1998)].
b) Supersymmetry requires the same behaviour of the real parts of S and T as that of the imaginary part; i.e., ReS and ReT couple differently to E_6 and E_8'. Since the vevs of these fields define the gauge coupling constants, g_6 and g_8' need no longer be equal. This might have consequences for the condensation scale of E_8'.
c) There exist now two axion-dilaton pairs, and this might generalize the relaxation of the cosmological constant to the observable sector in the same way as is appears in the hidden sector [compare Eq. (40)].
d) Imposition of supersymmetry also requires new terms in the Kähler potential at the one-loop level. We shall discuss this later.
e) As expected, these effects at the one-loop level lead to an induced breakdown of supersymmetry in the observable sector once it is broken in the hidden sector. Remember our discussion in Section 5, where the observable sector remained supersymmetric. Gaugino masses are given by

$$m_0 \sim F_T f_T + F_S f_S \tag{56}$$

and vanish because $F_S = f_T = 0$. But here we now have $f = S + \epsilon T$, and f_T no longer vanishes. As a result, non-trivial gaugino masses (and also non-trivial scalar masses and A-parameters) of order $\epsilon m_{3/2}$ are transmitted to the observable sector. This shows again that in a theory with $m_{3/2}$ of the order of the Planck mass, no sign of supersymmetry can survive in the TeV region. In addition we see that gaugino masses of order $\epsilon m_{3/2}$ tend to be small compared to the gravitino mass.

According to the classical symmetries, the new terms scale as

$$\mathcal{L}_{1-loop} \to t^0 r^{1/2} \mathcal{L}_{1-loop} \tag{57}$$

consistent with the expectation of a g^2-expansion. New counterterms at the n-loop level would therefore scale as

$$\mathcal{L}_{n-loop} \rightarrow t^{4(1-n)} r^{2n-1/2} \mathcal{L}_{n-loop}. \quad (58)$$

We can now try to extract the restrictions on f and G within this framework. Since, following (47), f scales like the action we write:

$$f = \sum_{n=o}^{\infty} f_n \quad (59)$$

with

$$f_n \rightarrow t^{4(1-n)} r^{2n-1/2} f_n \quad (60)$$

We know that $f_0 = S$. f_1 cannot contain S since this is the only field transforming non-trivially under the T-symmetry, leaving $f_1 = T + C^2$. For $n > 2$, the analyticity of f and the fact that Re f should be independent of ImS, T forces f_n to vanish. Thus we have a non-renormalization theorem for f beyond the one-loop level [Nilles (1986)] and

$$f = S + \epsilon(T + \alpha C^2). \quad (61)$$

Observe that the presence of supersymmetry allows us to obtain such a restrictive result from broken symmetries, since supersymmetry forces f to be analytic in the chiral superfields. This is not in contradiction with the logarithmic variation of gauge coupling constants defined through the vev of f, since this involves a discussion of the potential where non-analytic pieces appear.

G can be examined in a similar way. At the n-th level, we obtain

$$\exp(G/2) \rightarrow t^{-4n-2} r^{n+1/4} \exp(G/2) \quad (62)$$

and this leaves us with

$$\begin{aligned} G = & -\log(S + S^*) - 3\log(T + T^* + g(C, C^*)) \\ & + \log|W|^2 + 2\log\left(1 + \sum_{n=1}^{\infty} a_n \epsilon^n \left(\frac{S + S^*}{T + T^*}\right)^n\right) \\ & + \bar{K}\left(\frac{C}{\sqrt{T + T^*}}, \frac{C^*}{\sqrt{T + T^*}}\right) \end{aligned} \quad (63)$$

where we expect the non-renormalization theorem for W to hold in any finite order of perturbation theory, and the a_n are unknown coefficients. A discussion of the resulting scalar potential is very complicated and has not yet been attempted. As long as supersymmetry remains unbroken (due to the non-renormalization theorems), we expect this potential still to have flat directions at $E_{\text{vacuum}} = 0$, but with redefined S and T fields. However, once a breakdown of supersymmetry and the corresponding gaugino bilinears are included, this potential might fix $< t >$ as well as $< s >$ and that would allow to determine values of some of the parameters of the low-energy effective theory.

So far our naive discussion of the low energy limit of the $d = 10$ heterotic string theory as one would have discussed it already 12 years ago. Although we have used very crude approximations, a posteriori, it turned out that this picture was quite useful. Of course, meanwhile many of the calculations have been done in a more rigorous way in the framework of full string theory. An example are so-called string threshold corrections that allow a full determination of the gauge kinetic function as a function of the moduli field S and T, the calculation of Yukawa-couplings etc. It turned out through the whole discussion, that the mechanism of anomaly cancellation is a very useful tool to determine properties of low energy string theory, even in cases where a full string calculation is not possible.

8 More recent developments: M-Theory

Recently, with the discovery of string dualities, there has been a revival of the study of string theories that might also eventually become relevant for our discussion of the low-energy effective supergravity theories. From all the new and interesting results in string dualities, it is the heterotic M–theory of Hořava and Witten [Hořava and Witten (1996)] (that in $d = 11$ could be regarded as the strong coupling limit of $d = 10$ $E_8 \times E_8$ heterotic string theory) which seems to have immediate impact on the discussion of the phenomenological aspects of these theories. One of the results concerns the question of the unification of all fundamental coupling constants [Witten (1996)] and the second one the properties of the soft terms (especially the gaugino masses) once supersymmetry is broken [Nilles et al. (1997)], [Nilles et al. (1998)]. As we shall see in both cases, results that appear problematic in the weakly coupled case (as the formerly discussed heterotic string case will be called from now on) get modified in a satisfactory way, while the overall qualitative picture remains essentially unchanged. In these lectures we shall therefore concentrate on these aspects of the new picture.

The heterotic M–theory is an 11–dimensional theory with the $E_8 \times E_8$ gauge fields living on two 10–dimensional boundaries (walls), respectively, while the gravitational fields can propagate in the bulk as well. A $d = 4$ dimensional theory with $N = 1$ supersymmetry emerges at low energies when 6 dimensions are compactified on a Calabi–Yau manifold. The scales of that theory are M_{11}, the $d = 11$ Planck scale, R_{11} the size of the x^{11} interval, and $V \sim R^6$ the volume of the Calabi–Yau manifold. The quantities of interest in $d = 4$, the Planck mass, the GUT–scale and the unified gauge coupling constant α_{GUT} should be determined through these higher dimensional quantities. The fit of ref. [Witten (1996)] identifies $M_{GUT} \sim 3 \cdot 10^{16}$ GeV with the inverse Calabi–Yau radius R^{-1}. Adjusting $\alpha_{GUT} = 1/25$ gives M_{11} to be a few times larger than M_{GUT}. On the other hand, the fit of the actual value of the Planck scale can be achieved by the choice of R_{11} and, interestingly enough, R_{11} turns out to be an order of magnitude larger than the fundamental length scale M_{11}^{-1}. A satisfactory fit of the $d = 4$ scales is thus possible, in contrast to the case of the weakly coupled heterotic string where, as we shall discuss in the next section, naively the string scale seems to be a factor 20 larger than M_{GUT}.

Otherwise the heterotic $E_8 \times E_8$ string looks rather attractive from the point of view of phenomenological applications. One seems to be able to accommodate the correct gauge group and particle spectrum. The mechanism of hidden sector gaugino condensation leads to a breakdown of supersymmetry with vanishing cosmological constant to leading order. With a condensate scale $\Lambda \sim 10^{13}$ GeV, one obtains a gravitino mass in the TeV range and soft scalar masses in that range as well. In the simplest models [Derendinger et al. (1985)], [Dine et al. (1985)], [Derendinger et al. (1986)] this type of supersymmetry breakdown is characterized through the vacuum expectation value of moduli fields other than the dilaton, giving a small problem with the soft gaugino masses in the observable sector: they turn out to be too small, generically about two orders of magnitude smaller than the soft scalar masses. It is again in the framework of heterotic M–theory that this problem is solved [Nilles et al. (1997)]; gaugino masses are of the same size as (or even larger than) the soft scalar masses.

The mechanism of hidden sector gaugino condensation itself can be realized in a way very similar to the weakly coupled case. This includes the mechanism of cancellation of the vacuum energy, which in the weakly coupled case arises because of a cancellation of the gaugino condensate with a vacuum expectation value of the three index tensor field H of $d = 10$ supergravity. This cancellation is at the origin of the fact that supersymmetry breakdown is dominated by a T modulus field rather than the dilaton (S). Hořava [Hořava (1996)] observed that this compensation of the vacuum expectation values of the condensate and H carries over to the M–theory case. In [Nilles et al. (1997)] this has been explicitly worked out for the mechanism of gaugino condensation in the heterotic M–theory and the similarity to the weakly coupled case was shown. Now the gaugino condensate forms at the hidden 4–dimensional wall and is canceled locally at that wall by the vev of a Chern–Simons term. This also clarifies some questions concerning the nature of the vev of H that arose in the weakly coupled case.

In the remainder of these lectures we want to discuss the phenomenological properties of the heterotic M–theory. This includes a presentation of the full effective four–dimensional $N = 1$ supergravity action in leading and next–to–leading order, the mechanism of hidden sector gaugino condensation and its explicit consequences for supersymmetry breaking and the scalar potential and finally the resulting soft breaking terms in the 4–dimensional theory. Although some of the issues have already been discussed earlier, we shall at each step first explain the situation again for the weakly coupled theory and then compare it to the results obtained in the M–theory case.

These results are obtained using the method of reduction and truncation that has been successfully applied to the weakly coupled case [Witten (1985)], [Derendinger et al. (1986)], [Nilles (1986)]. It is a simplified prescription that shows the main qualitative features of the effective $d = 4$ effective theory. In orbifold compactification it would represent the fields and interactions in the untwisted sector. We compute Kähler potential (K), superpotential (W) and gauge kinetic function (f) both in the weakly and strongly coupled regime and explain similarities and differences.

The remainder of the lectures will proceed as follows. First we discuss the scales and the question of unification as suggested in [Witten (1996)] and compare the two cases. Then we derive the effective $d = 4$ action of M–theory using the method of reduction and truncation. In this case we have to deal with a nontrivial obstruction first encountered in [Witten (1996)]. It leads to an explicit x^{11} dependence of certain fields, which is induced by vevs of antisymmetric tensor fields at the walls. To obtain the effective action in $d = 4$ we have to integrate out this dependence. This then leads to corrections to K and f in next to leading order, which are very similar to those in the weakly coupled case. We also discuss the appearance and the size of a critical radius for R_{11}. The phenomenological fit presented in our discussion of unification implies that we are not too far from that critical radius. We then turn again to the question of supersymmetry breakdown. We start with the weakly coupled case and investigate the nature of the vev of the H–field (concerning some quantization conditions) and the cancellation of the vacuum energy. In the strongly coupled case we shall see that such a cancellation appears locally at one wall. This supports the interpretation that the gaugino condensate is matched by a nontrivial vev of a Chern–Simons term. We then explicitly identify the mechanism of supersymmetry breakdown and the nature of the gravitino. The goldstino turns out to be the fermionic component of the T superfield that represents essentially the radius of the 11th dimension. It is a bulk field, with a vev of its auxiliary component on one wall. Integrating out the 11th dimension we then obtain explicitly the mass of the gravitino.

The remainder deals with the induced soft breaking terms in the observable sector: scalar and gaugino masses. We shall see a strong model dependence of the scalar masses and argue that they are not too different from the gravitino mass. This all is very similar to the situation in the weakly coupled case. We then compute the soft gaugino masses and see that in the strongly coupled case they are of the order of the gravitino mass. This comes from the fact that we are quite close to the critical radius and represents a decisive difference to the weakly coupled regime.

9 Scales and unification

One of the expectations of string theory is the possibility that it might ultimately lead to an explanation of the unification of all fundamental coupling constants. In contrast to usual grand unified models describing exclusively gauge interactions we have here a unification with the gravitational interaction as well. One therefore expects the grand unified scale M_{GUT} to be connected to the Planck scale. We shall first discuss the situation in the framework of the (weakly coupled) heterotic string theory and then compare it to the case of heterotic M-theory.

9.1 Weakly coupled $E_8 \times E_8$ heterotic string

Models of particle physics that are derived as the low energy limit of the $E_8 \times E_8$ heterotic string are able to accommodate the correct gauge group and particle spectrum to lead to the supersymmetric extension to the $SU(3) \times SU(2) \times U(1)$ standard model. It is exactly in this framework that a unification of the gauge coupling constants is expected to appear at a scale $M_{GUT} = 3 \cdot 10^{16}$ GeV. This heterotic string theory (weakly coupled at the string scale) in fact gives a prediction for the relation between gauge and gravitational coupling constants. To see this explicitly let us have again a look at the low energy effective action of the $d = 10$–dimensional field theory. In order to be able to compare these results with heterotic M-theory in a better way, we shall use a somewhat different notation here than that used previously in these lectures. The interested reader might try to figure out the correct transcription from one to the other notation. The more patient reader might wait till later sections where such a transcription will be given explicitly. We now write:

$$L = -\frac{4}{(\alpha')^3} \int d^{10}x \sqrt{g} \exp(-2\phi) \left(\frac{1}{(\alpha')} R + \frac{1}{4} \mathrm{tr} F^2 + \ldots \right), \tag{64}$$

where α' is the string tension and ϕ the dilaton field in $d = 10$. A definite relation between gauge and gravitational coupling appears because of the universal behaviour of the dilaton term in eq. (64). The effective $d = 4$–dimensional theory is obtained after compactification on a Calabi–Yau manifold with volume V:

$$L = -\frac{4}{(\alpha')^3} \int d^4x \sqrt{g} \exp(-2\phi) V \left(\frac{1}{(\alpha')} R + \frac{1}{4} \mathrm{tr} F^2 + \ldots \right). \tag{65}$$

Thus a universal factor $V \exp(-2\phi)$ multiplies both the R and F^2 terms. Newton's and Einstein's gravitational coupling constants are related as

$$G_N = \frac{1}{8\pi} \kappa_4^2 = \frac{1}{M_{\mathrm{Planck}}^2}, \tag{66}$$

with $M_{\mathrm{Planck}} \approx 1.2 \cdot 10^{19}$ GeV. From eq. (65) we then deduce:

$$G_N = \frac{\exp(2\phi)(\alpha')^4}{64\pi V}, \tag{67}$$

as well as

$$\alpha_{GUT} = \frac{\exp(2\phi)(\alpha')^3}{16\pi V}, \tag{68}$$

leading to the relation

$$G_N = \frac{\alpha_{GUT} \alpha'}{4}. \tag{69}$$

Putting in the value for M_{Planck} and $\alpha_{GUT} \approx 1/25$ one obtains a value for the string scale $M_{\mathrm{string}} = (\alpha')^{-1/2}$ that is in the region of 10^{18} GeV. This is apparently much larger than the GUT–scale of $3 \cdot 10^{16}$ GeV, while naively one would like to identify $M_{\rm string}$ with M_{GUT}. The discrepancy of the scales is sometimes called the unification problem in the framework of the weakly coupled heterotic string. Of course, the above argumentation is rather simple and more sophisticated (threshold) calculations are needed to settle this issue. In any case, the natural appearance of $M_{\rm string} \sim M_{GUT}$ would have been desirable. Let us now see how the situation looks in the case of heterotic string theory at stronger coupling.

9.2 $E_8 \times E_8$ M–theory

The effective action of the strongly coupled $E_8 \times E_8$ - M–theory in the "downstairs" approach is given by [Hořava and Witten (1996)] (we take into account the numerical corrections found in [Conrad (1997)])

$$\begin{aligned} L=&\frac{1}{\kappa^2}\int_{M^{11}} d^{11}x\sqrt{g}\left[-\frac{1}{2}R-\frac{1}{2}\bar{\psi}_I\Gamma^{IJK}D_J\left(\frac{\Omega+\hat{\Omega}}{2}\right)\psi_K-\frac{1}{48}G_{IJKL}G^{IJKL}\right.\\ &-\frac{\sqrt{2}}{384}\left(\bar{\psi}_I\Gamma^{IJKLMN}\psi_N+12\bar{\psi}^J\Gamma^{KL}\psi^M\right)\left(G_{JKLM}+\hat{G}_{JKLM}\right)\\ &\left.-\frac{\sqrt{2}}{3456}\epsilon^{I_1I_2\ldots I_{11}}C_{I_1I_2I_3}G_{I_4\ldots I_7}G_{I_8\ldots I_{11}}\right] \\ &+\frac{1}{4\pi(4\pi\kappa^2)^{2/3}}\int_{M_i^{10}} d^{10}x\sqrt{g}\left[-\frac{1}{4}F^a_{iAB}F_i^{aAB}-\frac{1}{2}\bar{\chi}_i^a\Gamma^A D_A(\hat{\Omega})\chi_i^a\right.\\ &\left.-\frac{1}{8}\bar{\psi}_A\Gamma^{BC}\Gamma^A\left(F^a_{iBC}+\hat{F}^a_{iBC}\right)\chi_i^a+\frac{\sqrt{2}}{48}\left(\bar{\chi}_i^a\Gamma^{ABC}\chi_i^a\right)\hat{G}_{ABC11}\right] \end{aligned} \tag{70}$$

where M^{11} is the "downstairs" manifold while M_i^{10} are its 10–dimensional boundaries. In the lowest approximation M^{11} is just a product $M^4 \times X^6 \times S^1/Z_2$. Compactifying to $d = 4$ in such an approximation we obtain [Witten (1996)], [Conrad (1997)]

$$G_N = \frac{\kappa_4^2}{8\pi} = \frac{\kappa^2}{8\pi R_{11} V}, \tag{71}$$

$$\alpha_{GUT} = \frac{(4\pi\kappa^2)^{2/3}}{V} \tag{72}$$

with V the volume of the Calabi–Yau manifold X^6 and $R_{11} = \pi\rho$ the S^1/Z_2 length.

The fundamental mass scale of the 11–dimensional theory is given by $M_{11} = \kappa^{-2/9}$. Let us see which value of M_{11} is favoured in a phenomenological application. For that purpose we identify the Calabi–Yau volume V with the GUT–scale: $V \sim (M_{GUT})^{-6}$. From (72) and the value of $\alpha_{GUT} = 1/25$ at the grand unified scale, we can then deduce the value of M_{11}

$$V^{1/6}M_{11} = (4\pi)^{1/9}\alpha_{GUT}^{-1/6} \approx 2.3\,, \tag{73}$$

to be a few times larger than the GUT–scale. In a next step we can now adjust the gravitational coupling constant by choosing the appropriate value of R_{11} using (71). This leads to

$$R_{11}M_{11} = \left(\frac{M_{Planck}}{M_{11}}\right)^2 \frac{\alpha_{GUT}}{8\pi(4\pi)^{2/3}} \approx 2.9 \cdot 10^{-4} \left(\frac{M_{Planck}}{M_{11}}\right)^2 . \tag{74}$$

This simple analysis tells us the following:

- In contrast to the weakly coupled case (where we had a prediction (69)), the correct value of M_{Planck} can be fitted by adjusting the value of R_{11}.
- The numerical value of R_{11}^{-1} turns out to be approximately an order of magnitude smaller than M_{11}.
- Thus the 11th dimension appears to be larger than the dimensions compactified on the Calabi–Yau manifold, and at an intermediate stage the world appears 5–dimensional with two 4–dimensional boundaries (walls).

We thus have the following picture of the evolution and unification of coupling constants. At low energies the world is 4–dimensional and the couplings evolve accordingly with energy: a logarithmic variation of gauge coupling constants and the usual power law behaviour for the gravitational coupling. Around R_{11}^{-1} we have an additional 5th dimension and the power law evolution of the gravitational interactions changes. Gauge couplings are not effected at that scale since the gauge fields live on the walls and do not feel the existence of the 5th dimension. Finally at M_{GUT} the theory becomes 11–dimensional and both gravitational and gauge couplings show a power law behaviour and meet at the scale M_{11}, the fundamental scale of the theory. It is obvious that the correct choice of R_{11} is needed to achieve unification. We also see that, although the theory is weakly coupled at M_{GUT}, this is no longer true at M_{11}. The naive estimate for the evolution of the gauge coupling constants between M_{GUT} and M_{11} goes with the sixth power of the scale. At M_{11} we thus expect unification of the couplings at $\alpha \sim O(1)$. In that sense, the M–theoretic description of the heterotic string gives an interpolation between weak coupling and moderate coupling. In $d = 4$ this is not strong–weak coupling duality in the usual sense. We shall later come back to these questions when we discuss the appearance of a critical limit on the size of R_{11}. A value of $\alpha \sim O(1)$ (and thus $S \sim O(1)$) at M_{11} might also be favoured in view of the question of the dynamic determination of the vev of the dilaton field [Lalak et al. (1995)].

These are, of course, rather qualitative results. In order to get a more quantitative feeling for the range of M_{11} and R_{11}, let us be a bit more specific and write the relation of the unification scale M_{GUT} to the characteristic size of the Calabi–Yau space as:

$$V^{1/6} = aM_{GUT}^{-1} \, . \tag{75}$$

The above formula corresponds to the situation in which we identify the unification scale with the radius, R, of X^6 which volume is given by $V = (aR)^6$. We expect the parameter a to be somewhere in the range from 1 to 2π. Using the above identification and the value of $M_{GUT} = 3 \cdot 10^{16}$ GeV we obtain:

$$M_{11} \approx \frac{2.3}{a} M_{GUT} . \tag{76}$$

As said before, the scale M_{11} occurs to be of the order of the unification scale M_{GUT}. However, we do not expect M_{11} to be smaller than M_{GUT} because we need the ordinary logarithmic evolution of the gauge coupling constants up to M_{GUT}. In fact, M_{11} should be somewhat bigger in order to allow for the evolution of α from its unification value 1/25 to the strong regime. Thus, we expect the parameter a to be quite close to 1. Putting the above value of M_{11} into eq. (74) we get the length of S^1/Z_2:

$$R_{11} \approx 9.2 a^2 M_{11}^{-1} \approx 4 a^3 M_{GUT}^{-1} . \tag{77}$$

It is about one order of magnitude bigger than the scale characteristic for the 11–dimensional theory. This is the reason for the relatively large value of the $d = 4$ Planck Mass. Of course R_{11} can not be too large. For a between 1 and 2.3 (values corresponding to $M_{11} > M_{GUT}$) we obtain R_{11}^{-1} in the range $(6.2 \cdot 10^{14} - 7.4 \cdot 10^{15})$ GeV (as we discussed, the parameter a should not be too different from 1 so the upper part of the above range is favoured). Smaller values of R_{11}^{-1} seem to be very unnatural. Trying to push R_{11}^{-1} to smaller values would need a redefinition of M_{11}. For that purpose in [Antoniadis et al. (1997)] a definition $m_{11} = 2\pi(4\pi\kappa^2)^{-1/9}$ was used. This allows then to push a to the extreme limit of 2π. With these extreme choices of both a and m_{11} one would then be able to obtain R_{11}^{-1} as small as $3 \cdot 10^{13}$ GeV. Values smaller than that (like values of 10^{12} GeV as sometimes quoted in the literature) cannot be obtained. In any case, even values in the lower 10^{13} GeV range seem to be in conflict with the critical value of R_{11}, as we shall see later.

10 The effective action in $d = 4$

We now want to repeat the construction of the effective action in $d = 4$ as obtained with the method of reduction and truncation, using the new notation. We shall first consider the $d = 10$ effective field theory for the heterotic string:

$$L = -\frac{4}{(\alpha')^3} \int d^{10}x \sqrt{g} \exp(-2\phi) \left(\frac{1}{(\alpha')} R + \frac{1}{4} \mathrm{tr} F^2 + \frac{1}{12} \alpha' H^2 + \ldots \right) , \tag{78}$$

where we have included the three index tensor field strength

$$H = dB + \omega^{YM} - \omega^L . \tag{79}$$

B is the two–index antisymmetric tensor while

$$\omega^{YM} = \mathrm{Tr}(AF - \frac{2}{3} A^3) \tag{80}$$

and

$$\omega^L = \mathrm{Tr}(\omega R - \frac{2}{3}\omega^3) \tag{81}$$

are the Yang–Mills and Lorentz–Chern–Simons terms, respectively. The addition of these terms in the definition of H is needed for supersymmetry and anomaly freedom of the theory.

To obtain the effective theory in $d = 4$ dimensions we use as an approximation the method of reduction and truncation explained in ref. [Witten (1985)]. It essentially corresponds to a torus compactification, with truncation of states to arrive at a $d = 4$ theory with $N = 1$ supersymmetry. In string theory compactified on an orbifold this would describe the dynamics of the untwisted sector. We retain the usual moduli fields S and T as well as matter fields C_i that transform nontrivially under the observable sector gauge group. In this approximation, the Kähler potential is given by [Witten (1985)], [Derendinger et al. (1986)]

$$G = -\log(S + S^*) - 3\log(T + T^* - 2C_i^* C_i) + \log|W|^2 \tag{82}$$

with the superpotential $W(C_i)$ originating from the Chern–Simons terms ω^{YM} [Derendinger et al. (1985)]

$$W(C) = d_{ijk} C_i C_j C_k \tag{83}$$

and the gauge kinetic function is given by the dilaton field

$$f = S\,. \tag{84}$$

For a detailed discussion of this method and the explicit definition of the fields see the review [Nilles (1986)]. These expressions for the $d = 4$ effective action look quite simple and it remains to be seen whether this simplicity is true in general or whether it is an artifact of the approximation. Our experience with supergravity models tells us that the holomorphic functions W and f might be protected by nonrenormalization theorems, while the Kähler potential is strongly modified in perturbation theory. In addition we have to be aware of the fact that the expressions given above are at best representing a subsector of the theory. In orbifold compactification this would be the untwisted sector, and we know that the Kähler potential for twisted sector fields will look quite different. Nonetheless the used approximation turned out to be useful for the discussion of those aspects of the theory that determine the dynamics of the T– and S–moduli. When trying to extract, however, detailed masses and other properties of the fields one should be aware of the fact, that some results might not be true in general and only appear as a result of the simplicity of the approximation.

So far the classical action. What about loop corrections? Not much can be said about the details of the corrections to the Kähler potential. This has to be discussed on a model by model basis. The situation with the superpotential is quite easy. There we expect a nonrenormalization theorem to be at work. The inclusion of other sectors of the theory will lead to new terms in the superpotential that in general have T–dependent coefficients. Such terms can be computed in simple cases by using e.g. methods of conformal field theory [Lauer et al. (1991)].

The situation for f, the gauge–kinetic function, is more interesting. Symmetries and holomorphicity lead us to believe, that although there are nontrivial corrections at one–loop, no more perturbative corrections are allowed at higher orders [Shifman and Vainshtein (1986)], [Nilles (1986)]. The existence of such corrections at one loop seems to be intimately connected to the mechanism of anomaly cancellation in the $d = 10$ theory [Choi and Kim (1985)], [Ibáñez and Nilles (1986)]. To see this consider one of the anomaly cancellation counter–terms introduced by Green and Schwarz [Green and Schwarz (1984)]:

$$\epsilon^{NEGAWSKLOV} B_{VO} \mathrm{Tr} F^2_{LKSW} F^2_{AGEN} \,. \tag{85}$$

We are interested in a $d = 4$ theory with $N = 1$ supersymmetry, and thus expect nontrivial vacuum expectation values for the curvature terms $\mathrm{Tr}R^2$ and field strengths $\mathrm{Tr}F^2$ in the extra six dimensions. Consistency of the theory requires a condition for the 3–index tensor field strength. For H to be well defined, the quantity

$$dH = \mathrm{Tr}F^2 - \mathrm{Tr}R^2 \tag{86}$$

has to vanish cohomologically [Candelas et al. (1985)]. In the simplest case (the so–called standard embedding leading to gauge group $E_6 \times E_8$) one chooses equality pointwise $\mathrm{Tr}R^2 = \mathrm{Tr}F^2$. Let us now assume that $\mathrm{Tr}F^2_{agen}$ is nonzero. The Green–Schwarz term given above by eq. (85) then leads to

$$\epsilon^{mn} B_{mn} \epsilon^{\mu\nu\rho\sigma} \mathrm{Tr} F_{\mu\nu} F_{\rho\sigma} \tag{87}$$

in the four–dimensional theory. An explicit inspection of the fields tells us that $\epsilon^{mn} B_{mn}$ is the pseudoscalar axion that belongs to the T–superfield. Upon supersymmetrization the term in eq. (87) will then correspond to a one–loop correction to the holomorphic f–function (84) that is proportional to T with the coefficient fixed entirely by the anomaly considerations. This is, of course, nothing else than a threshold correction. In the simple case of the standard embedding with gauge group $E_6 \times E_8$ one obtains e.g.

$$f_6 = S + \epsilon T \,, \qquad f_8 = S - \epsilon T \,, \tag{88}$$

respectively, where ϵ is the constant fixed by the anomaly. These results can be backed up by explicit calculations in string theory. In cases where such an explicit calculation is feasible, many more details about these corrections can be deduced. The above result (88) obtained in $d = 10$ field theory represents an approximation of the exact result in the large T–limit. For a detailed discussion of these calculations and the limiting procedure see [Nilles and Stieberger (1997)]. We have here mainly concentrated on that limit, because it represents a rather model independent statement.

Thus we have seen that there are corrections to the gauge–kinetic function at one loop. Their existence is found to be intimately related to the mechanism of anomaly cancellation. The corrections found are exactly those that are expected by general symmetry considerations [Nilles (1986)]. In (88) we have given the result for the standard embedding. Coefficients might vary for more general cases, but the fact that they have opposite sign for the two separate groups is true in all known cases.

Superpotential and f–function should not receive further perturbative corrections beyond one loop. This implies that the knowledge of f at one loop represents the full perturbative result. Combined with the fact that the coefficients are fixed by anomaly considerations one would then expect that this result for the f–function might be valid even beyond the weakly coupled limit. Not much can be said about the Kähler potential beyond one loop.

We now turn to the calculation in the M-theory case [Nilles et al. (1997)] [Lukas et al. (1997)]. In the strongly coupled case we have to perform a compactification from $d = 11$ to $d = 4$. Again we use the method of reduction and truncation. For the metric we write

$$g^{(11)}_{MN} = \begin{pmatrix} c_4 e^{-\gamma} e^{-2\sigma} g_{\mu\nu} & & \\ & e^{\sigma} g_{mn} & \\ & & e^{2\gamma} e^{-2\sigma} \end{pmatrix} \tag{89}$$

with $M, N = 1 \dots 11$; $\mu, \nu = 1 \dots 4$; $m, n = 5 \dots 10$ and $\det(g_{mn})=1$. This is the frame in which the 11–dimensional Einstein action gives the ordinary Einstein action after the reduction do $d = 4$:

$$-\frac{1}{2\kappa^2} \int d^{11}x \sqrt{g^{(11)}} R^{(11)} = -\frac{c_4 \hat{V}_7}{2\kappa^2} \int d^4x \sqrt{g} R + \dots \tag{90}$$

where $\hat{V}_7 = \int d^7x$ is the coordinate volume of the compact 7–manifold and the scaling factor c_4 describes our freedom to choose the units in $d = 4$. The most popular choice in the literature is $c_4 = 1$. This, however, corresponds to the unphysical situation in which the 4–dimensional Planck mass is determined by the choice of $\hat{V}_7$ which is just a convention. With $c_4 = 1$ one needs further rescaling of the 4–dimensional metric. We instead prefer the choice

$$c_4 = V_7 / \hat{V}_7 \tag{91}$$

where $V_7 = \int d^7x \sqrt{g^{(7)}}$ is the physical volume of the compact 7–manifold. This way we recover eq. (71) in which the 4–dimensional Planck mass depends on the physical (and not coordinate) volume of the manifold on which we compactify. As a result, if we start from the product of the 4–dimensional Minkowski space and some 7–dimensional compact space (in the leading order of the expansion in $\kappa^{2/3}$) as a ground state in $d = 11$ we obtain the Minkowski space with the standard normalization as the vacuum in $d = 4$.

To find a more explicit formula for c_4 we have to discuss the fields σ and γ in some detail. In the leading approximation σ is the overall modulus of the Calabi–Yau 6–manifold. We can divide it into a sum of the vacuum expectation value, $\langle\sigma\rangle$, and the fluctuation $\tilde{\sigma}$. In general both parts could depend on all 11 coordinates but in practice we have to impose some restrictions. The vacuum expectation value can not depend on x^μ if the 4-dimensional theory is to be Lorentz-invariant. In the fluctuations we drop the dependence on the compact coordinates corresponding to the higher Kaluza–Klein modes. Furthermore, we know that in the leading approximation $\langle\sigma\rangle$ is just a constant, σ_0 , while corrections depending on the internal coordinates, σ_1, are of the next order in $\kappa^{2/3}$. Thus, we obtain

$$\sigma(x^\mu, x^m, x^{11}) = \langle\sigma\rangle\,(x^m, x^{11}) + \tilde{\sigma}(x^\mu) = \sigma_0 + \sigma_1(x^m, x^{11}) + \tilde{\sigma}(x^\mu)\,. \qquad (92)$$

To make the above decomposition unique we define σ_0 by requiring that the integral of σ_1 over the internal space vanishes. The analogous decomposition can be also done for γ. With the above definitions the physical volume of the compact space is

$$V_7 = \int d^7x \left\langle e^{2\sigma} e^{\gamma} \right\rangle = e^{2\sigma_0} e^{\gamma_0} \hat{V}_7 \qquad (93)$$

up to corrections of order $\kappa^{4/3}$. Thus, the parameter c_4 can be written as

$$c_4 = e^{2\sigma_0} e^{\gamma_0}\,. \qquad (94)$$

The choice of the coordinate volumes is just a convention. For example in the case of the Calabi–Yau 6-manifold only the product $e^{3\sigma}\hat{V}_6$ has physical meaning. For definiteness we will use the convention that the coordinate volumes are equal 1 in M_{11} units. Thus, $\left\langle e^{3\sigma}\right\rangle$ describes the Calabi–Yau volume in these units. Using eqs. (73,74) we obtain $e^{3\sigma_0} = VM_{11}^6 \approx (2.3)^6$, $e^{\gamma_0}e^{-\sigma_0} = R_{11}M_{11} \approx 9.2a^2$. The parameter c_4 is equal to the square of the 4-dimensional Planck mass in these units and numerically $c_4 \approx (35a)^2$.

At the classical level we compactify on $M^4 \times X^6 \times S^1/Z_2$. This means that the vacuum expectation values $\langle\sigma\rangle$ and $\langle\gamma\rangle$ are just constants and eq. (92) reduces to

$$\sigma = \sigma_0 + \tilde{\sigma}(x^\mu), \qquad \gamma = \gamma_0 + \tilde{\gamma}(x^\mu)\,. \qquad (95)$$

In such a situation σ and γ are 4-dimensional fields. We introduce two other 4-dimensional fields by the relations

$$\frac{1}{4!c_4} e^{6\sigma} G_{11\lambda\mu\nu} = \epsilon_{\lambda\mu\nu\rho}\,(\partial^\rho D)\,, \qquad (96)$$

$$C_{11a\bar{b}} = C_{11}\delta_{a\bar{b}} \qquad (97)$$

where x^a ($x^{\bar{b}}$) is the holomorphic (antiholomorphic) coordinate of the Calabi–Yau manifold. Now we can define the dilaton and the modulus fields by

$$\mathcal{S} = \frac{1}{(4\pi)^{2/3}}\left(e^{3\sigma} + i24\sqrt{2}D\right)\,, \qquad (98)$$

$$\mathcal{T} = \frac{1}{(4\pi)^{2/3}}\left(e^{\gamma} + i6\sqrt{2}C_{11} + C_i^* C_i\right) \qquad (99)$$

where the observable sector matter fields C_i originate from the gauge fields A_M on the 10–dimensional observable wall (and M is an index in the compactified six dimensions). The Kähler potential takes its standard form as in eq. (82)

$$K = -\log(\mathcal{S}+\mathcal{S}^*) - 3\log(\mathcal{T}+\mathcal{T}^* - 2C_i^*C_i)\,. \tag{100}$$

The imaginary part of $\mathcal{S}$ (Im$\mathcal{S}$) corresponds to the model independent axion, and with the above normalization the gauge kinetic function is $f = \mathcal{S}$. We have also

$$W(C) = d_{ijk}C_iC_jC_k \tag{101}$$

Thus the action to leading order is very similar to the weakly coupled case.

Before drawing any conclusion from the formulae obtained above we have to discuss a possible obstruction at the next to leading order. For the 3–index tensor field H in $d = 10$ supergravity to be well defined one has to satisfy $dH = \mathrm{tr}F_1^2 + \mathrm{tr}F_2^2 - \mathrm{tr}R^2 = 0$ cohomologically. In the simplest case of the standard embedding one assumes $\mathrm{tr}F_1^2 = \mathrm{tr}R^2$ locally and the gauge group is broken to $E_6 \times E_8$. Since in the M–theory case the two different gauge groups live on the two different boundaries (walls) of space–time, such a cancellation point by point is no longer possible [Witten (1996)]. We expect nontrivial vevs of

$$(dG) \propto \sum_i \delta(x^{11} - x_i^{11})\left(\mathrm{tr}F_i^2 - \frac{1}{2}\mathrm{tr}R^2\right) \tag{102}$$

at least on one boundary (x_i^{11} is the position of i–th boundary). In the case of the standard embedding we would have $\mathrm{tr}F_1^2 - \frac{1}{2}\mathrm{tr}R^2 = \frac{1}{2}\mathrm{tr}R^2$ on one and $\mathrm{tr}F_2^2 - \frac{1}{2}\mathrm{tr}R^2 = -\frac{1}{2}\mathrm{tr}R^2$ on the other boundary. This might pose a severe problem since a nontrivial vev of G might be in conflict with supersymmetry ($G_{11ABC} = H_{ABC}$). The supersymmetry transformation law in $d = 11$ reads

$$\delta\psi_M = D_M\eta + \frac{\sqrt{2}}{288}G_{IJKL}\left(\Gamma_M^{IJKL} - 8\delta_M^I\Gamma^{JKL}\right)\eta + \ldots \tag{103}$$

Supersymmetry will be broken unless e.g. the derivative term $D_M\eta$ compensates the nontrivial vev of G. Witten has shown [Witten (1996)] that such a cancellation can occur and constructed the solution in the linearized approximation (linear in the expansion parameter $\kappa^{2/3}$). This solution requires some modification of the metric on M^{11}:

$$g_{MN}^{(11)} = \begin{pmatrix} (1+b)\eta_{\mu\nu} & & \\ & (g_{ij}+h_{ij}) & \\ & & (1+\gamma') \end{pmatrix}. \tag{104}$$

M^{11} is no longer a direct product $M^4 \times X^6 \times S^1/Z_2$ because b, h_{ij} and γ' depend now on the compactified coordinates. The volume of X^6 depends on x^{11} [Witten (1996)]:

$$\frac{\partial}{\partial x^{11}}V = -\frac{\sqrt{2}}{8}\int d^6x\sqrt{g}\,\omega^{AB}\omega^{CD}G_{ABCD} \tag{105}$$

where the integral is over the Calabi–Yau manifold X^6 and ω is the corresponding Kähler form. The parameter $(1+b)$ is the scale factor of the Minkowski 4-manifold and depends on x^{11} in the following way

$$\frac{\partial}{\partial x^{11}} b = \frac{1}{2} \frac{\partial}{\partial x^{11}} \log v_4 = \frac{\sqrt{2}}{24} \omega^{AB} \omega^{CD} G_{ABCD} \tag{106}$$

where v_4 is the physical volume for some fixed coordinate volume in M^4. In our simple reduction and truncation method with the metric $g^{(11)}_{MN}$ given by eq. (89) we can reproduce the x^{11} dependence of V and v_4. The volume of X^6 is determined by σ:

$$\frac{\partial}{\partial x^{11}} \log V = \frac{\partial}{\partial x^{11}} (3 \langle \sigma \rangle) = 3 \frac{\partial}{\partial x^{11}} \sigma \tag{107}$$

while the scale factor of M^4 can be similarly expressed in terms of σ and γ fields:

$$\frac{\partial}{\partial x^{11}} \log v_4 = -\frac{\partial}{\partial x^{11}} (2 \langle \gamma \rangle + 4 \langle \sigma \rangle) = -\frac{\partial}{\partial x^{11}} (2\gamma + 4\sigma)\,. \tag{108}$$

Substituting $\langle \sigma \rangle$ with σ in the above two equations is allowed because, due to our decomposition (92), only the vev of σ depends on the internal coordinates (the same is true for γ). The scale factor b calculated in ref. [Witten (1996)] depends also on the Calabi–Yau coordinates. Such a dependence can not be reproduced in our simple reduction and truncation compactification so we have to average eq. (106) over X^6. Using equations (105–108) after such an averaging we obtain (to leading order in the expansion parameter $\kappa^{2/3}$) [Nilles et al. (1997)]

$$\frac{\partial \gamma}{\partial x^{11}} = -\frac{\partial \sigma}{\partial x^{11}} = \frac{\sqrt{2}}{24} \frac{\int d^6 x \sqrt{g} \omega^{AB} \omega^{CD} G_{ABCD}}{\int d^6 x \sqrt{g}}\,. \tag{109}$$

Substituting the vacuum expectation value of G found in [Witten (1996)] we can rewrite it in the form

$$\frac{\partial \gamma}{\partial x^{11}} = -\frac{\partial \sigma}{\partial x^{11}} = \frac{2}{3} \alpha \kappa^{2/3} V^{-2/3} \tag{110}$$

where

$$\alpha = \frac{\pi c}{2(4\pi)^{2/3}} \tag{111}$$

and c is a constant of order unity given for the standard embedding of the spin connection by

$$c = V^{-1/3} \left| \int \frac{\omega \wedge \mathrm{tr}(R \wedge R)}{8\pi^2} \right|\,. \tag{112}$$

Our calculations, as those of Witten, are valid only in the leading nontrivial order in the $\kappa^{2/3}$ expansion. The expression (110) for the derivatives of σ and γ has an explicit factor $\kappa^{2/3}$. This means that we should take the lowest order value for the Calabi–Yau volume in that expression. An analogous procedure has been used in obtaining all formulae presented in this paper. We always expand in

$\kappa^{2/3}$ and drop all terms which are of higher order. Taking the above into account and using our units in which $M_{11} = 1$ we can rewrite eq. (110) in the simple form:

$$\frac{\partial \gamma}{\partial x^{11}} = -\frac{\partial \sigma}{\partial x^{11}} = \frac{2}{3}\alpha e^{-2\sigma_0}\,. \tag{113}$$

Eqs. (109–113) as derived in ref. [Nilles et al. (1997)] contain all the information to deduce the effective action, i.e. Kähler potential, superpotential and gauge kinetic function of the 4–dimensional effective supergravity theory.

It is the above dependence of σ and γ on x^{11} that leads to these consequences. One has to be careful in defining the fields in $d = 4$. It is obvious, that the 4–dimensional fields $\mathcal{S}$ and $\mathcal{T}$ can not be any longer defined by eqs. (98, 99) because now σ and γ are 5–dimensional fields. We have to integrate out the dependence on the 11th coordinate. In the present approximation, this procedure is quite simple: we have to replace σ and γ in the definitions of $\mathcal{S}$ and $\mathcal{T}$ with their averages over the S^1/Z_2 interval [Nilles et al. (1997)]. With the linear dependence of σ and γ on x^{11} their average values coincide with the values taken at the middle of the S^1/Z_2 interval

$$\bar{\sigma} = \sigma\left(\frac{\pi\rho}{2}\right) = \sigma_0 + \tilde{\sigma}(x^{\mu})\,, \tag{114}$$

$$\bar{\gamma} = \gamma\left(\frac{\pi\rho}{2}\right) = \gamma_0 + \tilde{\gamma}(x^{\mu})\,. \tag{115}$$

When we reduce the boundary part of the Lagrangian of M–theory to 4 dimensions we find exponents of σ and γ fields evaluated at the boundaries. Using eqs. (92) and (113) we get

$$e^{-\gamma}\big|_{M_i^{10}} = e^{-\gamma_0} \pm \frac{1}{3}\alpha e^{-3\sigma_0}\,, \tag{116}$$

$$e^{3\sigma}\big|_{M_i^{10}} = e^{3\sigma_0} \pm \alpha e^{\gamma_0}\,. \tag{117}$$

The above formulae have very important consequences for the definitions of the Kähler potential and the gauge kinetic functions. For example, the coefficient in front of the $D_\mu C_i^* D^\mu C_i$ kinetic term is proportional to $e^{-\gamma}$ evaluated at the E_6 wall where the matter fields propagate. At the lowest order this was just $e^{-\gamma_0}$ or $\langle\mathcal{T}\rangle^{-1}$ up to some numerical factor. From eq. (116) we see that at the next to leading order also $\langle\mathcal{S}\rangle^{-1}$ is involved with relative coefficient $\alpha/3$. Taking such corrections into account we find that at this order the Kähler potential is given by

$$K = -\log(\mathcal{S} + \mathcal{S}^*) + \frac{2\alpha C_i^* C_i}{\mathcal{S} + \mathcal{S}^*} - 3\log(\mathcal{T} + \mathcal{T}^* - 2C_i^* C_i) \tag{118}$$

with $\mathcal{S}$ and $\mathcal{T}$ now defined by

$$\mathcal{S} = \frac{1}{(4\pi)^{2/3}}\left(e^{3\bar{\sigma}} + i24\sqrt{2}\bar{D} + \alpha C_i^* C_i\right)\,, \tag{119}$$

$$\mathcal{T} = \frac{1}{(4\pi)^{2/3}}\left(e^{\bar{\gamma}} + i6\sqrt{2}\bar{C}_{11} + C_i^* C_i\right) \tag{120}$$

where bars denote averaging over the 11th dimension. It might be of some interest to note that the combination $\mathcal{S}\mathcal{T}^3$ is independent of x^{11} even before this averaging procedure took place. The solution above is valid only for terms at most linear in α. Keeping this in mind we could write the Kähler potential also in the form

$$K = -\log(\mathcal{S} + \mathcal{S}^* - 2\alpha C_i^* C_i) - 3\log(\mathcal{T} + \mathcal{T}^* - 2C_i^* C_i). \tag{121}$$

Equipped with this definition the calculation of the gauge kinetic function(s) from eqs. (113, 117) becomes a trivial exercise [Nilles et al. (1997)]. In the five–dimensional theory f depends on the 11–dimensional coordinate as well, thus the gauge kinetic function takes different values at the two walls. The averaging procedure allows us to deduce these functions directly. For the simple case at hand (the so–called standard embedding) eq. (117) gives [Nilles et al. (1997)]

$$f_6 = \mathcal{S} + \alpha\mathcal{T}\,; \qquad\qquad f_8 = \mathcal{S} - \alpha\mathcal{T}\,. \tag{122}$$

It is a special property of the standard embedding that the coefficients are equal and opposite. The coefficients might vary for more general cases. This completes the discussion of the $d = 4$ effective action in next to leading order, noting that the superpotential does not receive corrections at this level.

The nontrivial dependence of σ and γ on x^{11} can also enter definitions and/or interactions of other 4–dimensional fields. Let us next consider the gravitino. After all we have to show that this field is massless to give the final proof that the given solution respects supersymmetry. Its 11–dimensional kinetic term

$$-\frac{1}{2}\sqrt{g}\bar{\psi}_I \Gamma^{IJK} D_J \psi_K \tag{123}$$

remains diagonal after compactification to $d = 4$ if we define the 4–dimensional gravitino, $\psi_\mu^{(4)}$, and dilatino, $\psi_{11}^{(4)}$, fields by the relations

$$\psi_\mu = e^{-(\sigma-\sigma_0)/2} e^{-(\gamma-\gamma_0)/4} \left(\psi_\mu^{(4)} + \frac{1}{\sqrt{6}}\Gamma_\mu \psi_{11}^{(4)}\right), \tag{124}$$

$$\psi_{11} = -\frac{2}{\sqrt{6}} e^{(\sigma-\sigma_0)/2} e^{(\gamma-\gamma_0)/4} \Gamma^{11} \psi_{11}^{(4)}\,. \tag{125}$$

The $d = 11$ kinetic term (123) gives after the compactification also a mass term for the $d = 4$ gravitino of the form

$$\frac{3}{8} e^{\sigma_0} e^{-\gamma_0} \frac{\partial\gamma}{\partial x^{11}} = \frac{\sqrt{2}}{64} e^{\sigma_0} e^{-\gamma_0} \frac{\int d^6x \sqrt{g}\,\omega^{AB}\omega^{CD} G_{ABCD}}{\int d^6x \sqrt{g}} = \frac{1}{4}\alpha e^{-\sigma_0} e^{-\gamma_0}\,. \tag{126}$$

The sources of such a term are nonzero values of the spin connection components $\omega_\mu^{\alpha 11}$ and ω_m^{a11} resulting from the x^{11} dependence of the metric. It is a constant mass term from the 4–dimensional point of view. This, however, does not mean that the gravitino mass is nonzero. There is another contribution from the 11–dimensional term

$$-\frac{\sqrt{2}}{384}\sqrt{g}\bar{\psi}_I \Gamma^{IJKLMN}\psi_N\left(G_{JKLM}+\hat{G}_{JKLM}\right). \tag{127}$$

After redefining fields according to (124,125) and averaging the nontrivial vacuum expectation value of G over X^6 we get from eq. (127) a mass term which exactly cancels the previous contribution (126). The gravitino is massless – the result which we expect in a model with unbroken supersymmetry and vanishing cosmological constant. Thus, we find that our simple reduction and truncation method (including the correct x^{11} dependence in next to leading order) reproduces the main features of the model.

The factor $\langle\exp(3\sigma)\rangle$ represents the volume of the six–dimensional compact space in units of M_{11}^{-6}. The x^{11} dependence of σ then leads to the geometrical picture that the volume of this space varies with x^{11} and differs at the two boundaries:

$$V_{E_8}=V_{E_6}-2\pi^2\rho\left(\frac{\kappa}{4\pi}\right)^{2/3}\left|\int\omega\wedge\frac{\mathrm{tr}(F\wedge F)-\frac{1}{2}\mathrm{tr}(R\wedge R)}{8\pi^2}\right| \tag{128}$$

where the integral is over X^6 at the E_6 boundary. In the given approximation, this variation is linear, and for growing ρ the volume on the E_8 side becomes smaller and smaller. At a critical value of ρ the volume will thus vanish and this will provide us with an upper limit on ρ:

$$\rho<\rho_{crit}=\frac{(4\pi)^{2/3}}{c\pi^2}M_{11}^3V_{E_6}^{2/3} \tag{129}$$

where c was defined in eq. (112). To estimate the numerical value of ρ_{crit} we first recall that from eq. (71) we obtained *

$$M_{11}V_{E_6}^{1/6}=\left(\alpha_{GUT}(4\pi)^{-2/3}\right)^{-1/6}\approx 2.3\,. \tag{130}$$

Thus, we get

$$\rho^{-1}>\rho_{crit}^{-1}\approx 0.16cV_{E_6}^{-1/6}\,. \tag{131}$$

The numerical value of V at the E_6 boundary depends on what we identify with the unification scale M_{GUT} via eq. (75):

$$V_{E_6}^{-1/6}=aM_{GUT}^{-1} \tag{132}$$

with a somewhere between 1 and about 2. Thus, the bound (131) can be written in the form

$$R_{11}^{-1}>0.05\frac{c}{a}M_{GUT}\,. \tag{133}$$

* With V depending on x^{11} we have to specify which values should be used in eqs. (71,72,75). The appropriate choice in the expression for G_N is the average value of V while in the expressions for α_{GUT} and for the V–M_{GUT} relation we have to use V evaluated at the E_6 wall.

For the phenomenological applications we have to check whether our preferred choice of $6.2 \cdot 10^{14}$ GeV $< R_{11}^{-1} < 7.4 \cdot 10^{15}$ GeV that fits the correct value of the $d = 4$ Planck mass satisfies the bound (133). In a rather extreme case of $c = 1$ and $a = 2.3$ we find that the upper bound on R_{11}^{-1} is of the order of $6.5 \cdot 10^{14}$ GeV. Even for $c = 1$ this bound goes up to about $1.5 \cdot 10^{15}$ GeV if we identify $V^{-1/6}$ with M_{GUT}. Although some coefficients are model dependent we find in general that the bound can be satisfied, but that R_{11} is quite close to its critical value. Values of R_{11}^{-1} about 10^{12} GeV as necessary in [Antoniadis et al. (1997)] seem to be beyond the critical value, even with the modifications discussed before. In any case, models where supersymmetry is broken by a Scherk–Schwarz mechanism seem to require the absence of the next to leading order corrections in (122), i.e. $\alpha = 0$. It remains to be seen whether such a possibility can be realized.

Inspection of (88) and (122) reveals a close connection between the strongly and weakly coupled case [Banks et al. (1996)], [Nilles and Stieberger (1997)]. The variation of the Calabi–Yau manifold volume as discussed above is the analogue of the one loop correction of the gauge kinetic function (88) in the weakly coupled case and has the same origin, namely a Green–Schwarz anomaly cancellation counterterm. In fact, also in the strongly coupled case this leads to a correction for the gauge coupling constants at the E_6 and E_8 side. As seen, gauge couplings are no longer given by the (averaged) $\mathcal{S}$–field, but by that combination of the (averaged) $\mathcal{S}$ and $\mathcal{T}$ fields which corresponds to the $\mathcal{S}$–field before averaging at the given boundary leading to

$$f_{6,8} = \mathcal{S} \pm \alpha \mathcal{T} \tag{134}$$

at the E_6 or E_8 side, respectively. The critical value of R_{11} will correspond to infinitely strong coupling at the E_8 side $\mathcal{S} - \alpha\mathcal{T} = 0$. Since we are here close to criticality a correct phenomenological fit of $\alpha_{\text{GUT}} = 1/25$ should include this correction $\alpha_{\text{GUT}}^{-1} = \mathcal{S} + \alpha\mathcal{T}$ where $\mathcal{S}$ and $\alpha\mathcal{T}$ give comparable contributions. This is a difference to the weakly coupled case, where in $f = S + \epsilon T$ the latter contribution was small compared to S. The stability of this result for the corrections to f when going from weak coupling to strong coupling is only possible because of the rather special properties of f. f does not receive further perturbative corrections beyond one loop [Shifman and Vainshtein (1986)], [Nilles (1986)], and the one loop corrections are determined by the anomaly considerations. The formal expressions for the corrections are identical, the difference being only that in the strongly coupled case these corrections are as important as the classical value.

11 Supersymmetry breaking at the hidden wall

We shall now discuss the question of supersymmetry breakdown within this framework. We consider the breakdown of supersymmetry in a hidden sector, transmitted to the observable sector via gravitational interactions. Such a scenario was suggested in [Nilles (1982)] after having observed that gaugino condensation can break supersymmetry in $d = 4$ supergravity models. As we have seen, a nontrivial gauge kinetic function f seems to be necessary for such a mechanism to work [Ferrara et al. (1983)]. In the heterotic string both ingredients, a hidden sector E_8 and a nontrivial f, were present in a natural way and a coherent picture of supersymmetry breakdown via gaugino condensation emerged [Derendinger et al. (1985)], [Dine et al. (1985)], [Derendinger et al. (1986)]. In the strongly coupled case, such a mechanism can be realized as well [Hořava (1996)], [Nilles et al. (1997)]. In fact the notion of the hidden sector acquires a geometrical interpretation: the gaugino condensate forms at one boundary (the hidden wall) of spacetime. We shall now discuss this mechanism in detail. First we remind you of some relevant formulae in the weakly coupled case. Our aim then is to compare the strong coupling regime with the weak coupling regime and clarify similarities as well as differences. For the weakly coupled case we start with the action of $d = 10$ supergravity. Supersymmetry transformation laws for the $d = 10$ gravitino fields ψ_M and the dilatino field λ are written **

$$\begin{aligned}\delta\lambda &= \frac{1}{8}\varphi^{-3/4}\Gamma^{MNP}H_{MNP} + \frac{\sqrt{2}}{384}\Gamma^{MNP}\bar{\chi}^a\Gamma_{MNP}\chi^a + \ldots, \\ \delta\psi_M &= \frac{\sqrt{2}}{32}\varphi^{-3/4}(\Gamma_M^{NPQ} - 9\delta_M^N\Gamma^{PQ})H_{NPQ} \\ &\quad + \frac{1}{256}(\Gamma_{MNPQ} - 5g_{MN}\Gamma_{PQ})\bar{\chi}^a\Gamma^{NPQ}\chi^a + \ldots,\end{aligned} \tag{135}$$

implying that a condensate of gauginos $\bar{\chi}\chi$ and/or non–vanishing vevs of the H fields may break supersymmetry. Here we assume the appearance of the gaugino condensate in the hidden sector

$$\langle\bar{\chi}^a\Gamma_{mnp}\chi^a\rangle = \Lambda^3\epsilon_{mnp}, \tag{136}$$

with Λ being the gaugino condensation scale and ϵ_{mnp} the covariantly constant holomorphic three–form. The perfect square structure seen in the Lagrangian [Dine et al. (1985)]

$$-\frac{3}{4}\varphi^{-3/2}(H_{MNP} - \sqrt{2}\varphi^{3/4}\bar{\chi}^a\Gamma_{MNP}\chi^a)^2 \tag{137}$$

will be a very important ingredient to discuss the quantitative properties of the mechanism. When reducing to the $d = 4$ effective action we will find a cancellation of the vevs of the H field and the gaugino condensate at the minimum of the potential such that the term in eq. (137) vanishes. Before we look at this in detail, let us first comment on such a possible vev of H and a possible quantization condition of the antisymmetric tensor. In [Rohm and Witten (1986)] it was shown, that an antisymmetric tensor field $H = dB$ has a quantized vacuum expectation value. In many subsequent papers this has been incorrectly taken

** Here we use again the conventions of [Chamseddine (1981)] used in the first part of these lectures, where the Lagrangian is given in the Einstein frame. To recover the effective action (78) in the string frame, one has to make a proper Weyl transformation and identify $\varphi = \exp(\phi/3)$.

as an argument for the quantization of the vev of $H = dB + \omega^{YM} - \omega^L$ as given in eq. (79). The correct way to interpret this situation is to have a cancellation of the gaugino condensate with the vev of a Chern–Simons term [Derendinger et al. (1986)], for which such a quantization condition does not hold. After all the Chern–Simons term ω^{YM} contains the superpotential of the $d = 4$ effective theory [Derendinger et al. (1985)]. This cancellation leads to a certain combination of ψ_M and λ as the candidate goldstino that will provide the longitudinal component of the gravitino. While in $d = 10$ this looks rather complicated, it simplifies tremendously once one reduces to $d = 4$. Qualitatively the scalar potential takes the following form at the classical level (for the detailed factors see [Nilles (1990)]):

$$V = \frac{1}{ST^3}\left[\mid W - 2(ST)^{3/2}(\bar{\chi}\chi) \mid^2 + \frac{T}{3} \mid \frac{\partial W}{\partial C} \mid^2\right]. \tag{138}$$

We observe the important fact that the potential is positive and vanishes at the minimum. Thus we have broken supersymmetry with a vanishing cosmological constant at the classical level. The first term in the brackets of eq. (138) corresponds to the contribution from eq. (137) once reduced to $d = 4$ and vanishes at the minimum. In the $d = 4$ theory it represents the auxiliary component F_S of the dilaton superfield S. Thus we have $F_S = 0$ and supersymmetry is broken by a nonvanishing vev of F_T [Derendinger et al. (1986)]. The goldstino is then the fermion in the T–multiplet and we are dealing with a situation that has later been named moduli–dominated supersymmetry breakdown. This fact has its origin in the special properties of the $d = 10$ action (the term in eq. (137)) and seems to be of rather general validity. The statement $F_S = 0$ is, of course, strictly valid only in the classical theory. The corrections discussed earlier, eq. (88), will slightly change these results as we shall see later.

Having minimized the potential and identified the goldstino we can now compute the gravitino mass according to the standard procedure. The result has a direct physical meaning because we are dealing with a theory with vanishing vacuum energy. We obtain

$$m_{3/2} \sim \frac{F_T}{M_{Planck}} \sim \frac{\Lambda^3}{M^2_{Planck}}. \tag{139}$$

A value of $\Lambda \sim 10^{13}$ GeV will thus lead to a gravitino mass in the TeV region.

Next we turn to supersymmetry breaking in the strongly coupled case ($d = 11$ M–theory picture) and start with the $d = 11$ action. Supersymmetry transformation laws for the gravitino fields in this case are given by

$$\begin{aligned}\delta\psi_A &= D_A\eta + \frac{\sqrt{2}}{288}G_{IJKL}\left(\Gamma_A^{IJKL} - 8\delta_A^I\Gamma^{JKL}\right)\eta \qquad (140)\\ &\quad - \frac{1}{1152\pi}\left(\frac{\kappa}{4\pi}\right)^{2/3}\delta(x^{11})\left(\bar{\chi}^a\Gamma_{BCD}\chi^a\right)\left(\Gamma_A^{BCD} - 6\delta_A^B\Gamma^{CD}\right)\eta + \ldots\\ \delta\psi_{11} &= D_{11}\eta + \frac{\sqrt{2}}{288}G_{IJKL}\left(\Gamma_{11}^{IJKL} - 8\delta_{11}^I\Gamma^{JKL}\right)\eta\end{aligned}$$

$$+\frac{1}{1152\pi}\left(\frac{\kappa}{4\pi}\right)^{2/3}\delta(x^{11})\left(\bar{\chi}^a\Gamma_{ABC}\chi^a\right)\Gamma^{ABC}\eta+\dots \tag{141}$$

where gaugino bilinears appear in the right hand side of both expressions. Again we consider gaugino condensation at the hidden E_8 boundary

$$\langle\bar{\chi}^a\Gamma_{ijk}\chi^a\rangle = g_8^2\Lambda^3\epsilon_{ijk}. \tag{142}$$

The E_8 gauge coupling constant appears in this equation because the straightforward reduction and truncation leaves a non–canonical normalization for the gaugino kinetic term. An important property of the weakly coupled case (d=10 Lagrangian) was the fact that the gaugino condensate and the three–index tensor field H contributed to the scalar potential in a full square. Hořava made the important observation that a similar structure appears in the M–theory Lagrangian as well [Hořava (1996)]:

$$-\frac{1}{12\kappa^2}\int_{M^{11}} d^{11}x\sqrt{g}\left(G_{ABC11}-\frac{\sqrt{2}}{32\pi}\left(\frac{\kappa}{4\pi}\right)^{2/3}\delta(x^{11})\bar{\chi}^a\Gamma_{ABC}\chi^a\right)^2 \tag{143}$$

with the obvious relation between H and G. Let us now have a closer look at the form of G. At the next to leading order we have

$$\begin{aligned}G_{11ABC} &= (\partial_{11}C_{ABC}+\text{permutations})\\ &\quad+\frac{1}{4\pi\sqrt{2}}\left(\frac{\kappa}{4\pi}\right)^{2/3}\sum_i\delta(x^{11}-x_i^{11})(\omega^{YM}_{ABC}-\frac{1}{2}\omega^L_{ABC}).\end{aligned} \tag{144}$$

Observe, that in the bulk we have $G=dC$ with the Chern–Simons contributions confined to the boundaries. Formula (143) suggests a cancellation between the gaugino condensate and the G–field in a way very similar to the weakly coupled case, but the nature of the cancellation of the terms becomes much more transparent now. In the former case we had to argue via the quantization condition for dB that the gaugino condensate is canceled by one of the Chern–Simons terms. Here this becomes obvious. The condensate is located at the wall as are the Chern–Simons terms, so this cancellation has to happen locally at the wall and dC should vanish for G not to have a vev in the bulk. In any case there is a quantization condition for dC as well [Witten (1997)].

So this cancellation is very similar to the one in the weakly coupled case. At the minimum of the potential we obtain $G_{ABCD}=0$ everywhere and

$$G_{ABC11}=\frac{\sqrt{2}}{32\pi}\left(\frac{\kappa}{4\pi}\right)^{2/3}\delta(x^{11})\bar{\chi}^a\Gamma_{ABC}\chi^a \tag{145}$$

at the hidden wall. Eqs. (141) and (141) then become

$$\delta\psi_A = D_A\eta+\dots \tag{146}$$

$$\delta\psi_{11} = D_{11}\eta+\frac{1}{384\pi}\left(\frac{\kappa}{4\pi}\right)^{2/3}\delta(x^{11})\left(\bar{\chi}^a\Gamma_{ABC}\chi^a\right)\Gamma^{ABC}\eta+\dots. \tag{147}$$

An inspection of the potential shows that $\delta\psi_{11}$ is nonvanishing and supersymmetry is spontaneously broken. Because of the cancellation in eq. (143), the cosmological constant vanishes to leading order. Recalling the supersymmetry transformation law for the elfbein

$$\delta e_I^m = \frac{1}{2}\bar{\eta}\Gamma^m\psi_I, \tag{148}$$

one finds that the superpartner of the $\mathcal{T}$ field plays the role of the goldstino. Again we have a situation where $F_{\mathcal{S}} = 0$ (due to the cancellation in (143)) with nonvanishing $F_{\mathcal{T}}$. But here we find the novel and interesting situation that $F_{\mathcal{T}}$ differs from zero only at the hidden wall, although the field itself is a bulk field. In general, it would be interesting to consider also situations where the goldstino is not a bulk but a wall field.

At that wall our discussion is completely 4–dimensional although we are still dealing effectively with a $d = 5$ theory. To reach the effective theory in $d = 4$ we have to integrate out the dependence of the x^{11} coordinate. As in the previous section this can be performed by the averaging procedure explained there. With the gaugino condensation scale Λ sufficiently small compared to the compactification scale M_{GUT}, the low–energy effective theory is well described by four dimensional $N = 1$ supergravity in which supersymmetry is spontaneously broken. In this case, the modes which remain at low energies will be well approximated by constant modes along the x^{11} direction. This observation justifies our averaging procedure to obtain four dimensional quantities. Averaging $\delta\psi_{11}$ over x^{11}, we thus obtain the vev of the auxiliary field $F_{\mathcal{T}}$

$$F_{\mathcal{T}} = \frac{1}{2}\mathcal{T}\frac{\int dx^{11}\sqrt{g_{11\,11}}\delta\psi_{11}}{\int dx^{11}\sqrt{g_{11\,11}}}. \tag{149}$$

Note that this procedure allows for a nonlocal cancellation of the vev of the auxiliary field in $d = 4$. A condensate with equal size and opposite sign at the observable wall could cancel the effect and restore supersymmetry. Using $\int dx^{11}\sqrt{g_{11\,11}}\delta(x^{11}) = 1$, the auxiliary field is found to be

$$F_{\mathcal{T}} = \mathcal{T}\frac{1}{32\pi(4\pi)^{2/3}}\frac{g_8^2\Lambda^3}{R_{11}M_{11}^3} \tag{150}$$

Similarly one can easily show that $F_{\mathcal{S}}$ as well as the vacuum energy vanish. This allows us then to unambiguously determine the gravitino mass, which is related to the auxiliary field in the following way:

$$m_{3/2} = \frac{F_{\mathcal{T}}}{\mathcal{T}+\mathcal{T}^*} = \frac{1}{64\pi(4\pi)^{2/3}}\frac{g_8^2\Lambda^3}{R_{11}M_{11}^3} = \frac{\pi}{2}\frac{\Lambda^3}{M_{Planck}^2}. \tag{151}$$

As a nontrivial check one may calculate the gravitino mass in a different way. A term in the Lagrangian

$$-\frac{\sqrt{2}}{192\kappa^2}\int dx^{11}\sqrt{g}\bar{\psi}_I\Gamma^{IJKLMN}\psi_N G_{JKLM}, \tag{152}$$

becomes the gravitino mass term when compactified to four dimensions. Using the vevs of the G_{IJK11} given by eq. (145), one can obtain the same result as eq. (151). This is a consistency check of our approach and the fact that the vacuum energy vanishes in the given approximation.

It follows from eq. (151), that the gravitino mass tends to zero when the radius of the eleventh dimension goes to infinity. When the four–dimensional Planck scale is fixed to be the measured value, however, the gravitino mass in the strongly coupled case is expressed in a standard manner, similar to the weakly coupled case as can be seen by inspecting (151) and (139). To obtain the gravitino mass of the order of 1 TeV, one has to adjust Λ to be of the order of 10^{13} GeV when one constructs a realistic model by appropriately breaking the E_8 gauge group at the hidden wall.

In the minimization of the potential we have implicitly used the leading order approximation. As was explained in a previous section, the next to leading order correction gives the non–trivial dependence of the background metric on x^{11}. Then the Einstein–Hilbert action in eleven dimensions gives additional contribution to the scalar potential in the four–dimensional effective theory, which shifts the vevs of the G_{IJKL}. As a consequence, F_S will no longer vanish. Though this may be significant when we discuss soft masses, it does not drastically change our estimate of the gravitino mass (151) and our main conclusion drawn here is still valid after the higher order corrections are taken into account.

12 Soft supersymmetry breaking terms

In the previous section, we have shown that gaugino condensation breaks supersymmetry both in the weakly coupled heterotic string and in heterotic M–theory. We chose Λ in such a way that the gravitino mass appeared in the TeV–range. In this section we shall discuss the soft supersymmetry breaking terms that appear in the low–energy effective theory as a consequence of this nonzero gravitino mass.

We first give the relevant formulae for gaugino and scalar masses in the observable sector. Given the gauge kinetic function f_6 in the observable sector, the gaugino mass is calculated to be

$$m_{1/2} = \frac{\partial f_6}{\partial \phi^i} \frac{F^i}{2\mathrm{Re} f_6}, \tag{153}$$

where ϕ^i symbolically denote hidden sector fields responsible for supersymmetry breakdown. Writing the Kähler potential

$$K = \hat{K}(\phi^i, \phi^*_i) + Z(\phi^i, \phi^*_i) C^* C + (\text{higher orders in } C, C^*), \tag{154}$$

one can also calculate the mass of a matter field C [Kaplunovsky and Louis (1993)], [Brignole et al. (1994)]

$$m_0^2 = m_{3/2}^2 - F^i F^*_j \frac{Z_i^j - Z_i Z^{-1} Z^j}{Z}. \tag{155}$$

Here a vanishing cosmological constant is assumed.

Using the classical approximation naively, these formulae lead to a surprising result. All soft masses vanish. At the basis of this fact it had been suggested that the gravitino mass could be arbitrarily high, still leading to softly broken supersymmetry in the TeV range. It has been observed meanwhile that this surprising result is an artifact of the approximation and it is now commonly accepted that generically the soft masses tend to be of the order of the gravitino mass or at least not arbitrarily small compared to it. In general, the result for the soft scalar masses is strongly model dependent. We shall see in the following that the situation concerning the gaugino mass is less model dependent but varies when we go from the weakly to the strongly coupled case [Nilles et al. (1997)].

We start again with the weakly coupled case. At the leading order (tree level), the gauge kinetic function for the observable sector is simply $f_6 = S$, whereas the gaugino condensation gives $F_S = 0$, $F_T = m_{3/2}(T + T^*)$. Thus, at this level, the gaugino mass vanishes. As was discussed earlier in these lectures, the gauge kinetic function receives corrections at one–loop order. Using eq. (88), the gaugino mass is explicitly written as

$$m_{1/2} = \frac{F_S + \epsilon F_T}{2\mathrm{Re}(S + \epsilon T)}. \tag{156}$$

Note that $F_T/(T+T^*) \sim m_{3/2}$. Also we expect F_S to be of the order of $\epsilon T m_{3/2}$ due to the one–loop corrections. Plugging them into the above expression, we obtain

$$m_{1/2} \sim \frac{\epsilon T}{S} m_{3/2}. \tag{157}$$

Since in the weakly coupled case the ratio $\epsilon T/S$ is small, the gaugino becomes much lighter than the gravitino.

Let us now consider the scalar masses. At the tree level, the Kähler potential is

$$K = -\ln(S+S^*) - 3\ln(T+T^*) + (T+T^*)^n C^* C + (\text{higher orders in } C^*C), \tag{158}$$

where n denotes the modular weight of a field C. For a field with $n = -1$ (untwisted sector in an orbifold construction), which naturally appears in the simple truncation procedure, we recover the previous formula (82). From eq. (155), it follows that

$$m_0^2 = m_{3/2}^2 + \frac{|F_T|^2}{(T+T^*)^2} = (1+n) m_{3/2}^2. \tag{159}$$

A scalar field with the modular weight -1 has a vanishing supersymmetry breaking mass at the leading order. It is a special property of the approximation of reduction and truncation (i.e. torus compactification) that the fields have modular weight -1. A field whose modular weight is different from -1 has a mass comparable to the gravitino mass. Though, as discussed in section 3, corrections at the one–loop level are model dependent, one expects them to be of the order of $\epsilon T/S m_{3/2}^2$. Summarizing these contributions, one obtains

$$m_0^2 = (1+n)m_{3/2}^2 + O(\frac{\epsilon T}{S} m_{3/2}^2), \tag{160}$$

where the actual value of the second term depends on the model one considers. A conclusion we can draw from eqs. (157) and (160) is that the gaugino masses tend to be much smaller than the scalar masses:

$$m_{1/2} << m_0 \leq O(m_{3/2}). \tag{161}$$

Phenomenologically this relation might be problematic. Requiring that the gaugino masses are at the electro–weak scale, eq. (161) would then imply that the masses of the squarks and sleptons should be well above the 1 TeV region, which raises the fine–tuning problem to reproduce the Fermi scale. Another potential problem is the relic abundance of the lightest superparticles (LSPs) which are likely the lightest neutralinos in the present case. With the parameters characterized by (161), the standard computation of the relic abundances shows that too many LSPs would (if stable) still be around today, resulting in the overclosure of the Universe.

Thus in the weak coupling regime, one can conclude that, though the gaugino condensation realizes supersymmetry breaking, it tends to lead to a picture where gaugino masses are generically smaller than gravitino and scalar masses. A satisfactory situation might only be achieved, if one fine–tunes the scalar masses in a way that they become comparable to the gaugino masses.

Next we want to discuss how the situation changes when one considers the strongly coupled case (heterotic M–theory). As in the weakly coupled heterotic string theory, the gaugino mass vanishes at the leading order of the $\kappa^{2/3}$ expansions, because $f_6 = \mathcal{S}$ and $F_{\mathcal{S}} = 0$. Again the next to the leading order is important. The analogue of eq. (156) in the strongly coupled case is

$$m_{1/2} = \frac{F_{\mathcal{S}} + \alpha F_{\mathcal{T}}}{2\mathrm{Re}(\mathcal{S} + \alpha\mathcal{T})}. \tag{162}$$

Thus we obtain, as before

$$m_{1/2} \sim \frac{\alpha\mathcal{T}}{\mathcal{S}} m_{3/2}. \tag{163}$$

A crucial difference in this case, however, is the fact that the ratio $\alpha\mathcal{T}/\mathcal{S}$ is not a small number, but can be as large as unity. This is because the values of $\mathcal{S}$ and $\mathcal{T}$ inferred from our input variables suggests that we are rather close to criticality (in which case the ratio becomes unity). Thus we can conclude that, unlike the weakly coupled case, the gaugino mass in the strongly coupled regime is comparable to the gravitino mass. This observation confirms the expectation that the gravitino mass should be in the TeV–region and the gaugino condensation scale $\Lambda \sim 10^{13}$ GeV. Because of the simplicity of the mass formula (153) and the fact that the gauge–kinetic function f is stable in higher order perturbation theory, the statement concerning the soft gaugino masses is rather model independent.

The situation is more complicated in the case of the scalar masses which we consider now in the framework of heterotic M–theory. At the leading order we arrive at the same conclusions as in the weak coupling case, since the Kähler potential is identical in both cases. Earlier, we calculated the corrections to the Kähler potential at the next to leading order, which read

$$\hat{K} = -\ln(\mathcal{S}+\mathcal{S}^*) - 3\ln(\mathcal{T}+\mathcal{T}^*) \tag{164}$$

$$Z = \frac{6}{\mathcal{T}+\mathcal{T}^*} + \frac{2\alpha}{\mathcal{S}+\mathcal{S}^*} \tag{165}$$

where the latter is valid for a field with the modular weight -1. Now using the formula (155) one may be able to calculate the scalar masses, with the result

$$m_0^2 = m_{3/2}^2 - \frac{2-\frac{1}{1+\delta}}{1+\delta}\frac{|F_{\mathcal{T}}|^2}{(\mathcal{T}+\mathcal{T}^*)^2} - \frac{\delta(2-\delta\frac{1}{1+\delta})}{1+\delta}\frac{|F_{\mathcal{S}}|^2}{(\mathcal{S}+\mathcal{S}^*)^2} - \frac{\delta}{(1+\delta)^2}(F_{\mathcal{S}}F_{\mathcal{T}}^* + F_{\mathcal{S}}^*F_{\mathcal{T}}) \tag{166}$$

where

$$\delta \equiv \frac{\alpha}{3}\frac{\mathcal{T}+\mathcal{T}^*}{\mathcal{S}+\mathcal{S}^*}. \tag{167}$$

We can clearly see from this expression that the structure obtained in the leading order is badly violated. Given the fact that the expansion parameter $\alpha(\mathcal{T}+\mathcal{T}^*)/(\mathcal{S}+\mathcal{S}^*)$ is of order unity it is no longer possible to fine tune the scalar masses (by choosing modular weight -1 for all of them) to a small value and then hope that the corrections respect this fine tuning. In addition, the scalar masses depend strongly on the form of the Kähler potential which, in contrast to the gauge kinetic function, receives further corrections in higher order. Thus detailed statements about the scalar masses are very model dependent. It remains to be seen whether any sensible quantitative statement can be made about the scalar masses with the formulae given above. The results for the gaugino masses are more reliable since f does not receive corrections in higher order.

In summary we can, however, conclude with the qualitative statement that in the strong coupling regime,

$$m_{1/2} \sim m_0 \sim m_{3/2}. \tag{168}$$

This contrasts with the relation (161) for the weak coupling regime and represents an important improvement concerning phenomenological applications. In the strongly coupled case, the difference between dilaton– and moduli–dominated supersymmetry breakdown seems less pronounced than it is in the weakly coupled case.

13 Some phenomenological consequences

We have presented a consistent framework of supersymmetry breaking and soft breaking terms triggered by the gaugino condensate at the hidden wall. In the strongly coupled case, in complete analogy to the weakly coupled case, the gravitino mass $m_{3/2}$ is related to the gaugino condensation scale Λ as

$$m_{3/2} \approx \frac{\Lambda^3}{M_{Planck}^2}. \tag{169}$$

Furthermore, as explained in detail, the soft masses are of the order of the gravitino mass. This implies that these masses should be in the TeV range in order to solve the naturalness problem of the Higgs boson mass in the supersymmetric framework. This requires that Λ should be around 10^{13} GeV, three orders of magnitude smaller than the GUT scale (the compactification scale) and thus the 11D Planck scale as well. The gauge coupling constant at the E_8 wall, where the gaugino condensate is supposed to occur, is larger than the one at the E_6 wall. If the eleventh dimensional radius ρ approaches the critical radius ρ_{crit}, the E_8 gauge coupling constant becomes strong at a scale as large as the GUT scale, and the running coupling constant will blow up at that scale already. Then the gaugino condensation scale Λ, which is approximately identified with the blow–up energy scale, would become too large. For a value of $\Lambda \sim 10^{13}$ GeV, ρ should (although close) not be too close to the critical value so that the gauge coupling constant does not blow up immediately. This gives a constraint on the constant α (defined in (111)), which depends on the detailed properties of the Calabi–Yau manifold under consideration. In any case it is probably necessary to break the hidden E_8 to a smaller group to obtain a smaller coefficient of the β–function. One might also consider the situation, where the role of hidden and observable sector are interchanged, with the observable sector more strongly coupled than the hidden sector at the GUT-scale. These considerations should be kept in mind when one attempts to construct a realistic model.

The fact that the gravitino mass cannot be arbitrarily large, but should lie in the TeV range in the heterotic M–theory regime suggests that the theory might share a problem already encountered in the weakly coupled case [Pagels and Primack (1982)], [Weinberg (1983)], [Ellis et al. (1984)]. Late time decay of the gravitinos would upset the success of the standard big–bang nucleosynthesis scenario. This problem is rather universal in most of the supergravity models where breakdown of supersymmetry is mediated through gravity. Indeed, this is not really a serious difficulty, but just implies that the universe underwent inflationary expansion followed by reheating at a relatively low temperature ($T < 10^9$ GeV for $m_{3/2} = 1$ TeV [Kawasaki and Moroi (1995)]), in which the gravitino number density is diluted by the inflation and the low reheat temperature suppresses gravitino production after that.

A main difference between the weakly and the strongly coupled case manifests itself when we consider phenomenological issues associated with the soft masses. In the weakly coupled string case, the gaugino condensation scenario gives a very small gaugino mass compared to the scalar masses. For a typical size of the compactification radius of the 6D manifold, the gaugino mass is shown to be more than one order of magnitude smaller than the scalar mass (see for example eqs. (7.20) and (7.24) (with $\sin\theta \to 0$ limit) of ref. [Brignole et al. (1994)] for more detail). This hierarchy among the soft masses obviously raises a naturalness problem. With gaugino masses of the order of 100 GeV, the scalar masses would be far above 1 TeV, requiring fine tuning to obtain the electroweak symmetry breaking scale. This causes problems for explicit model building. Another phenomenological difficulty caused by the small gaugino mass arises in the context of relic abundances of the lightest superparticles (LSPs). Under the assumption of R-parity conservation, the LSP is stable and remains today as a dark matter candidate. Given the superparticle spectrum in the weak coupling regime, the bino, the superpartner of the $U(1)_Y$ gauge boson, is most likely to be the LSP. To evaluate the relic abundances of the bino, one has to know its annihilation cross section (see ref. [Jungman et al. (1996)] and references therein). In our case, the bino pair annihilates into fermion (quarks and leptons) pairs via t-channel scalar (squarks and sleptons) exchange. The cross section is roughly proportional to

$$\sigma \propto \frac{m_{\tilde{B}}^2}{m_{\tilde{f}}^4} \tag{170}$$

where $m_{\tilde{B}}$ is the bino mass and $m_{\tilde{f}}$ represents a scalar mass. As the scalar becomes heavier, the cross section is suppressed, yielding a larger relic abundance. Indeed, when the scalar mass is more than an order of magnitude larger than the gaugino mass, a standard calculation shows that the relic abundance exceeds the critical value of the universe. This overclosure is a serious problem in the weakly coupled case.

In the strong coupling regime, the gaugino acquires a mass comparable to the gravitino mass and the scalar masses. Thus the above two problems do not appear. All the soft masses are in the same range. If this is not far from the electroweak scale, one can naturally realize the electroweak symmetry breaking at the correct scale without fine tuning. Moreover in this scenario, the annihilation cross section of the bino becomes larger, and thus we can obtain a relic abundance compatible with the observations. In some regions of parameter space we may even realize a situation where the LSP is the dominant component of the dark matter of the universe.

A characteristic of the mechanism of gaugino condensation is the fact that it is the T field that plays the dominant role in the breakdown of supersymmetry. In this scenario scalar fields with different modular weight will have different masses, which may cause problems with flavor changing neutral currents (FCNC). In the strong coupling case, the situation may be improved through the presence of a large gaugino mass which contributes to the scalar masses at low energies through radiative corrections that can be computed via renormalization group methods. In a situation where scalar masses at the GUT scale are small enough, this universal radiative contribution might wash out nonuniversalities and avoid problems with FCNC. Details of the superparticle phenomenology in the strongly coupled case, including the issues outlined above, will be discussed elsewhere [Kawamura et al. (1998)].

Eqs. (88) (in the weak coupling case) and (122) (in the strong coupling case) show that the imaginary part of the complex scalar fields, S and T, has an axion–like coupling to the gluon fields. In the weakly coupled case, world–sheet instanton effects [Dine et al. (1987)] and possibly other non–perturbative effects give non–negligible contributions to the potential. Then the axion candidates receive masses comparable to the gravitino mass, and they do not solve the strong CP problem. However, in the strongly coupled case, it has been argued that these non–perturbative contributions originated at high energy physics might be suppressed to a negligible level [Banks et al. (1996)], [Banks et al. (1997)], [Choi (1997)]. If this is the case, a linear combination of the Im$\mathcal{S}$ and Im$\mathcal{T}$ will play a role of the axion, whose potential is dominated by the QCD contribution. Then this axion, referred to as the M–theory axion, will be able to solve the strong CP problem. A word of caution should be added here, since a reliable calculation of these world sheet nonperturbative effects has only been performed in the weakly coupled case [Lauer et al. (1991)]. The above argumentation in the M–theory framework uses the implicit assumption that those Yukawa couplings remain as weak as in the case of the weakly coupled string, an assumption that might not be necessarily correct. Apart from that, the axion decay constant in this case becomes as large as 10^{16} GeV, which leads to the potential problem that the energy density of the coherent oscillation of the axion field exceeds the critical energy density of the universe. This problem could be solved if the entropy production occurs after the QCD phase transition when the axion gets massive, or if this world is almost CP conserving and the initial displacement of the axion field is very small. The direct detection of the relic axions with such a large decay constant would be extremely difficult. However the M–theory axion may give a significant contribution to the isocurvature density fluctuations during the inflationary epoch, which may be detectable in future satellite observations [Kawasaki and Yanagida (1997)]. It remains to be seen whether this mechanism leads to a satisfactory solution of the strong CP–problem.

14 Summary and outlook

In any case we have seen that the M–theoretic version of the heterotic string shows some highly satisfactory phenomenological properties concerning the unification of fundamental coupling constants as well as the nature of the soft supersymmetry breaking parameters.

Still there remain some problems that still resist attempts for a satisfactory solution. Certainly one of them is the question of fixing the vev of the dilaton. One would like to see whether the M-theoretic approach to the problem might give us some new hints in that direction.

In the last years there has been revolutionary progress in the understanding of nonperturbative aspects of string theory. Here we have discussed the first consequences of phenomenological interest that could be derived from this new insights. Let us hope that other aspects of that field might also be of relevance for this questions and increase our understanding of the low-energy effective actions that could be derived from string theory.

Acknowledgements

I would like to thank J. Conrad, Z. Lalak, A. Niemeyer, M. Olechowski, S. Stieberger, and M. Yamaguchi for useful discussions and collaboration. This work was partially supported by the European Commission programs ERB FMRX-CT96-0045 and CT96-0090. Thanks to the organizers of the school for their hospitality.

References

Aldazabal, G., Font, A., Ibáñez, L. E. and Uranga, A.M., *String GUTs*, Madrid FTUAM-94/28 (1994); (hep-th/9410206)

Amaldi, U., de Boer, W., Fürstenau, H., *Phys. Lett.* **B 281** (1992) 374; Antoniadis, I., Ellis, J., Kelley, S., and Nanopoulos, D. V., *Phys. Lett.* **B 272** (1991) 31

Antoniadis, I., Ellis, J., Lacaze, R. and Nanopoulos, D. V., *Phys. Lett.* **B 268** (1991) 188;
Dolan, L. and Liu, J. T., *Nucl. Phys.* **B 387** (1992) 86;
Chemtob, M., Saclay T95/086 (hep–th/9506178)

Antoniadis, I., Gava, E., Narain K. S. and Taylor, T. R., *Nucl. Phys.* **B 432** (1994) 187

Antoniadis, I. , Narain, K. S. and Taylor, T. R., *Phys. Lett.* **B 267** (1991) 37;
Antoniadis, I., Gava, E. and Narain, K. S., *Nucl. Phys.* **B 383** (1992) 93; *Phys. Lett.* **B 283** (1992) 209;
Mayr, P., and Stieberger, S., *Nucl. Phys.* **B 412** (1994) 502

Antoniadis, I. and Quirós, M., Phys. Lett. **B392** (1997) 61; hep–th/9705037

Banks, T. and Dine, M., Nucl. Phys. **B479** (1996) 173

Banks, T. and Dine, M., Nucl. Phys. **B505** (1997) 445

Bershadsky, M., Cecotti, Ooguri, S. H. and Vafa, C., *Nucl. Phys.* **B 405** (1993) 279; *Comm. Math. Phys.* **165** (1994) 311;
Hosono, S., Klemm, A., Theisen, S. and Yau, S. T., *Nucl. Phys.* **B 433** (1995) 501

Binetruy, P. and Gaillard, M.K., Phys. Lett. **232B** (1989) 83

Brignole, A., Ibáñez, L. E. and Muñoz, C., Nucl. Phys. **B422** (1994) 125 and references therein

Candelas, P., Horowitz, G., Strominger A. and Witten, E., Nucl. Phys. **B258** (1985) 46

Chamseddine, A. H., Phys. Rev. **D24** (1981) 3065;
Chapline, G. F. and Manton, N. S., Phys. Lett. **B120** (1983) 105

Choi, K., Phys. Rev. **D56** (1997) 6588

Choi, K. and Kim. J. E., Phys. Lett. **B165** (1985) 71

Choi, K., Kim, H. B. and Muñoz, C., hep-th/9711158
Conrad, J. O., hep-th/9708031
Conrad, J. O. and Nilles, H. P., to appear
Cremmer, E., Ferrara, S., Girardello, L. and van Proeyen, A., *Nucl. Phys.* **B 212** (1983) 413
Cvetič, M., Font, A., Ibáñez, L. E., Lüst, D. and Quevedo, F., *Nucl. Phys.* **B 361** (1991) 194
For an extended list of references see: Carlos, B. de, Casas, J. A., Muñoz, C., Nucl.Phys. **B399** (1993) 623
Font, A., Ibáñez, L. E., Lüst, D., Quevedo, F., Phys.Lett. **249B** (1990) 35
Derendinger, J. P., Ferrara, S., Kounnas, C. and Zwirner, F., *Nucl. Phys.* **B 372** (1992) 145;
Antoniadis, I., Gava, E., Narain, K. S. and Taylor, T. R., *Nucl. Phys.* **B 407** (1993) 706
Derendinger, J. P., Ibáñez, L. E. and Nilles, H. P., Phys. Lett. **B155** (1985) 65
Derendinger, J. P., Ibáñez, L. E. and Nilles, H. P., Nucl. Phys. **B267** (1986) 365
Dienes, K. R. and Faraggi, A. E., *Making ends meet: string unification and low-energy data*, Princeton IASSNS-HEP-95/24 (hep-th/9505018); *Gauge coupling unification in realistic free-fermionic string models*, Princeton IASSNS-HEP-94/113 (hep-th/9505046)
Dine, M., Fischler, W., Srednicki, M., Nucl.Phys. **B189** (1981) 575
Dine, M., Rohm, R., Seiberg, N. and Witten, E., Phys. Lett. **B156** (1985) 55
Dine, M., Seiberg, N., Wen, X. G. and Witten, E., Nucl. Phys. **B289** (1987) 319; Nucl. Phys. **B278** (1986) 769
Dixon, L., Harvey, J., Vafa, C. and Witten, E., *Nucl. Phys.* **B 261** (1985) 678; **B 274** (1986) 285;
Ibáñez, L. E., Mas, J., Nilles, H. P. and Quevedo, F., *Nucl. Phys.* **B 301** (1988) 157
Dixon, L., Kaplunovsky, V. and Louis, J. *Nucl. Phys.* **B 355** (1991) 649
Dudas, E., hep-th/9709043
Dudas, E. and Grojean, C., hep-th/9704177
Ellis, J., Kim. J. E. and Nanopoulos, D. V., Phys. Lett. **B145** (1984) 181
Ellis, J., Kelley, S. and Nanopoulos, D. V., *Phys. Lett.* **B 249** (1990) 441;
Amaldi, U., Boer, W. de and Fürstenau, H., *Phys. Lett.* **B 260** (1991) 447;
Langacker, P. and Luo, M. X., *Phys. Rev.* **D 44** (1991) 817
Ferrara, S., Girardello, L. and Nilles, H. P., Phys. Lett. **B125** (1983) 457
Ferrara, S., Kounnas, C., Lüst, D. and Zwirner, F., *Nucl. Phys.* **B 365** (1991) 431
Font, A., Ibáñez, L. E., D. Lüst, Quevedo, F., Phys.Lett. **245B** (1990) 401;
Ferrara, S., Magnoli, N., Taylor, T. R., Veneziano, G., Phys.Lett. **245B** (1990) 409;
Nilles, H. P., M. Olechowski, Phys.Lett. **248B** (1990) 268;
Binetruy, P., Gaillard, M. K., Phys.Lett. **253B** (1991) 119;
Cvetič, M., Font, A., Ibáñez, L. E., Lüst, D., Quevedo, F., Nucl.Phys. **B361** (1991) 194
Georgi, H., Kim, Jihn E., Nilles, H. P., hep-ph/9805510, to appear in Physics Letters B
Ginsparg, P. *Phys. Lett.* **B 197** (1987) 139
Horne, J. H., Moore, G., Nucl.Phys. **B432** (1994) 109
Ibáñez, L. E., Nilles, H. P., Phys.Lett. **169B** (1986) 354;
Dixon, L., Kaplunovsky, V., Louis, J., Nucl.Phys. **B355** (1991) 649
Green, M. and Schwarz, J., Phys. Lett. **B149** (1984) 117

Green, M., Schwarz, J. and Witten, E., Superstring Theory, Cambridge University Press, 1987
Hořava, P., Phys. Rev. **D54** (1996) 7561
Hořava, P. and Witten, E., Nucl. Phys. **B460** (1996) 506; Nucl. Phys. **B475** (1996) 94.
Ibáñez, L. E., *Phys. Lett.* **B 318** (1993) 73
Ibáñez, L. E. and Lüst, D., *Nucl. Phys.* **B 382** (1992) 305
Ibáñez, L. E., Lüst, D. and Ross, G. G., *Phys. Lett.* **B 272** (1991) 251
Ibáñez, L. E. and Nilles, H. P., Phys. Lett. **B169** (1986) 354
Ibáñez, L. E., Nilles, H. P. and Quevedo, F., *Phys. Lett.* **B 187** (1987) 25; Ibáñez, L. E., Kim. J. E., Nilles, H. P. and Quevedo, F., *Phys. Lett.* **B 191** (1987) 283
Ibáñez, L. E., Nilles, H. P. and Quevedo, F., *Phys. Lett.* **B 192** (1987) 332
Jungman, G., Kamionkowski, M. and Griest, K., Phys. Rep. **267** (1996) 195
Kaplunovsky, V. S., *Nucl. Phys.* **B 307** (1988) 145, Erratum: *Nucl. Phys.* **B 382** (1992) 436
Kaplunovsky, V. S. and Louis, J., Phys. Lett. **B306** (1993) 269
Kaplunovsky, V. S. and Louis, J., *Nucl. Phys.* **B 444** (1995) 191
Kawamura, Y., Nilles, H. P., Olechowski, M. and Yamaguchi, M., hep-ph/9805397, to appear in JHEP
Kawasaki, M. and Moroi, T., Prog. Theor. Phys. **93** (1995) 879
Kawasaki, M. and Yanagida, T., Prog. Theor. Phys. **97** (1997) 809
Kiritsis, E. and Kounnas, C., *Nucl. Phys.* **B 41** [Proceedings Sup.] (1995) 331; *Nucl. Phys.* **B 442** (1995) 472; *Infrared-regulated string theory and loop corrections to coupling constants*, hep-th/9507051
Krasnikov, N. V., Phys.Lett. **193B** (1987) 37;
Casas, J. A., Lalak, Z., Muñoz, C., Ross, G. G., Nucl.Phys. **B347** (1990) 243
Lalak, Z., Niemeyer, A., Nilles, H. P., Phys.Lett. **349B** (1995) 99
Lalak, Z., Niemeyer, A., Nilles, H. P., hep-th/9503170, Nucl.Phys. **B453** (1995) 100
Lalak, Z. and Thomas, S., htp-th/9707223
For a review see: Langacker, P., *Grand Unification and the Standard Model*, hep-ph/9411247
Langacker, P. and Polonsky, N., *Phys. Rev.* **D 47** (1993) 4028 and references therein
Lauer, J., Mas, J. and Nilles, H. P., Nucl. Phys. **B351** (1991) 353
Li, T., Lopez, J. L. and Nanopoulos, D. V., hep-ph/9702237; hep-ph/9704247
Lopes Cardoso, G., Lüst, D. and Mohaupt, T., *Nucl. Phys.* **B 450** (1995) 115
Lukas, A., Ovrut, B. A. and Waldram, D., hep-th/9710208
Lukas, A., Ovrut, B. A. and Waldram, D., hep-th/9711197
Macorra, A. de la, Ross, G.G., Nucl.Phys. **B404** (1993) 321
Matalliatakis, D., Nilles, H. P., Theisen, S., hep-th/9710247; *Phys. Lett.* **B 421** (1998) 169
Mayr, P., Nilles, H. P. and Stieberger, S., *Phys. Lett.* **B 317** (1993) 53
Mayr, P. and Stieberger, S., *Nucl. Phys.* **B 407** (1993) 725;
Bailin, D. , Love, A., Sabra, W. A. and Thomas, S., Mod. Phys. Lett. **A9** (1994) 67; **A10** (1995) 337
Mayr, P. and Stieberger, S., *Phys. Lett.* **B 355** (1995) 107
Mayr, P. and Stieberger, S., TUM-HEP-212/95 to appear;
Stieberger, S., *One-loop corrections and gauge coupling unification in superstring theory*, Ph.D. thesis, TUM-HEP-220/95

For a review see: Mayr, P. and Stieberger, S., Proceedings *28th International Symposium on Particle Theory*, p. 72–79, Wendisch–Rietz (1994) (hep–th/9412196, DESY 95–027
Montonen, C., Olive, D., Phys.Lett. **72B** (1977) 117;
Seiberg, N., Witten, E., Nucl.Phys. **B426** (1994) 19
Nilles, H. P., Phys.Lett. **112B** (1982) 455
Nilles, H. P., Phys. Lett. **B115** (1982) 193
Nilles, H. P., Nucl.Phys. **B217** (1983) 366
Nilles, H. P., Physics Reports **110** (1984) 1
Nilles, H. P., Phys. Lett. **B180** (1986) 240
Nilles, H. P., Lectures at the Trieste Spring School on Supersymmetry, Supergravity and Superstrings 1986, Eds. B. de Wit et al., World Scientific (1986), page 37
Nilles, H. P., Tortured Tori, International Workshop on Superstrings, Composite Structures and Cosmology, Univ. of Maryland, March 1987, Ed. S.J. Gates et al., World Scientific 1987, page 312
Nilles, H. P., Strings on Orbifolds: An Introduction, Lectures given at the International Summer School on Conformal Invariance and String Theory, Poiona Brasov, Roumania, September 1987, Ed. P. Dita and v. Georgescu, Academic Press 1989, page 305
Nilles, H. P., Int. Journ. of Modern Physics **A5** (1990) 4199
Nilles, H. P., TASI lectures 1990, Testing the Standard Model, Ed. M. Cvetic and P. Langacker, World Scientific 1991, page 633
Nilles, H. P., TASI lectures 1993, The Building Blocks of Creation, Ed. S. Raby and T. Walker, World Scientific 1994, page 291
Nilles, H. P., Olechowski. M. and Yamaguchi, M., hep–th/9707143, Phys. Lett. **B415** (1997) 24.
Nilles, H. P., Olechowski. M. and Yamaguchi, M., hep–th/9801030, to appear in Nuclear Physics B
Nilles, H. P. and Stieberger, S., *How to reach the correct* $\sin^2\theta_W$ *and* α_s *in string theory*, hep-th/9510009, Phys.Lett. **B367** (1996) 126
Nilles, H. P. and Stieberger, S., hep–th/9702110, Nucl. Phys. **B499** (1997) 3
Nilles, H. P., Phys.Lett. **115B** (1982) 193;
Nilles, H. P., Nucl.Phys. **B217** (1983) 366;
Chamseddine, A. H., Arnowitt, R. and Nath, P., Phys.Rev.Lett. **49** (1982) 970;
Barbieri, R., Ferrara, S. and Savoy, S., Phys.Lett. **119B** (1982) 343;
Nilles, H. P., Srednicki, M. and Wyler, D., Phys.Lett. **120B** (1983) 346;
Hall, L., Lykken, J. and Weinberg, S., Phys.Rev. **D27** (1983) 2359
Pagels, H. and Primack, J. R., Phys. Rev. Lett **48** (1982) 223
Rohm, R. and Witten, E., Ann. Physics **170** (1986) 454
Shifman, M. and Vainshtein, A., Nucl. Phys. **B277** (1986) 456
Taylor, T. R., Phys.Lett. **164B** (1985) 43
G. 't Hooft, *'Naturalness, chiral symmetry and spontaneous chiral symmetry breaking'*, in 'Recent Developments in Gauge Theories', Cargèse 1979, G. 't Hooft et al, New York 1980, *Plenum Press*
Veneziano, G., Yankielowicz, S., Phys.Lett. **113B** (1982) 231
Weinberg, S., Phys. Rev. Lett. **48** (1982) 1303
Weinberg, S., *Phys. Lett.* **B 91** (1980) 51
Witten, E., Nucl.Phys. **B202** (1982) 253
Witten, E., Phys. Lett. **B155** (1985) 151

Witten, E., Nucl. Phys. **B471** (1996) 135
Witten, E., hep-th/9609122, J. Geom. Phys. **22** (1997) 1
Witten, E., Nucl.Phys. **B188** (1981) 513;
Dimopoulos, S., Raby, S. and Wilzcek, F., Phys.Rev. **D24** (1981) 1681;
Nilles, H. P. and Raby, S., Nucl.Phys. **B198** (1982) 102;
Ibáñez, L. E. and Ross, G. G.,Phys.Lett **105B** (1982) 439
Witten, E., *Nucl. Phys.* **B 258** (1985) 75
Witten, E., *Nucl. Phys.* **B 269** (1986) 79

Abstracts of the Seminars

CP Violation in Angular and Energy Distributions of b and $\bar{b}$ Quarks from Top Decays in $e^+e^- \to t\bar{t}$

A. Bartl[1], E. Christova[2], Th. Gajdosik[3], W. Majerotto[3]

[1]Institut für Theoretische Physik, Universität Wien, A-1090 Vienna, Austria
[2] Institute of Nuclear Research and Nuclear Energy, Boul. Tzarigradsko Chaussee 72, Sofia 1784, Bulgaria
[3] Institut für Hochenergiephysik der Österreichischen Akademie der Wissenschaften, A-1050 Vienna, Austria

Abstract. We obtain analytic formulae for the angular and energy distributions of the secondary b and $\bar{b}$ quarks in the processes:

$$e^+ + e^- \to t + \bar{t} \to b + X \quad (1)$$
$$e^+ + e^- \to t + \bar{t} \to \bar{b} + \bar{X} \ . \quad (2)$$

Here X, $\bar{X}$ stand for $\bar{t}W^+$ and tW^-, irrespectively how the W's are identified. CP violation is assumed in the $\gamma t\bar{t}$ and $Zt\bar{t}$ vertices. The obtained distributions are sensitive to two different combinations of the imaginary parts of the electroweak dipole moment form factors of the top quark, $d^\gamma(s)$ and $d^Z(s)$, that determine the CP violating contribution to the top–quark polarization vector in the production plane. Suitable energy and angular CP violating asymmetries both for unpolarized and polarized beams are defined. All phase space integrations are performed analytically and rather simple expressions are obtained.

The real parts of $d^\gamma(s)$ and $d^Z(s)$ determine the component of the top–quark polarization perpendicular to the production plane and can be determined by measuring the triple–product correlations $(\mathbf{q_e} \times \mathbf{p_t}) \cdot \mathbf{p_{b,\bar{b}}}$. We derive simple analytic expressions for the corresponding asymmetries.

Our expressions are general and model independent. Numerical estimates are presented in the Minimal Supersymmetric Standard Model (MSSM) with complex parameters. The asymmetries are rather sensitive to the beam polarizations. The effects are of the order 10^{-3}.

For more details see A. Bartl, E. Christova, T. Gajdosik, W. Majerotto, hep-ph/9802352 and hep-ph/9803426 and the references therein.

Can the Higgs Mechanism Favour an Electron-Pair Condensation in Three Dimensions?

O.M. Del Cima

Institut für Theoretische Physik, Technische Universität Wien,
Wiedner Hauptstraße 8-10, A-1040 Vienna, Austria

Abstract. The main purpose of this talk is to show that electrons scattered in $D = 1 + 2$ can experience a mutual net attractive interaction, not depending on their spin states [De Andrade et al. (1996)]. This attractive scattering potential comes from processes in which the electrons are correlated in momentum space with opposite spin polarisations (s-wave state). Also, in the case of equal spin polarisations (p-wave state), a net attraction may appear, as due to the Higgs interaction, if some special conditions are set up on the parameters. The latter possibility should be investigated for the cases in which very high external magnetic fields are applied, since it is suspected that the resistance of the superconducting state in the presence of high magnetic fields, in the re-entrant superconductivity effect, could be explained by p-wave states, p-electron pairing [Boebinger (1996)]. The intermediate bosons involved in such scatterings are a massive vector meson and a Higgs scalar, both resulting from the breaking of a local $U(1)$-symmetry. The breaking-down is accomplished by a sixth-power potential. The conditions on the parameters are in order to avoid metastable vacuum states. The behaviour at the quantum level of this model [De Andrade et al. (1996)], in the symmetric and broken regimes, is analysed [Del Cima et al. (1997) and (1998)] by using the algebraic renormalisation method, which is independent of any kind of regularisation scheme.

Now the low-energy s and p-wave bound states are under investigation by using the Schrödinger equations associated to both cases [Carvalho et al. (1998)].

Note. Work supported by the *Fonds zur Förderung der Wissenschaftlichen Forschung (FWF)* under the contract number P11654-PHY.

References

Boebinger, G. (1996): *Correlated Electrons in a Million Gauss*, Physics Today **49**, 36.

De Andrade, M.A., Del Cima, O.M. and Helayël-Neto, J.A. (1996): *Electron-pair condensation in parity-preserving QED*$_3$, hep-th/9603054, Il Nuovo Cimento A (to appear). Talk given at *QS96 - Quantum Systems: New Trends and Methods*, Minsk, Belarus, published in the proceedings.

Del Cima, O.M., Franco, D.H.T., Helayël-Neto, J.A. and Piguet, O. (1997): *Algebraic renormalization of parity-preserving QED*$_3$ *coupled to scalar matter I: unbroken case*, Phys. Lett. **B410**, 250 and (1998): *Algebraic renormalization of parity-preserving QED*$_3$ *coupled to scalar matter II: broken case*, Phys. Lett. **B416**, 402.

Carvalho, H.S., Colatto, L.P., De Andrade, M.A., Del Cima, O.M. and Helayël-Neto, J.A. (1998): *Low-energy electron-electron bound-states in parity-preserving QED*$_3$, work in progress.

Towards a Common Origin of Supersymmetry Breaking, Compositeness, and Gauge Mediation[1]

S. Dubovsky, D. Gorbunov, S. Troitsky

Institute for Nuclear Research of the Russian Academy of Sciences, 60th October Anniversary prospect, 7a, 117312, Moscow, Russia.

Abstract. We propose an approach that incorporates in an economical way both supersymmetry breaking and its mediation to the visible sector. We present a toy model in which the Standard Model matter fields of one generation are composite and appear as low energy effective degrees of freedom of another theory which breaks supersymmetry at strong coupling. A subgroup of the flavor symmetry group of the strongly coupled theory is gauged and identified with the gauge group of the Standard Model. Effects of supersymmetry breaking are transferred to the visible sector by means of gauge interactions. It is desirable that apparently different phenomena – supersymmetry breaking, appearance of both matter content of the Standard Model and messenger superfields of direct gauge mediation – are manifestations of one and the same mechanism. Here we propose how such a mechanism may emerge due to strong coupling dynamics of supersymmetric gauge theories. Though our toy model is far from being viable, we hope that its main features are common to realistic models exploiting the same mechanism.

The simplest version of our model (Dubovsky et al. (1998)) deals with only one generation of the Standard Model matter. Its main ingredient is $SU(5)$ gauge theory with matter in one chiral (antisymmetric tensor A plus antifundamental $\bar{Q}_6$) and five vector-like (fundamental Q_i plus antifundamental $\bar{Q}_i$) generations. The global symmetry group besides $U(1)$ factors contains $SU(5) \times SU(6)$ flavor symmetry. We embed the Standard Model gauge group into the vector-like $SU(5)_W$ subgroup of this flavor symmetry group. The effective theory for this model (Pouliot (1996)) breaks supersymmetry at tree level and has charged under the Standard Model degrees of freedom which correspond to one generation of matter and two sets of messenger fields. The generalization to three generations deals with gauge group $SU(9)$ with essentially the same matter content.

References

Dubovsky, S., Gorbunov, D., Troitsky, S. (1998): Phys. Lett. **B**, to appear; hep-ph/9712397.

Pouliot, P. (1996): Phys. Lett. **B367**, 151.

[1] This work was supported in part by RFBR grant 96-02-17449a and (for S.T.) CRDF Award No. RP1-187. S.T. acknowledges support by the organizers of the Schladming Winter School and by RFBR travel grant.

The Fixed-Point Action for Lattice Gauge Theories

F. Farchioni, C.B. Lang, M. Wohlgenannt

Institut für Theoretische Physik, Universität Graz, A-8010 Graz, Austria

Abstract. We review a recent theoretical development concerning the chiral properties of the fixed point (FP) action for lattice gauge theories. The FP action is the action associated to the FP, in the space of couplings of the lattice theory, of a renormalization group transformation. The corresponding lattice theory reproduces all the classical properties of the theory of the continuum. We concentrate in particular on the chiral properties. As a matter of fact, lattice fermionic theories are forced to break the chiral invariance explicitly (Nielsen and Ninomiya (1981)). In the case of the FP action this obligatory breaking is so mild, that all the relevant chiral properties of the continuum theory are preserved. In formulae, this is guaranteed by the Ginsparg-Wilson relation (Ginsparg and Wilson (1982)) satisfied by the FP Dirac operator. We point out the main consequences of this relation, following the recent works by Hasenfratz and collaborators (Hasenfratz (1997), Hasenfratz, Laliena and Niedermayer (1998), Hasenfratz (1998)). At the classical level, the zero modes of the Dirac operator have definite chirality, a lattice version of the Atiyah-Singer Theorem (Atiyah and Singer (1971)) holds, no exceptional configurations are allowed: the fermion determinant is always positive, except on zero modes of topological origin, where it vanishes; at the quantum level, no additive renormalization of the quark mass occurs (i.e. the chiral limit is realized in the limit of *zero* bare quark mass), the chiral currents do not renormalize, and operators with different chiral properties do not mix. We point out how these theoretical expectations can be checked in the case of the Schwinger model, regarded as a laboratory for QCD (cfr. the contribution by M. Wohlgenannt in this same volume; see also Farchioni, Lang and Wohlgenannt (1998)).

References

M. Atiyah and I.M. Singer (1971): Ann. Math. **93**, 139.
F. Farchioni, C.B. Lang and M. Wohlgenannt (1998): Phys. Lett. B. **433**, 377.
P.H. Ginsparg and K.G. Wilson (1982): Phys. Rev. D **25**, 2649.
P. Hasenfratz (1998): Nucl. Phys. B **401**, 525.
P. Hasenfratz (1997): Nucl. Phys. B (Proc. Suppl.) **63A-C**, 53.
P. Hasenfratz, V. Laliena and F. Niedermayer (1998): Phys. Lett. B **125**, 427.
H. Nielsen and M. Ninomiya (1981): Nucl. Phys. B **185**, 20; *ibid.* **193**, 173.

Spectral Properties of the Dirac Operator with Fixed-Point Action

F. Farchioni, C. B. Lang, M. Wohlgenannt

Karl-Franzens Universität Graz, Universitätsplatz 5, A-8010 Graz, Austria

Abstract. Discretization of space and time leads to severe problems, such as cut-off dependence of observables and explicit breaking of chiral symmetry (Nielsen and Ninomiya (1981)). One can cope with these problems by introducing **perfect actions**. The fixed-point action defined by Lang and Pany (1998) is (numerically close to) a **classical perfect action**. Its classical predictions agree with those of the continuum (considering the same physical volume), no matter how coarse the lattice. **Quantum perfect actions** agree with all the continuum predictions (if one considers quantum perfect operators or observables).

We studied this fixed-point action for the massless one-flavour Schwinger model and compared the results with the theoretical predictions found by Hasenfratz, Laliena and Niedermayer (1998) and by Farchioni and Laliena (1998). The numerical results agree nicely with the predicted circular shape of the spectrum. The distribution roughens at low values of β due to the necessary truncation of the couplings and to numerical errors. The mean deviation $|\lambda - 1|$ from the unit circle in the region close to $\lambda = 0$, in an angular window of $|\arg(1-\lambda)| < \pi/4$, exhibits a scaling behaviour $\propto 1/\beta^{2.41} \simeq a^5$. The parametrized action pA_{FP} is truncated in a finite range (7x7 in our case). Heuristically this implies an error for the eigenvalues in the form of some operator of higher dimension k. From the observed deviation we estimate an effective value $k \simeq 5$.

We note, that there are configurations with real eigenvalues. We checked the eigenvectors v_i for those and confirm that these modes have definite chirality $\langle v_i \gamma_5 v_i \rangle$. Also, we can clearly distinguish the real values around zero from those around 2 (right-hand part of the spectrum). We may identify these real eigenvalues (around zero) with zero-modes and relate their number n_0 with the geometrically (i.e. from the gauge field configuration) defined topological charge Q_G. We find agreement in the following sense: The ratio of the number of configurations, where these numbers coincide over all configurations approaches unity in the limit $\beta \to \infty$. These results and results on the chiral condensate are presented in more detail by Farchioni, Lang and Wohlgenannt (1998).

References

F. Farchioni and V. Laliena, Phys. Rev.**D 58** (1998) 054501.

F. Farchioni, C. B. Lang, and M. Wohlgenannt, Phys. Lett. **B 433** (1988) 377.

P. Hasenfratz, V. Laliena, and F. Niedermayer, Phys. Lett. **B 125** (1998) 427.

C. B. Lang and T. K. Pany, Nucl. Phys. **B** (Proc. Suppl.) **63A-C** (1998) 898; Nucl. Phys. **B 513** (1998) 645.

H. Nielsen and M. Ninomiya, Nucl. Phys. **B 185** (1981) 20; ibid. **193** (1981) 173.

Sources of CP Violation in the Minimal Supersymmetric Standard Model

Th. Gajdosik

Institut für Hochenergiephysik der Österreichischen Akademie der Wissenschaften, A-1050 Vienna, Austria

Abstract. In the Minimal Supersymmetric Standard Model (MSSM) with complex parameters one has additional phases compared to the Standard Model (SM), which induce CP violation. It is explicitly shown that only two phases can be rotated away. The other phases, i.e. the phases of μ, M_1, M_3, $A_{\tilde{q}}$, are physical quantities. μ is the Higgs mixing parameter in the superpotential, M_1 and M_3 are the mass parameters of the U(1) and SU(3) gauge group, respectively, and $A_{\tilde{q}}$ are the trilinear Higgs–sfermion parameters. These phases have an impact on the masses of the supersymmetric particles at tree level. The dependence of the masses of charginos and neutralinos on the phase of μ is smaller than 20% and the dependence on the phase of M_1 is smaller than 10%. Mass formulae and mixing matrices are given. Interesting plots of the chargino and neutralino masses depending on the phases are shown.

Quantum Equivalence of Dual σ Models

R.L. Karp

Institute for Theoretical Physics, Roland Eötvös University,
H-1088 Budapest, Puskin u. 5-7, Hungary

Abstract. In this paper we review the new developments about the perturbative quantum equivalence of dual σ models. We start with Buscher's formula for T-duality, state the problem of quantum equivalence, study several examples with different amount of isometry, conclude that Buscher's formula has to be modified at two loop order, and point to some proposals for resolution.

The original Buscher's formula for T-duality of σ models [1] gives a prescription of defining a new σ model out of one that contains an isometry, modifying the metric, antisymmetric tensor and the dilaton fields. The easiest way to derive it is through functional integral manipulations. However the question arises whether it should be implemented on the bare or renormalized quantities.

As a first example we consider the principal $SU(2)$ σ model, with a perturbation that breaks the global SU(2)×SU(2) to SU(2)×U(1) [2] having two abelian isometries. One of the duals has the same symmetry as the original σ model, and it is shown that the two loop β-functions, expressed in terms of the renormalization group quantities agree. The other possible dual has less symmetry, (U(1)×U(1)), and it is shown to be non-renormalizable at two loop order, although at one loop order there is agreement. In this case the perturbative quantum equivalence cannot be an issue.

One of the differences between the two models is that the g_{00} metric component in the isometry adopted coordinates is constant in the first case and field dependent in the second. One might argue that this is the source of failure. But this is not the case, as it can be seen on the case of a properly deformed principal $SL(3)$ σ model [3]. We have also shown that it is possible to further deform the model in such a way to regain the two loop equivalence, at least in the vicinity of the so called fixed points. This way, we obtained an infinitesimal modification of the original Buscher's formula.

A consistent analysis of the necessary modification of Buscher's formula at two loop order was considered for example in [4]. Unfortunately, it was impossible to solve the corresponding equations in general, even though there has been major advances in this matter from the low energy effective action point of view of the σ model [5]. Due to lack of time this will be reviewed elsewhere.

References

[1] T. Buscher, Phys.Lett. **B**194 (1987) 51; **B**201 (1988) 466.
[2] J. Balog, P. Forgacs, Z. Horvath, L. Palla Nucl.Phys. **B**(PS)49 (1996)16.
[3] Z. Horvath, R. L. Karp, L. Palla Nucl.Phys. **B**490 (1997) 435.
[4] J. Balog, P. Forgacs, Z. Horvath, L. Palla Phys.Lett. **B**388 (1996) 121.
[5] N. Kaloper, K. Meissner (1997), Phys. Rev. **D**56 (1997) 7940.

Renormalizations in Softly Broken SUSY Gauge Theories

D. Kazakov

Bogoliubov Laboratory of Theoretical Physics,
Joint Institute for Nuclear Research,
141980 Dubna, Russia

Abstract. It is shown that a softly broken theory is equivalent to a rigid theory in an external spurion superfield. This enables one to get the singular part of the effective action in a broken theory from a rigid one by a simple redefinition of the couplings. This way one can reproduce all known results on the renormalization of soft couplings and masses in a softly broken theory. As an example, the renormalization group functions for soft couplings and masses in the Minimal Supersymmetric Standard Model up to the three-loop level are calculated. The method opens a possibility to construct a totally all loop finite N=1 SUSY gauge theory, including the soft SUSY breaking terms. Explicit relations between the soft terms, which lead to a completely finite theory in any loop order, are given.

Standard-Model-Like Chiral Spectra and the Origin of the Families

O.C.W. Kong

Department of Physics and Astronomy, University of Rochester
Rochester NY 14627, USA

Abstract. Each family of the standard model (SM) fermions composes of a chiral spectrum tightly bounded by gauge anomaly cancellaton constraints. However, the three family structure and the hierarchy among the masses is a major puzzle.

The fermions in one SM family can be uniquely derived by assuming one multiplet transforming nontrivially under each component gauge group and requiring the minimal chiral spectrum canceling all the anomalies. We seek to understand the family structure through the idea of a SM-like chiral fermion spectrum, one with the same feature as the one family SM under an extended symmetry, which after breaking to the SM symmetry yields naturally the three families as the residual chiral content. The natural choice of the gauge group is $SU(N)\otimes SU(3)\otimes SU(2)\otimes U(1)$. For instance, start with $(\mathbf{4},\mathbf{3},\mathbf{2},\mathbf{1})$ for $N=4$, the strategy leads to the spectrum

$$(\mathbf{4},\mathbf{3},\mathbf{2},\mathbf{1}),(\bar{\mathbf{4}},\bar{\mathbf{3}},\mathbf{1},\mathbf{x}),(\bar{\mathbf{4}},\mathbf{1},\mathbf{2},\mathbf{y}),(\bar{\mathbf{4}},\mathbf{1},\mathbf{1},\mathbf{z}),$$
$$(\mathbf{1},\bar{\mathbf{3}},\mathbf{2},\mathbf{a}),(\mathbf{1},\bar{\mathbf{3}},\mathbf{1},\mathbf{b}),(\mathbf{1},\bar{\mathbf{3}},\mathbf{1},\mathbf{c}),(\mathbf{1},\mathbf{1},\mathbf{2},\mathbf{k}),(\mathbf{1},\mathbf{1},\mathbf{1},\mathbf{s})\ .$$

Solution for the $U(1)$ charges canceling all anomalies exists but fails to give the correct SM embedding. However, analysis of the potentially successful embeddings of the three families suggests that $SU(3)_C\otimes SU(2)_L\otimes U(1)_Y \subset SU(4)_A\otimes SU(3)_C\otimes SU(2)_L\otimes U(1)_X$ works when the above spectrum is augmented with an anomaly-free $SU(4)_A$ multiplet charged under $U(1)_X$. The resulted models have nontrivial $U(1)_Y$ embeddings. Similar constructions with some other N values are also obtained.

These SM-like chiral models have interesting phenomenological predictions. In the case of a specific model with $N=4$, we also constructed a Higgs sector giving rise to a natural mass hierarchy $m_t, m_b > m_c > m_s > m_d, m_u$. The scalars multiplets are $\phi_0 = (\bar{\mathbf{4}},\mathbf{1},\mathbf{1},\mathbf{9})$ and ϕ_a ($a=1$ or 2), in $(\bar{\mathbf{4}},\mathbf{1},\mathbf{1},-\mathbf{3})$, together with $SU(2)_L$ doublets $\Phi = (\mathbf{15},\mathbf{1},\mathbf{2},-\mathbf{6})$. A $C_{ab}\phi_{ai}\phi_b^{\dagger j}\Phi_j^k\Phi_k^{\dagger i}$ mass term with natural VEVs for the ϕ_a's decouples twelve of the fifteen doublets from the EW-scale. There remain two EW Higgs doublets and an extra doublet of singly- and doubly-charged scalars. FCNC constraints can be easily satisfied and the quark mass hierarchy resulted.

References

Kong, O.C.W. (1996), Mod. Phys. Lett. **A 11**, 2547.
Kong, O.C.W. (1997), Phys. Rev. **D 55**, 383.

Duality in Supersymmetric Gauge Theories via Branes

K. Landsteiner

Institut für Theoretische Physik, TU Wien,
Wiedner Hauptstraße 8-10, A-1040 Wien, Austria

Abstract. Supersymmetric Gauge Theories show the remarkable property of duality. This means that two a priori different theories with different gauge group and matter content describe the same physics in the infrared limit. While in field theory only indirect arguments are known to establish a pair of dual gauge theories, string theory and in particular D-branes provide us with a tool to derive these dualities. In this stringy approach one stretches N_c D-4-branes in between two Neveu-Schwarz fivebranes. These branes have a common 3+1 dimensional worldvolume. At low energies the physics is described by a supersymmetric gauge theory. If one rotates the Neveu-Schwarz fivebranes such that the rotation lies in an $SU(2)$ subgroup of the rotation group in a four dimensional embedding space one effectively constructs a brane configuration with $N = 1$ supersymmetry. Chiral multiplets in the fundamental and antifundamental representation of $SU(N_c)$ can be obtained by including also D-6-branes in between the two fivebranes. The field theory duality can be derived by moving all D-6-branes to one side of the brane configuration and exchanging the two Neveu-Schwarz branes. In this process one carefully has to take into account that a D-4-brane is created every time a D-6-brane crosses a Neveu-Schwarz brane (Hanany-Witten transition). Using a configuration with three Neveu-Schwarz branes and an orientifold sixplane on top of the middle one we could construct an example of a chiral $N = 1$ gauge theory. In the brane configuration there is a jump of the RR 7-form charge where the orientifold sixplane is divided by the middle Neveu-Schwarz five brane. To compensate this jump one needs to introduce also 8 half D-6-branes. The model has $SU(N_c)$ gauge group, a chiral multiplet in the antisymmetric representation, a chiral multiplet in the conjugate symmetric representation, $N_f + 8$ chiral multiplets in the fundamental and N_f in the antifundamental representation. The dual model can be found by moving all D-6-branes (except the 8 half ones) to the sides and exchanging the outer two Neveu-Schwarz fivebranes. The gauge group of the dual model turns out to be $SU(3N_f + 4 - N_c)$[1].

References

Landsteiner, K., Lopez, E., Lowe, D.A. (1998a): Duality of Chiral $N = 1$ Supersymmetric Gauge Theories via Branes, JHEP02 (1998) 007, hep-th/9801002.

Landsteiner, K., Lopez, E., Lowe, D.A. (1998b): Supersymmetric Gauge Theories from Branes and Orientifold Six-planes, JHEP07 (1998) 011, hep-th/9805158.

[1] Work supported by the FWF under project Nr. P10268-PHY.

Curves for $SU(N)$ Theories with Matter in Two-Index Representations from Branes

E. Lopez [1]

Institut für Theoretische Physik, TU Wien,
Wiedner Hauptstraße 8-10, A-1040 Wien, Austria

Abstract. A variety of gauge theories with different number of supersymmetries can be induced on the world-volume of the Dirichlet branes by considering configurations of Dirichlet branes ending on Neveu-Schwarz fivebranes. We will be interested in configurations of Neveu-Schwarz fivebranes, Dirichlet fourbranes and an orientifold sixplane in type IIA string theory. Our configurations have 3+1 macroscopic dimensions shared by all the objects and preserve 1/4 of the initial type IIA supersymmetries. Thus they give raise to four dimensional gauge theories with $\mathcal{N} = 2$ extended supersymmetry. Dirichlet fourbranes and Neveu-Schwarz fivebranes derive from a single object in M-theory, the M-fivebrane. Therefore when lifted to M-theory the four and fivebrane intersections are smoothed out and we obtain instead a single M-fivebrane wrapped around a Riemann surface. This Riemann surface behaves as the Seiberg-Witten curve for the effective gauge theory living on the world-volume of the branes. In this way, upon lifting our configurations to M-theory and proposing a description of how to include the effects of the orientifold sixplane, we derive the Seiberg-Witten curves describing the Coulomb branch of $\mathcal{N} = 2$ gauge theories with orthogonal and symplectic gauge groups, product gauge groups of the form $\bigotimes_i SU(k_i) \otimes SO(N)$ and $\bigotimes_i SU(k_i) \otimes Sp(N)$. Of particular interest are configurations with a Neveu-Schwarz fivebrane on top of the orientifold sixplane. We concentrate in the case of three Neveu-Schwarz fivebranes. Such configurations induce a theory with $SU(N)$ gauge group and matter in the symmetric or antisymmetric representation. As before we lift the configurations to M-theory and obtain the associated curves. The curves pass several consistency checks. They reproduce Seiberg-Witten curves for previously known cases after appropriate scaling limits. For $N = 2$ the symmetric representation coincides with the adjoint and the antisymmetric with the singlet. For $N = 3$ the antisymmetric representation is equivalent to the antifundamental. The curves we obtain for these cases differ from those already known. Coincidence of the curves is however not necessary. We compare then their discriminants and find agreement.

References

Landsteiner, K., Lopez, E. (1998): New Curves from Branes. Nucl.Phys. **B516** (1998), 273-296, hep-th/9708118.

[1] Supported by Lise Meitner Fellowship M456-TPH.

Symmetries in Low Energy Pion Physics

H. Machner

Institut für Kernphysik, FZ Jülich, D-52425 Jülich, Germany

Abstract. We use a new value for the s-wave matrix element at threshold for the $pp \to d\pi^+$ reaction to determine the isoscalar πN scattering length. This is done by invoking symmetries and three ratios. Two of the ratios are measured and the third one is obtained from a new calculation. This leads to a new value for the πNN coupling constant $f^2/4\pi = 0.0763 \pm 0.0014$.

The s-wave matrix element at threshold α_0 can be related to the isovector scattering length b_1 through a chain of symmetries and ratios. The cross section $\sigma(pp \to d\pi^+)$ can be transformed into $\sigma(\pi^+d \to 2p)$ by making use of time reversal invariance. From this, one has $\sigma(\pi^-d \to 2n)$ by applying charge symmetry. The cross section is converted into a rate $w(\pi^-d \to 2n) \propto \lim_{k\to 0} \sigma(\pi^-d \to 2n)/k$ by extrapolation to zero energy. Then one has to apply three ratios: $S = w(\pi^-d \to 2n)/w(\pi^-d \to 2n\gamma)$, $T = w(\pi^-d \to 2n\gamma)/w(\pi^-p \to n\gamma)$, and the Panofsky ratio $P = w(\pi^-p \to n\gamma)/w(\pi^-p \to \pi^0 n)$. Then one can go back to a cross section via $w(\pi^-p \to \pi^0 n) \propto \lim_{k\to 0} \sigma(\pi^-p \to \pi^0 n)$. Finally, isospin symmetry yields from this cross section the one for elastic π scattering on the proton. This cross section is determined by the isovector scattering length.

Recently new cross section data for the $pp \to d\pi^+$ reaction close to threshold were published by ? and ? making the extraction of the s-wave partial cross section possible. Older data are dominated by the p-wave cross sections due to Δ excitation. We now make use of the new s-wave pion production matrix element at threshold derived by ? $\alpha_0 = 0.230 \pm 0.019$ (mb). The ratios S and P are often measured and we make use of the newest values. The ratio T can be taken from the individually calculated rates as given by ? However, if one is only interested in the ratio this can be calculated in an quasi free model (see ?) yielding $T = 0.78 \pm 0.04$. With these ingredients one gets $b_1 = -(87.3 \pm 4.4)10^{-3}/m_\pi$. Including this new value together with all recent values from different analysis from data with a real pion one gets a new mean value. This can be converted into a new value of the πNN coupling constant by making use of the GMO sum rule: $f^2/4\pi = 0.0763 \pm 0.0014$ which is smaller than the previously accepted value.

References

Drochner, M. et al., Phys. Rev.Lett. **77**, 454 (1996)
Gibbs, W. R. et al., Phys. Rev. **C 16**, 322 (1977) and ibid. 327
Heimberg, P. et al., Phys. Rev. Lett. **77**, 1012 (1996)
Machner, H. Nucl. Phys. A (in press)

Zero Modes and Vacuum Structure on the Light-Front

Ľ. Martinovič

Institute of Physics, Slovak Academy of Sciences,
Dúbravská cesta 9, SK-842 28 Bratislava, Slovakia

Abstract. Quantum field theory formulated in the light front coordinates has a unique property that the Fock vacuum $|0\rangle$ is, in the sector of normal modes, an eigenstate of the full – free plus interacting – Hamiltonian. This "trivial" vacuum can only mix with the dynamical zero modes (ZM), i.e. with the Fourier modes of quantum fields carrying vanishing light-front (LF) momentum k^+ and having non-vanishing conjugate momentum. The LF quantization in the compactified space ($-L \le x^- \le L$ in 1+1 dimensions) with periodic boundary conditions imposed on fields provides an IR regularized theory where the consequences of the gauge field zero-mode dynamics and residual gauge symmetry can conveniently be studied. Due to the new LF constraints the Dirac – Bergmann (or similar) constrained quantization has to be used. In the present work, the fermionic as well as bosonized formulation of the Schwinger model, which is known to possess non-perturbative properties like the θ-vacuum and spontaneous symmetry breaking expected in the realistic theories, is considered. In the fermionic formulation, the ZM part of the LF Hamiltonian contains only a term quadratic in the (rescaled) momentum $\hat{\pi}_0$ conjugate to the (rescaled) zero-mode ζ, which is the only gauge-field degree of freedom left in the finite-volume light cone gauge. The operators obey $[\zeta, \hat{\pi}_0] = i$. The residual gauge symmetry of the Hamiltonian $\zeta \to \zeta - n, n = \pm 1, \pm 2, \ldots$ is at the quantum level implemented by the unitary operator $\hat{T}_1 = \exp(-i\hat{\pi}_0)$. This leads to an infinite set of vacuum states $(\hat{T}_1)^n|0\rangle$ which are degenerate in LF energy. They have in the second-quantized picture the coherent-state form $\exp[\frac{\alpha}{\sqrt{2}}(a_0^\dagger - a_0)]$ (where $a = \frac{1}{\sqrt{2}}(\zeta + i\hat{\pi}_0))$ and can be understood as a condensate of massless bosons. A usual superposition of these vacuum states with a simple phase factor gives the θ-vacuum invariant under the operator of large transformations $\hat{T}_1$.

Next, the bosonic formulation of both the massless and massive Schwinger model is studied to obtain further insight into the ZM dynamics. The Dirac – Bergmann quantization yields a non-trivial commutator between the normal-mode part and ZM ϕ_0 of the equivalent boson field ϕ, and a complicated constraint relating conjugate momentum of the gauge ZM to the field ϕ. After a suitable change of variables one finds a quantum mechanical commutator between ζ and ϕ_0. The θ-vacuum of the massive model is then easily constructed in an analogy with the fermionic case. The non-trivial vacuum structure affects the physical quantities only through the vacuum angle θ present in the non-linear (fermion) mass term of the Hamiltonian. The $O(m^2)$ corrections to the Schwinger boson mass and the corresponding momentum densities are calculated within the mass perturbation theory. Finally, the chiral symmetry breaking is studied for $m = 0$. The conserved axial charge $\tilde{Q}_5$ (not invariant under $\hat{T}_1$) is proportional to ZM operator ζ. The θ-vacuum breaks chiral symmetry implemented by the unitary operator $\hat{V}[\beta] = \exp(-i\beta\tilde{Q}_5)$. Thus, non-perturbative vacuum properties of the usual formulation of the Schwinger model are reproduced in the light front theory.
Acknowledgements. This work was supported by the NSF grant INT-9515511.

Chiral Phase Transition and Quantum Chaos in Lattice QCD

R. Pullirsch[1], K. Rabitsch[1], T. Wettig[2], H. Markum[1]

[1] Institut für Kernphysik, Technische Universität Wien, A-1040 Wien, Austria
[2] Institut für Theoretische Physik, Technische Universität München
D-85747 Garching, Germany

Abstract. The properties of the spectrum of the Dirac operator are of great importance for the understanding of certain features of QCD. For example, the accumulation of small eigenvalues is, via the Banks-Casher formula, related to the spontaneous breaking of chiral symmetry. Recently, the fluctuation properties of the eigenvalues in the bulk of the spectrum have also attracted attention. In particular, it was shown that the nearest-neighbor spacing distribution $P(s)$, i.e., the distribution of spacings s between adjacent eigenvalues, agrees with predictions from random-matrix theory (RMT). According to the so-called Bohigas-conjecture, quantum systems whose classical analogs are chaotic have a nearest-neighbor spacing distribution given by RMT whereas systems whose classical counterparts are integrable obey a Poisson distribution, $P(s) = e^{-s}$. Therefore, the specific form of $P(s)$ indicates the presence or absence of quantum chaos.

We have worked on a lattice of size $6^3 \times 4$ with various values of the inverse gauge coupling $\beta = 6/g^2$ both in the confinement and in the deconfinement phase. We have studied full QCD with $N_f = 3$ degenerate flavors of staggered quarks with mass $ma = 0.05$. SU(3) with staggered fermions corresponds to the chiral unitary ensemble of RMT. A very good approximation to the nearest-neighbor spacing distribution of this ensemble is provided by the Wigner surmise, $P(s) = (32/\pi^2)\, s^2\, e^{-(4/\pi)\, s^2}$. We set the quark mass m in the fermionic matrix to zero and compare the nearest-neighbor spacing distribution $P(s)$ of full QCD with the RMT result. In the confinement as well as in the deconfinement phase we observe agreement with the Wigner surmise. No signs for a transition to Poisson regularity are found. Thus, the deconfinement phase transition does not seem to coincide with a transition in the spacing distribution. This means that the chiral phase transition which coincides with the deconfinement phase transition is not related to a chaos-to-order transition of the quark degrees of freedom. This provides evidence that quantum chaos persists in the quark-gluon plasma-phase. The reason is that the Dirac equation is non-linear due to the coupling to the gauge fields and, therefore, non-integrable in both phases. Only for extremely large values of β at fixed lattice size the spectrum of a free theory is approached.

References

Pullirsch, R., Rabitsch, K., Wettig, T., Markum, H. (1998): Evidence of quantum chaos in the plasma phase of QCD. Phys. Lett. **B** in print, hep-ph/9803285.

Fixing the Ambiguities in the Quantum Corrections to Soliton Masses

A. Rebhan

Institut für Theoretische Physik, Technische Universität Wien,
Wiedner Hauptstr. 8-10, A-1040 Vienna, Austria

Abstract. One-loop quantum corrections to the mass of two-dimensional solitons are well-known to be sensitive to the boundary conditions imposed on quantum fluctuations about the classical soliton. Recently Rebhan and van Nieuwenhuizen (1997) have shown that these quantum corrections also depend critically on the regularization method used for ultraviolet divergences and that the methods employed to confirm the conjectured saturation of the quantum Bogomolnyi bound in $N = 1$ supersymmetric theories are incompatible with those used in bosonic theories to confirm the exactness of the WKB result in the sine-Gordon model. We (Nastase et al. (1998)) propose to fix these ambiguities by adopting a set of boundary conditions which follow from the symmetries of the action and which depend only on the topology of the sector considered. Concerning the regularization dependence we invoke a physical principle that ought to hold generally in quantum field theories with a topological sector: for vanishing mass and other dimensionful constants, the vacuum energies in the trivial and topological sectors have to become equal. These requirements are found to lead to results that are consistent with the exact solution of both the bosonic and the $N = 1$ supersymmetric sine-Gordon model. They imply however that the quantum Bogomolnyi bound in $N = 1$ theories is violated. This is explained by the appearance of an additional mass-independent renormalization of the Hamiltonian in the topologically nontrivial sector. In theories with more symmetries ($N > 1$ and/or higher dimensions) none of these issues arise and the quantum Bogomolnyi bound remains saturated in accordance with the Witten-Olive theorem.

References

Rebhan, A., van Nieuwenhuizen, P. (1997): No saturation of the quantum Bogomolnyi bound by two-dimensional N=1 supersymmetric solitons. Nucl. Phys. **B 508**, 449–467

Naštase, H., Stephanov, M., van Nieuwenhuizen, P., Rebhan, A. (1998): Topological boundary conditions, the BPS bound, and elimination of ambiguities in the quantum mass of solitons. Preprint ITP-SB 98-9/TUW 98-01 (hep-th/9802074)

Pion-Kaon Scattering in Chiral $SU(2)$ Perturbation Theory

A. Roessl

Institut de Physique Théorique, Université de Lausanne,
CH-1015 Lausanne, Switzerland

Abstract. The appropriate modern approach to pion-kaon scattering near threshold is chiral perturbation theory [1, 2]. In chiral $SU(3)$ perturbation theory the scattering amplitude is expanded in powers of the light quark masses (m_u, m_d, m_s) and external momenta [3].

Both pions and kaons are pseudo-Goldstone bosons (their masses vanish in the chiral limit) and the largest expansion parameter is $M_K/4\pi F_\pi \sim 0.5$, where $M_K \sim$ 490 MeV is the kaon mass and $F_\pi \sim$ 93 MeV the pion constant. The motivation for treating π-K scattering in the framework of chiral $SU(2)$ perturbation theory is the elimination of this large parameter from the theory. In this approach, m_u and m_d remain parameters but m_s – like the other heavy quark masses – does not appear explicitly in the theory, but only implicitly in the low-energy constants. The term of lowest order is $\mathcal{L} = 3DD_\mu K^+ D^\mu K - M^2 K^+ K$, where the kaon field K is a complex isospin doublet and $D_\mu = 3D\partial_\mu + \Gamma_\mu$ the covariant derivative containing the pion field. At order p^3 we find 10 independent terms contributing to π-K scattering and respecting the symmetry properties of QCD (Lorentz-, parity- and charge-conjugation invariance and approximate $SU(2)_\mathrm{R} \times SU(2)_\mathrm{L}$ invariance). The chiral $SU(3)$ theory can be used to obtain the first coefficients in the expansion of the corresponding low-energy constants in powers of m_s. As in baryon chiral perturbation theory [4, 5], chiral power counting in the presence of loops is non-trivial because the kaon mass does not vanish in the chiral limit. In the second part of my thesis I shall try to use methods from heavy-baryon χPT (see [6] for a recent application) to obtain a straightforward chiral power counting procedure and to calculate all loop diagrams contributing to a given chiral order.

Note. First part of a doctoral thesis under the guidance of Heinrich Leutwyler, University of Bern, Switzerland.

References

[1] J. Gasser, H. Leutwyler, Ann. of Phys. **158** (1984) 142.
[2] J. Gasser, H. Leutwyler, Nucl. Phys. **B250** (1985) 465.
[3] V. Bernard, N. Kaiser, U. G. Meißner, Nucl. Phys. **B357** (1991) 129.
[4] J. Gasser, M. E. Sainio, A. Švarc, Nucl. Phys. **B307** (1988) 779.
[5] A. Krause, Helv. Phys. Acta **63** (1990) 3.
[6] G. Ecker, M. Mojžiš, Phys. Lett. **B410** (1997) 266.

The Chiral Deformation of the Dirac Sea

M. Rosina

Department of Physics, University of Ljubljana,
Jadranska 19, POB.2964, SI-1001 Ljubljana, Slovenia

Abstract. The spontaneous chiral symmetry breaking can be described as a chiral deformation of the Dirac sea of quarks. Pions (Goldstone bosons) and σ mesons then appear as chiral rotational states and chiral vibrational states of such a "deformed" system. Some new insight is obtained from the pictorial analogies with the rotational and vibrational spectra of quadrupolarly deformed atomic nuclei. The Nambu–Jona-Lasinio interaction is compared with the quadrupole-quadrupole interaction.

1 The Deformed Dirac Sea

Da Providência, Ruivo and de Sousa (1987) fruitfully describe the vacuum as a Slater determinant of negative energy quark states: $\mathcal{N}(-\frac{M}{p+E}, 0, 1, 0)$, $\mathcal{N}(0, 1, 0, -\frac{M}{p+E})$. Here the Weyl representation is used, $\mathcal{N}$ is normalization, $M = m_o + U$ is the "deformation parameter" (the "constituent mass") and U is a scalar mean field potential due to the NJL interaction (isospin is suppressed for simplicity) $V_{NJL} = -G \sum_{u \neq v} [\beta(u)\beta(v) + \mathrm{i}\beta(u)\gamma_5(u)\ \mathrm{i}\beta(v)\gamma_5(v)]$.

The Time-Dependent Hartree-Fock generates rotations and vibrations.

2 The Deformed Nucleus

In the example of ^{8}Be we describe four p-shell valence nucleons feeling a spin-orbit potential and interacting with a quadrupole-quadrupole interaction $V_{qq} = -G \sum [q_{xx}(u)\, q_{xx}(v) + q_{xy}(u) q_{xy}(v)]$. For simplicity we use here only four "sea levels" ($m_l m_s = +1 \uparrow$) or ($-1 \downarrow$), for a neutron or a proton, and four "positive energy levels" ($m_l m_s = +1 \downarrow$) or ($-1 \uparrow$). This looks like a 2-dimensional system.

Due to V_{qq}, these orbitals deform: $\phi_\uparrow = \mathrm{e}^{\mathrm{i}\varphi} \Uparrow - \frac{Q}{\eta+E} \mathrm{e}^{-\mathrm{i}\varphi} \Uparrow$, $\phi_\downarrow = \mathrm{e}^{-\mathrm{i}\varphi} \Downarrow - \frac{Q}{\eta+E} \mathrm{e}^{\mathrm{i}\varphi} \Downarrow$.
Here Q is a deformation parameter and $E = \sqrt{\eta^2 + Q^2}$. Both orbitals are prolate in the x-direction and such a correlated orientaton is favoured by V_{qq}. The analogy with the quark vacuum is obvious from the "vocabulary" in the table.

L_z	$\Longleftrightarrow$	γ_5
s_z	$\Longleftrightarrow$	$\sigma \cdot \mathbf{p}/\lvert\mathbf{p}\rvert$
$q_{xx} = -q_{yy}$	$\Longleftrightarrow$	β
q_{xy}	$\Longleftrightarrow$	$i\beta\gamma_5$
η	$\Longleftrightarrow$	*average* $\lvert\mathbf{p}\rvert$
$-\eta L_z s_z$	$\Longleftrightarrow$	$\gamma_5 \sigma \cdot \mathbf{p}$
V_{qq}	$\Longleftrightarrow$	V_{NJL}

References

da Providência, J., Ruivo, M.C. and de Sousa, C.A.(1987): Phys. Rev. D **36**, 1882–1896

Soft Supersymmetry Breaking as a Spontaneous Breaking in the Auxiliary Sector

A. Slavnov

Steclov Mathematical Institute, Gubkina st. 8, 117966 Moscow, Russia

Abstract. Physical applications of supersymmetry (SUSY) require some mechanism of SUSY breaking. It is desirable that this mechanism preserve the most important properties of SUSY theories such as nonrenormalization theorems and the possibility to use a superdiagram techniques. Spontaneous SUSY breaking would serve this goal the best, however it seems to be too restrictive to be used in realistic models. At present, most popular is the mechanism of soft SUSY breaking by introducing mass terms for component fields and some threelinear scalar vertices. Recently the studies of renormalization procedure in softly broken SUSY models were carried out (see Avdeev et al. (1998) and references therein) in the framework of the so called spurion mechanism (Girardello, Grisaru (1982)).

In the present talk I demonstrate that the spurion mechanism is a particular realization of a spontaneous SUSY breaking in the auxiliary sector mechanism, developed in our papers (Slavnov (1977)). In this method a SUSY model is extended by adding some additional auxiliary fields which acquire nonzero expectation values via spontaneous SUSY breaking. After shifting this fields to the stable minimum they decouple from the physical ones, and their sole effect is to produce soft SUSY breaking terms.

This procedure allows to use for an analysis of the renormalization procedure in softly broken SUSY theories the machinery of generalized SUSY Ward identities (see Slavnov (1975) and references therein). It is demonstrated that some results obtained in Avdeev et al. (1998) by means of superdiagram techniques may be easily derived in the framework of this approach using simple symmetry arguments (Slavnov (1998)).

References

Avdeev, L., Kazakov, D., Kondrashuk, I. (1998): Renormalization in Softly Broken SUSY Gauge Theories. Nucl. Phys. **B510**, 289–301

Girardello, L., Grisaru, M. (1982): Soft Breaking of Supersymmetry. Nucl. Phys. **B194**, 65–80

Slavnov, A. (1977): Spontaneous Breaking of Supersymmetry and It's Possible Application to Unified Models of Weak and Electromagnetic Interactions. Nucl. Phys. **B124**, 301–311

Slavnov, A. (1975): Generalized Ward Identities for Supersymmetric Gauge Theories. II.Nonabelian case. Nucl. Phys. **B96**, 134–146

Slavnov, A. (1998): Soft Supersymmetry Breaking as a Spontaneous Breaking in Auxiliary Sector. Theor. Math. Phys., in print.

Mixing and Decay Constants of Pseudoscalar Mesons

B. Stech

Institut für Theoretische Physik, Universität Heidelberg,
D-69120 Heidelberg, Germany

Abstract. In collaboration with Thorsten Feldmann and Peter Kroll a new approach to the $\eta - \eta'$ mixing problem has been developed (Feldmann et al. (1998)). We start from the quark flavour basis in which mixing is entirely due to the anomaly and assume that the decay constants taken in that basis follow the particle state mixing. On exploiting the divergencies of the axial vector currents – which embody the axial vector anomaly – all basic parameters are fixed to first order of flavour symmetry breaking in terms of f_π and f_K. To this order our method provides a parameter-free determination of the mixing angle and allows the calculation of the four decay constants of η and η'. The resulting values automatically satisfy the constraints from chiral perturbation theory (Leutwyler, Kaiser (1997)). One obtains a mass matrix, quadratic in the particle masses, with specified elements thus providing an answer to the old problem of quadratic versus linear mass matrices. The ratios of matrix elements of the anomaly for the two states of our basis turn out to be inversely proportional to the corresponding decay constants. We tested our scheme against several independent experiments and determined the corrections to the first-order values of the basic parameters from phenomenology. All results were consistent with each other. Thus, the weighted average value of the mixing angle is rather precise: the angle describing the deviation from ideal mixing turned out to be $39.3^0 \pm 1^0$.

Finally, we generalized the new mixing scheme to include the mixing with the η_c. Here the decay constant of η_c enters which we took equal to the one for the J/ψ particle. With this ingredient the $c\bar{c}$ admixture of η and η' could be determined in magnitude and sign. For the decay constant of the η' originating from the $c\bar{c}$ current we obtain $-(6.3 \pm 0.6)$ MeV.

I like to thank the organizers for all their efforts which made the Schladming meeting a very fruitful and enjoyable one.

References

Feldmann, Th., Kroll, P., and Stech, B. (1998), hep-ph/9802409.
Leutwyler, H. (1997), hep-ph/9709408; Kaiser, R., diploma thesis, Bern (1997).

A Theory-Driven Analysis of the Effective QED Coupling at M_Z

M. Steinhauser

Max-Planck-Institut für Physik, Werner-Heisenberg-Institut,
D-80805 Munich, Germany

Abstract. An evaluation of the effective QED coupling at the scale M_Z is presented. It employs the predictions of perturbative QCD for the cross section of electron positron annihilation into hadrons, respectively the ratio $R(s) = \sigma(e^+e^- \to \text{hadrons})/\sigma(e^+e^- \to \mu^+\mu^-)$, up to order α_s^2, including the full quark mass dependence, and of order α_s^3 in the high energy region. This allows to predict the input for the dispersion relations over a large part of the integration region. The perturbative piece is combined with data for the lower energies. The normalization of data from the heavy quark thresholds is deduced from a comparison between data and pQCD outside the threshold region. For the energy range between 3.7 GeV and 5.0 GeV data for $R(s)$ from the experiments DASP, PLUTO and MARK I are used. Two models are constructed which account for the differences between the normalization factors from above and below the resonance region. A remarkable consistency both between the two models and the three experiments is found. For the three lowest J/Ψ resonances and the six Υ resonances the narrow width approximation is employed. The result for the hadronic contribution to the running of the coupling reads $\Delta\alpha_{\text{had}}^{(5)}(M_Z^2) = (277.4 \pm 1.7) \times 10^{-4}$.

For the contribution from the top quark the polarization function, $\Pi(q^2)$, is evaluated up to three loops. Thereby it is possible to restrict to the first two terms in the expansion for large top quark mass. The small error of $\Delta\alpha_{\text{had}}^{(5)}(M_Z^2)$ makes it also necessary to consider besides the dominant leptonic one-loop term, which amounts to 314.19×10^{-4}, also two-loop corrections giving a contribution of 0.78×10^{-4}. The combination of the different parts leads after resummation of the leading logarithms to $(\alpha(M_Z^2))^{-1} = 128.928 \pm 0.023$. Compared to previous analyses the uncertainty is thus significantly reduced, albeit at the price of a more pronounced dependence on pQCD at relatively low energies.

Top Quark Pair Production at Threshold: Complete Next-to-Next-to-Leading Order Relativistic Corrections

Th. Teubner

Deutsches Elektronen-Synchrotron DESY, D-22603 Hamburg, Germany

Abstract. In order to get insight into the electroweak symmetry breaking sector of the Standard Model (SM) or physics beyond the SM it is mandatory to measure the mass and the couplings of the top quark with high accuracy. Of particular interest is the production of $t\bar{t}$ in the threshold region at a future e^+e^- Linear Collider. To determine the top quark mass through a threshold scan, precise theoretical predictions of the cross section are needed. Here we present the complete next-to-next-to-leading order (i.e. $\mathcal{O}(v^2)$, $\mathcal{O}(v\alpha_s)$ and $\mathcal{O}(\alpha_s^2)$) relativistic corrections to the total photon mediated $t\bar{t}$ production cross section at threshold. They are obtained in the framework of non-relativistic quantum chromodynamics (NRQCD). The cross section can be expressed as a sum of nonrelativistic (long distance) current-current correlators multiplied by short distance coefficients. We use semi-analytic methods to calculate the correlators and determine the coefficients by *direct matching* of the cross section in NRQCD to the analytical result in full QCD. Figure 1 shows that the size of the next-to-next-to-leading order relativistic corrections is comparable to the size of the next-to-leading order ones. The band of the curves demonstrates the main uncertainty coming from the scale ambiguity in the long distance correlators. For a more detailed description of the calculation, results and references we refer the interested reader to A. H. Hoang and T. Teubner, DESY Preprint 98-008 and hep-ph/9801397.

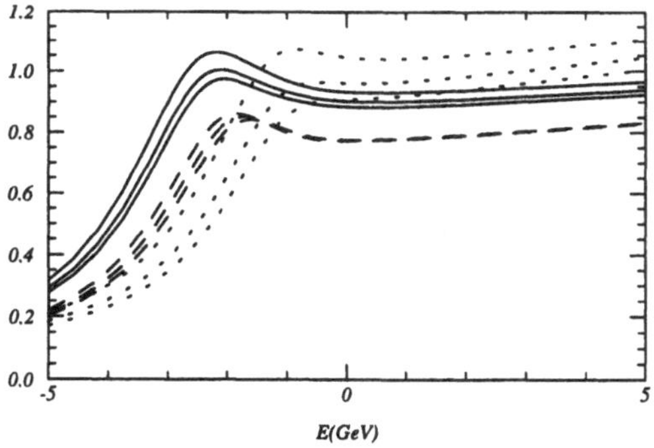

Fig. 1. The total normalized photon-mediated $t\bar{t}$ cross section at LO (dotted lines), NLO (dashed lines) and NNLO (solid lines) for the soft scales $\mu_{soft} = 50$ (upper lines), 75 and 100 GeV (lower lines).

Springer and the environment

At Springer we firmly believe that an international science publisher has a special obligation to the environment, and our corporate policies consistently reflect this conviction.
We also expect our business partners – paper mills, printers, packaging manufacturers, etc. – to commit themselves to using materials and production processes that do not harm the environment. The paper in this book is made from low- or no-chlorine pulp and is acid free, in conformance with international standards for paper permanency.

Springer

Lecture Notes in Physics

For information about Vols. 1–482 please contact your bookseller or Springer-Verlag

Vol. 483: G. Trottet (Ed.), Coronal Physics from Radio and Space Observations. Proceedings, 1996. XVII, 226 pages. 1997.

Vol. 484: L. Schimansky-Geier, T. Pöschel (Eds.), Stochastic Dynamics. XVIII, 386 pages. 1997.

Vol. 485: H. Friedrich, B. Eckhardt (Eds.), Classical, Semi-classical and Quantum Dynamics in Atoms. VIII, 341 pages. 1997.

Vol. 486: G. Chavent, P. C. Sabatier (Eds.), Inverse Problems of Wave Propagation and Diffraction. Proceedings, 1996. XV, 379 pages. 1997.

Vol. 487: E. Meyer-Hofmeister, H. Spruik (Eds.), Accretion Disks - New Aspects. Proceedings, 1996. XIII, 356 pages. 1997.

Vol. 488: B. Apagyi, G. Endrédi, P. Lévay (Eds.), Inverse and Algebraic Quantum Scattering Theory. Proceedings, 1996. XV, 385 pages. 1997.

Vol. 489: G. M. Simnett, C. E. Alissandrakis, L. Vlahos (Eds.), Solar and Heliospheric Plasma Physics. Proceedings, 1996. VIII, 278 pages. 1997.

Vol. 490: P. Kutler, J. Flores, J.-J. Chattot (Eds.), Fifteenth International Conference on Numerical Methods in Fluid Dynamics. Proceedings, 1996. XIV, 655 pages. 1997.

Vol. 491: O. Boratav, A. Eden, A. Erzan (Eds.), Turbulence Modeling and Vortex Dynamics. Proceedings, 1996. XII, 245 pages. 1997.

Vol. 492: M. Rubí, C. Pérez-Vicente (Eds.), Complex Behaviour of Glassy Systems. Proceedings, 1996. IX, 467 pages. 1997.

Vol. 493: P. L. Garrido, J. Marro (Eds.), Fourth Granada Lectures in Computational Physics. XIV, 316 pages. 1997.

Vol. 494: J. W. Clark, M. L. Ristig (Eds.), Theory of Spin Lattices and Lattice Gauge Models. Proceedings, 1996. XI, 194 pages. 1997.

Vol. 495: Y. Kosmann-Schwarzbach, B. Grammaticos, K.M. Tamizhmani (Eds.), Integrability of Nonlinear Systems. VII, 380 pages. 1997.

Vol. 496: F. Lenz, H. Grießhammer, D. Stoll (Eds.), Lectures on QCD. VII, 483 pages. 1997.

Vol. 497: J. P. Greve, R. Blomme, H. Hensberge (Eds.), Stellar Atmospheres: Theory and Observations. VIII, 352 pages. 1997

Vol. 498: Z. Horváth, L. Palla (Eds.), Conformal Field Theories and Integrable Models. Proceedings, 1996. X, 251 pages. 1997.

Vol. 499: K. Jungmann, J. Kowalski, I. Reinhard, F. Träger (Eds.), Atomic Physics Methods in Modern Research, IX, 448 pages. 1997.

Vol. 500: D. Joubert (Ed.), Density Functionals: Theory and Applications, XVI, 194 pages. 1998.

Vol. 501: J. Kertész, I. Kondor (Eds.), Advances in Computer Simulation. VIII, 166 pages. 1998.

Vol. 502: H. Aratyn, T. D. Imbo, W.-Y. Keung, U. Sukhatme (Eds.), Supersymmetry and Integrable Models. Proceedings, 1997. XI, 379 pages. 1998.

Vol. 503: J. Parisi, S. C. Müller, W. Zimmermann (Eds.), A Perspective Look at Nonlinear Media. From Physics to Biology and Social Sciences. VIII, 372 pages. 1998.

Vol. 504: A. Bohm, H.-D. Doebner, P. Kielanowski (Eds.), Irreversibility and Causality. Semigroups and Rigged Hilbert Spaces. XIX, 385 pages. 1998.

Vol. 505: D. Benest, C. Froeschlé (Eds.), Impacts on Earth. XVII, 223 pages. 1998.

Vol. 506: D. Breitschwerdt, M. J. Freyberg, J. Trümper (Eds.), The Local Bubble and Beyond. Proceedings, 1997. XXVIII, 603 pages. 1998.

Vol. 507: J. C. Vial, K. Bocchialini, P. Boumier (Eds.), Space Solar Physics. Proceedings, 1997. XIII, 296 pages. 1998.

Vol. 508: H. Meyer-Ortmanns, A. Klümper (Eds.), Field Theoretical Tools for Polymer and Particle Physics. XVI, 258 pages. 1998.

Vol. 509: J. Wess, V. P. Akulov (Eds.), Supersymmetry and Quantum Field Theory. Proceedings, 1997. XV, 405 pages. 1998.

Vol. 510: J. Navarro, A. Polls (Eds.), Microscopic Quantum Many-Body Theories and Their Applications. Proceedings, 1997. XIII, 379 pages. 1998.

Vol. 511: S. Benkadda, G. M. Zaslavsky (Eds.), Chaos, Kinetics and Nonlinear Dynamics in Fluids and Plasmas. Proceedings, 1997. VIII, 438 pages. 1998.

Vol. 512: H. Gausterer, C. Lang (Eds.), Computing Particle Properties. Proceedings, 1997. VII, 335 pages. 1998.

Vol. 513: A. Bernstein, D. Drechsel, T. Walcher (Eds.), Chiral Dynamics: Theory and Experiment. Proceedings, 1997. IX, 394 pages. 1998.

Vol. 514: F. W. Hehl, C. Kiefer, R.J.K. Metzler, Black Holes: Theory and Observation. Proceedings, 1997. XV, 519 pages. 1998.

Vol. 515: C.-H. Bruneau (Ed.), Sixteenth International Conference on Numerical Methods in Fluid Dynamics. Proceedings. XV, 568 pages. 1998.

Vol. 516: J. Cleymans, H. B. Geyer, F. G. Scholtz (Eds.), Hadrons in Dense Matter and Hadrosynthesis. Proceedings, 1998. XII, 253 pages. 1999.

Vol. 517: Ph. Blanchard, A. Jadczyk (Eds.), Quantum Future. Proceedings, 1997. X, 244 pages. 1999.

Vol. 518: P. G. L. Leach, S. E. Bouquet, J.-L. Rouet, E. Fijalkow (Eds.), Dynamical Systems, Plasmas and Gravitation. Proceedings, 1997. XII, 397 pages. 1999.

Vol. 519: A. Pękalski, K. Sznajd-Weron (Eds.), Anomalous Diffusion. From Basics to Applications. Proceedings, 1998. XVIII, 378 pages. 1999.

Vol. 520: J. A. van Paradijs, J. A. M. Bleeker (Eds.), X-Ray Spectroscopy in Astrophysics. EADN School X. Proceedings, 1997. XV, 530 pages. 1999.

Vol. 521: L. Mathelitsch, W. Plessas (Eds.), Broken Symmetries. Proceedings, 1998. VII, 299 pages. 1999.

Monographs

For information about Vols. 1–10 please contact your bookseller or Springer-Verlag